Edited by
Tharwat F. Tadros

Self-Organized Surfactant Structures

Related Titles

Tadros, T. F. (ed.)

Colloids and Interface Science Series

6 Volume Set

2010

ISBN: 978-3-527-31461-4

Tadros, T. F.

Rheology of Dispersions

Principles and Applications

2010

ISBN: 978-3-527-32003-5

Wilkinson, K. J., Lead, J. R. (eds.)

Environmental Colloids and Particles

Behaviour, Separation and Characterisation

2007

ISBN: 978-0-470-02432-4

Edited by Tharwat F. Tadros

Self-Organized Surfactant Structures

WILEY-VCH Verlag GmbH & Co. KGaA

The Editor

Prof. Dr. Tharwat F. Tadros
89 Nash Grove Lane
Wokingham, Berkshire RG40
4HE
United Kingdom

Library of Congress Card No.: applied for

British Library Cataloguing-in-Publication Data
A catalogue record for this book is available from the British Library.

Bibliographic information published by the Deutsche Nationalbibliothek
The Deutsche Nationalbibliothek lists this publication in the Deutsche Nationalbibliografie; detailed bibliographic data are available on the Internet at <http://dnb.d-nb.de>.

Typesetting Toppan Best-set Premedia Limited, Hong Kong

Printing and Binding Strauss GmbH, Mörlenbach

Cover Design Adam-Design, Weinheim

Printed in the Federal Republic of Germany
Printed on acid-free paper

ISBN: 978-3-527-31990-9

In memory of Professor Hironobu Kunieda.

Yokohema, November 17, 1948 – November 17, 2005

Dedicated to
his wife Akemi Kunieda and daughters
Yukiko Kunieda and Naoko Kunieda

Contents

Self-Organized Surfactant Structures. Edited by Tharwat F. Tadros

ISBN: 978-3-527-31990-9

Preface

This volume is dedicated to the late Professor H. Kunieda who did a great deal of research on self-organized surfactant structures. The scientific contributions of Prof. H. Kunieda are given in the first part of the book. The authors of the chapters in this volume have known Prof. H. Kunieda for many years and some of them were his students. The book addresses a variety of topics ranging from structure of lamellar liquid-crystalline phases to micellar systems both in aqueous and nonaqueous media. Aspects of the rheological behavior of micellar systems have also been addressed. The preparation of mesoporous materials using surfactant-organized structures has also been described.

This book, which deals with many diverse topics of self-organized surfactant structures, will be valuable to many research workers who deal with the phase behavior of surfactant systems. It can also be of much value for industrial researchers who are interested in application of these structures in their formulation, in particular in the area of cosmetics and pharmaceuticals.

The authors of the book address the topics at a fundamental level and try to relate the phase behavior to the surfactant structure. We are very grateful for the care the authors took in the preparation of the manuscripts.

Barcelona, March 2010

Tharwat Tadros
Conxita Solans

Self-Organized Surfactant Structures. Edited by Tharwat F. Tadros

ISBN: 978-3-527-31990-9

Scientific Contributions by Professor Hironobu Kunieda

Conxita Solans and Björn Lindman

The research interests of Prof. H. Kunieda focused on surfactant chemistry from the beginning of his scientific career. Understanding surfactant self-organizing structures by studying thermodynamic (i.e. phase equilibria), structural and dynamic aspects was his main scientific objective.

His earliest publications, dating from the early 1970s, dealt with the study of dissolution mechanisms in surfactant systems and the factors by which the mutual solubility of oil and water is increased. The reported results [1, 2] resolved the existing controversy about the nature of microemulsions allowing the conclusion to be drawn that microemulsions should not be regarded as emulsions but as solubilized systems. A particularly interesting outcome of his early research was to show that double-chain surfactants display a biomimetic behavior [3–5]. He reported in 1977 [3] that double-chain cationic surfactants form, in water, a rather stable two-phase dispersion consisting of water and lamellar liquid-crystalline phase (region II in Figure 1), the same phase behavior as lecithin. In the same year, Kunitake *et al.* reported for the first time that synthetic surfactants form normal vesicles [6]. He also found that the phase behavior of Aerosol OT is strongly influenced by the presence of salts and proposed a method to purify it [7–11]. This finding prompted other authors to revise previous papers related to the phase behavior of this surfactant.

Research on microemulsions was a major topic in his scientific activity, since the earlier work under Prof. Shinoda's supervision [1, 2], through his entire scientific career. First, the attention was focused to find the conditions to produce three-phase equilibria (balanced conditions) in both ionic [9–12] and nonionic [13–17] surfactant systems. In this context, it was shown that the effect of temperature in ionic surfactant systems is opposite to that in polyoxyethylene-type nonionic surfactants [10] and that both types of surfactant systems display similarities in phase behavior [18]. The most detailed phase equilibria of a water/ nonionic surfactant/aliphatic hydrocarbon system around the HLB temperature (Figure 2) was reported in 1982 [16].

In spite of the complexity in obtaining the phase behavior for the chosen surfactant, $C_{12}H_{25}(OCH_2CH_2)_5OH$), due to the formation of liquid-crystalline phases,

Self-Organized Surfactant Structures. Edited by Tharwat F. Tadros

ISBN: 978-3-527-31990-9

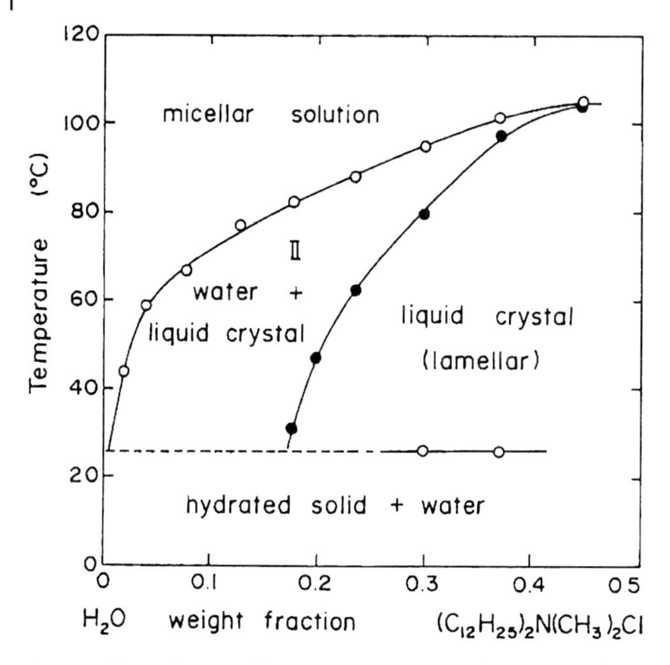

Figure 1 Phase diagram of water/$(C_{12}H_{25})_2N(CH_3)_2Cl$ system as a function of temperature (from Ref. [4], with permission).

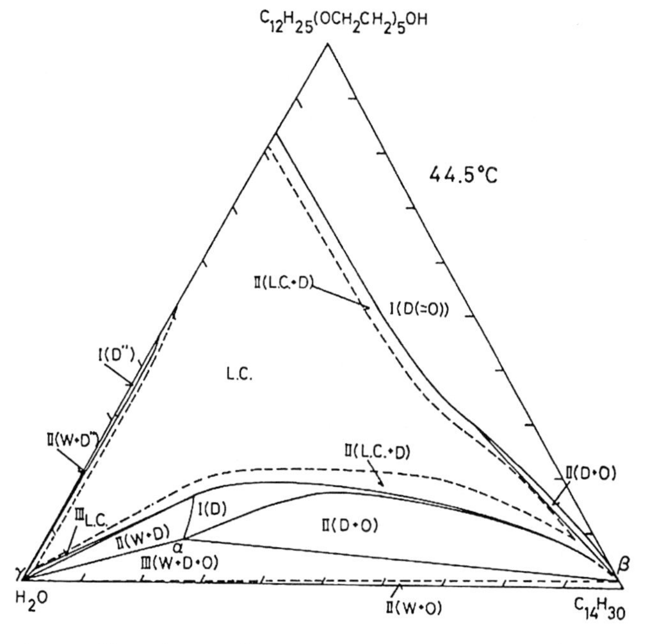

Figure 2 Phase diagram of water/$C_{12}H_{25}(OCH_2CH_2)_5OH$/tetradecane system at 44.5 °C (from Ref. [16], with permission).

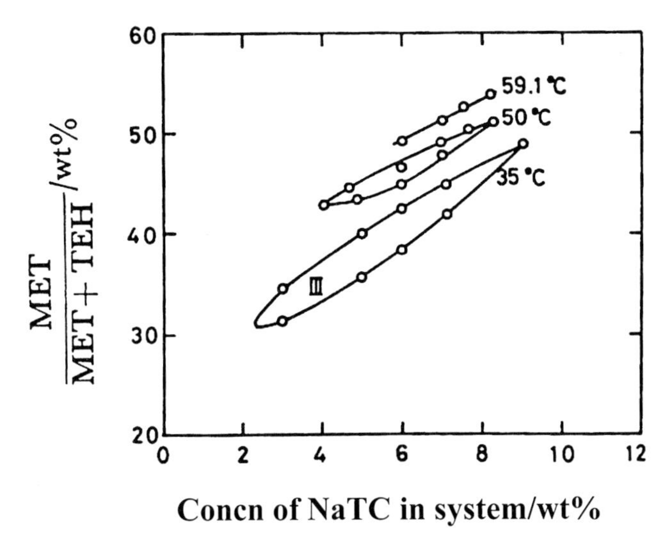

Figure 3 The effect of temperature on the three-phase region at constant salinity in a surfactant (sodium taurocholate NaTC)/ brine/cosurfactant (monoglycerideTEH)/oil (triglyceride THE) system. The three-phase region becomes narrower with the increase of temperature and vanishes above 59.1 °C, where a tricritical point exists, (adapted from Ref. [19], with permission).

the basic changes around the three-phase region were shown to be the same as those in short-chain nonionic surfactants [13].

A remarkable achievement was the discovery in 1984 of tricritical-point phenomena (Figure 3) in surfactant-containing systems [19]. He was the first researcher to report such behavior [19, 20] and to show that three-phase microemulsions are related to multicritical-solution phenomena (when three phases become identical and ultralow interfacial tension is attained) [21]. He reported tricritical phenomena in several surfactant systems [19–23].

Evaluation of the hydrophile–lipophile balance (HLB) properties of nonionic surfactants and understanding the phase behavior of surfactant mixtures in water and oil were the main objectives of his research from the early 1980s. In a series of papers published in 1985 [24–26] he clarified the concept of HLB temperature in mixed-surfactant systems by considering the monomeric solubility of surfactant in oil. In mixed nonionic surfactant or ionic surfactant–cosurfactant mixed systems, the monomeric solubility of lipophilic surfactant or cosurfactant is usually large. Therefore, the surfactant mixing ratio at water/oil interfaces deviates from the initial mixing ratio in the presence of a sufficient amount of oil. By geometrical calculations in a space of temperature and compositions, an equation of the HLB plane was developed by which the effect of temperature, oil/water ratio, weight ratio between the surfactants and surfactant concentration on the phase behavior of mixed-surfactant systems is explained. The HLB plane is expressed in a tetrahedral phase diagram in a four-component system at constant temperature as depicted in Figure 4.

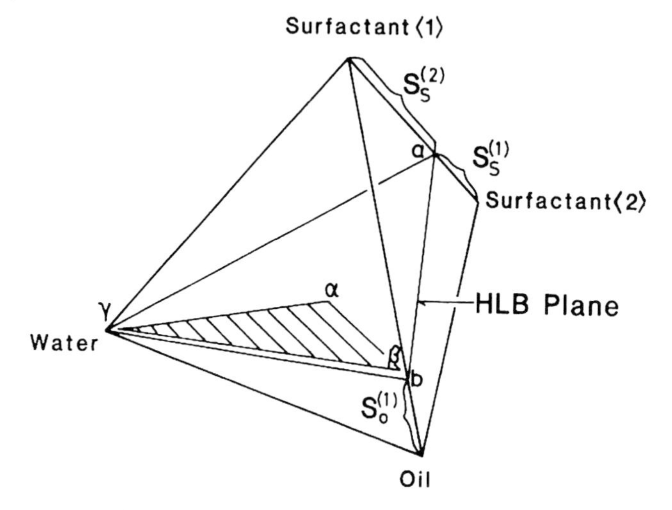

Figure 4 The HLB plane in a tetrahedral phase diagram of a four-component system at constant temperature. The shaded area means a three-phase triangle in the midst of a three-phase region, (from Ref. [24], with permission).

It was shown that the HLB temperature is shifted to higher temperature with decreasing surfactant concentration or increasing oil content even if the total mixing ratio of surfactant is fixed [24, 25]. HLB numbers for ionic surfactants could be estimated by the proposed equation [26]. Furthermore, although the conditions to produce a three-phase region in ionic surfactant systems had been studied, the effect of the oil/water ratio and surfactant concentration had not been taken into account. Kunieda investigated the effect of these parameters in brine/ionic surfactant/nonionic surfactant/oil systems. He showed that temperature-insensitive microemulsions are achieved in a dilute region of ionic–nonionic surfactant mixtures by depressing the solubility of the lipophilic surfactant in the oil phase of the three-phase region [26]. The important outcome of this research was that the phase-inversion temperature (PIT) can be predicted [24–26].

The deep knowledge in surfactant phase behavior led to the discovery of four-phase microemulsions containing excess of water and oil phases and two kinds of microemulsions, in a water/nonionic surfactant/two-oil system (Figure 5), in 1998 [27]. The mechanism for the formation of the four-phase region and the existence of four types of three-phase regions were discovered in carefully selected systems [28–31]. Of particular interest was the finding of a four-phase region in a single surfactant system (a ternary water/nonionic surfactant/triglyceride system) at constant temperature [31]. Moreover, the correlation between the formation of liquid crystal in dilute regions and the hydrophile–lipophile balance (HLB) of a surfactant as well as the transition from lamellar liquid crystal into microemulsion in both nonionic and ionic surfactant systems were described [32–37]. In this context, the effect of amphiphilic [33–36] and mixed [37] oils was thoroughly studied.

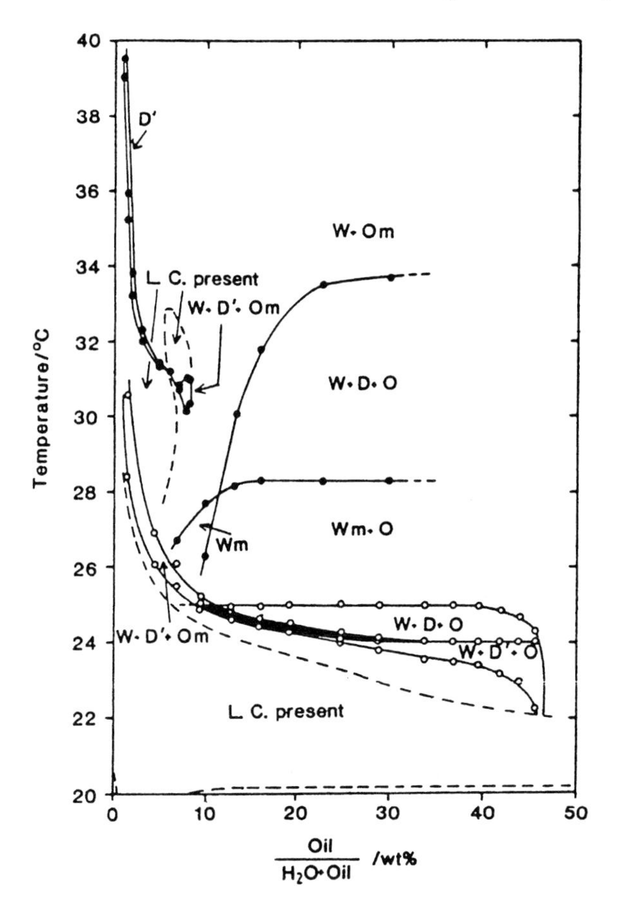

Figure 5 Phase diagrams for a water/$C_{12}H_{25}(OCH_2CH_2)_4OH$/mixed oil (triglyceride/hexadecane) system (○) and water/$C_{12}H_{25}(OCH_2CH_2)_4OH$/ single oil (hexadecane) system (●) as a function of oil/ (water+oil) and temperature. The black area is a four-phase region (from Ref. [27], with permission).

Another important finding by Kunieda was that of reverse vesicles, the counter structure to normal vesicles. He was the first to predict and report [38, 39], in 1991, the existence of this new class of molecular assembly, a manifestation of the symmetry between normal and reverse amphiphilic structures. Reverse vesicles consist of lamellar liquid crystals dispersions swollen with a large amount of oil (i.e. closed bimolecular layers in which the hydrophilic groups of surfactants orient inwards). The basis for the prediction was the observation that with long-chain surfactants, lamellar liquid crystals extend into both the water- and oil-rich regions of the phase diagrams of ternary water/surfactant/oil systems at the HLB temperature. As a consequence, if the lamellar liquid-crystal phase coexists with excess water and oil, normal vesicles should form in the water-rich region and reverse vesicles should form in the oil-rich-region at the HLB temperature (Figure 6).

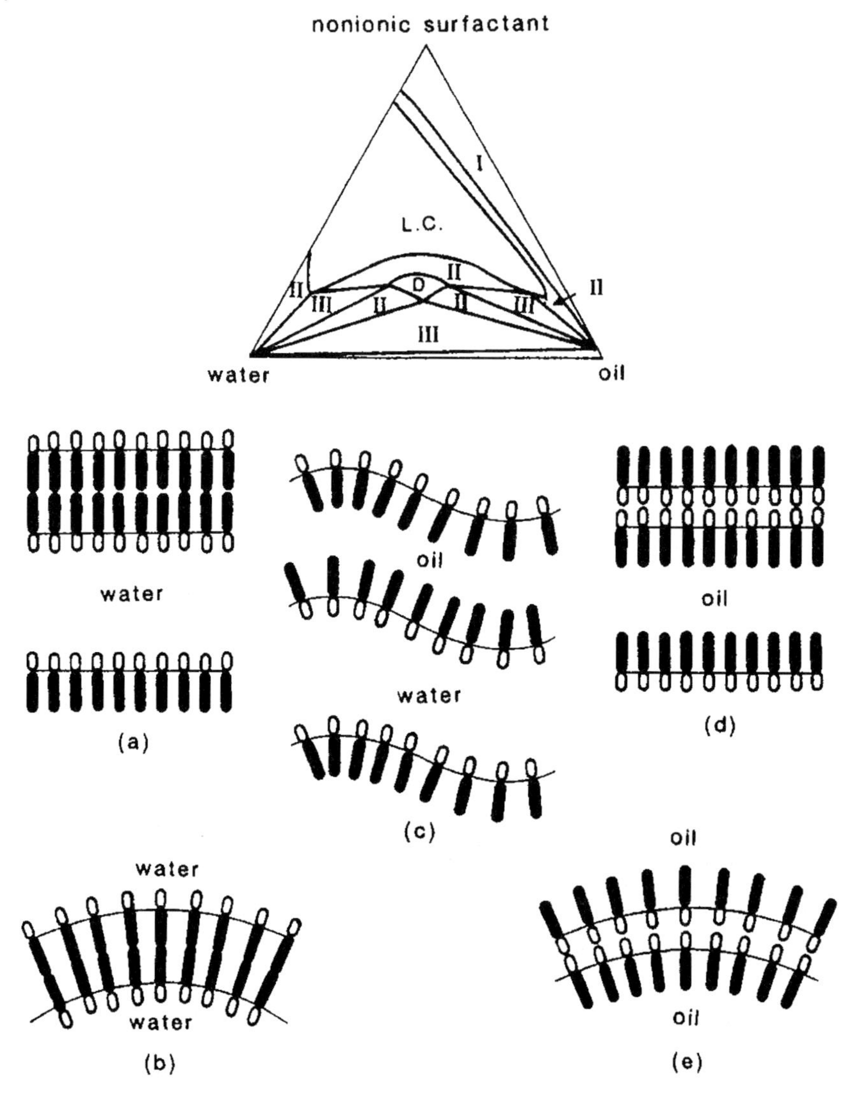

Figure 6 Schematic phase diagram of a water (W)/nonionic surfactant (S)/oil (O) system at the HLB temperature and self-organizing structures: (a) "normal" lamellar liquid crystal; (b) normal vesicles; (c) bicontinuous surfactant phase (middle-phase microemulsion; (d) "reversed" lamellar liquid crystal; (e) reversed vesicles. The lipophilic moiety of surfactant is represented by a solid rod and the hydrophilic part is represented by an open ellipse attached to the rod (from Ref. [39], with permission).

Indeed, these phase equilibria was found in several single surfactant systems (e.g., water/$R_{12}EO_4$/dodecane system, shown in Figure 7) that allowed him to establish a general correlation between phase behavior, structure of lamellar liquid crystals and the formation of both normal and reverse vesicles in single amphiphilic

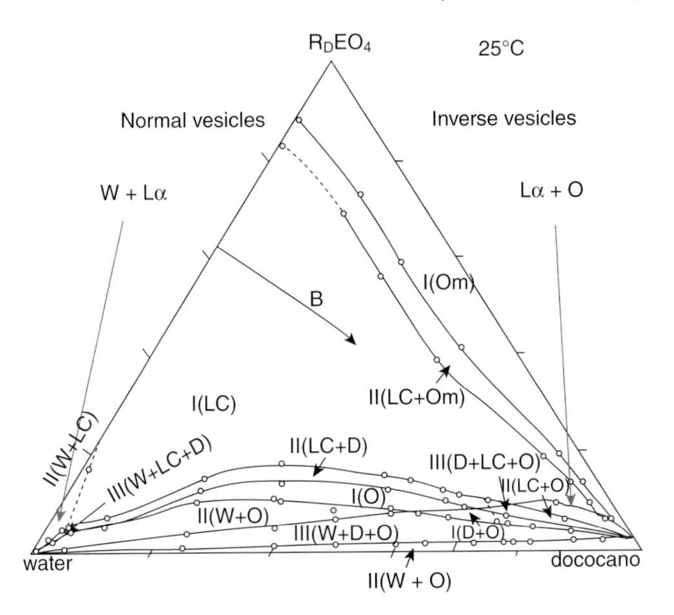

Figure 7 Phase diagram for a water/$C_{12}H_{25}(OCH_2CH_2)_4OH$/dodecane system at 25 °C. L.C. lamellar liquid crystal; D, isotropic surfactant phase; I, II, II, one-, two- and three-phase regions, respectively. W and O are excess water and oil phases. Om is a reversed micellar solution phase (adapted from Ref. [39], with permission).

systems near the HLB temperature [38–43]. Several aspects were studied systematically such as the conditions to produce reverse vesicles below the HLB temperature [40], the formation of reverse vesicles in different type of surfactants systems [42–47], including anionic [44], biocompatible [45–47] and long-polyoxyethylene-chain nonionic [48] surfactant systems. Studies of stability and size control [49] and membrane permeability [50], were also the object of research. His work on reverse vesicles was reviewed up to 1996 [51].

Kunieda was involved since 1985 in the study of highly concentrated emulsions (gel emulsions), an interesting class of emulsions characterized by an internal phase volume fraction exceeding 0.74, the critical value of the most compact arrangement of uniform, undistorted spherical droplets. Important outcomes in the early stages of this research were the relation found between stability and microstructure of the continuous phase and the explanation of their "spontaneous" formation through phase behavior [52–60]. The higher stability of W/O gel emulsions in water/ethoxylated nonionic surfactant/oil systems at 25–30 °C above the HLB temperature [55] was shown to be due to a change in the continuous phase structure from bicontinuous to droplet-type microemulsion with increase of temperature. The "spontaneous" formation of this type of emulsions by the phase-inversion temperature method, which leads to emulsions with smaller droplets and lower polydispersity than those obtained by conventional methods, was explained [58, 59] by the change in the spontaneous curvature of the surfactant in the formation process. Figure 8 shows the changes in self-organizing structures

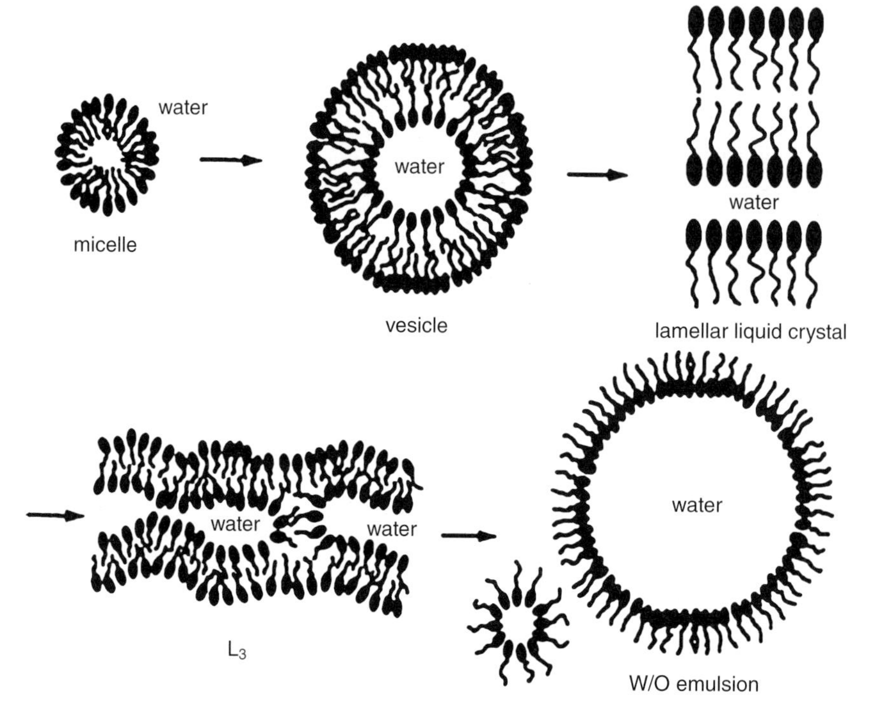

Figure 8 Schematic change in spontaneous curvature of surfactant layers in the process of spontaneous formation of highly concentrated W/O emulsions (from Ref. [58], with permission).

during the spontaneous formation process of highly concentrated W/O emulsions [58].

Another important achievement on this subject was the description for the first time of cubic-based highly concentrated O/W emulsions having a micellar cubic as the continuous phase [61]. It was earlier reported that the formation of a micellar (I_1) cubic liquid-crystal phase is promoted by adding long-chain alkanes to hydrophilic surfactant/water systems [48, 62]. Highly concentrated O/W, I_1-phase-based, emulsions were formed in the oil-rich region of water/hydrophilic surfactant/oil systems [61, 63]. The formation of W/O highly concentrated emulsions having a reverse micellar cubic (I_2) liquid-crystal phase as the continuous phase was later reported in a poly(oxyethylene) poly(dimethylsiloxane) surfactant system [64] and in systems containing a perfume, limonene, as the oil component [65]. These contributions established the basis for further progress, namely the use of highly concentrated emulsions as templates for the preparation of novel materials with dual meso/macroporous structure in single-step methods.

Although the general features of the phase behavior of water/mixed-surfactant/oil systems had been reported [24–26], the detailed behavior of mixed-surfactant systems was not completely elucidated until the 1990's. During this decade,

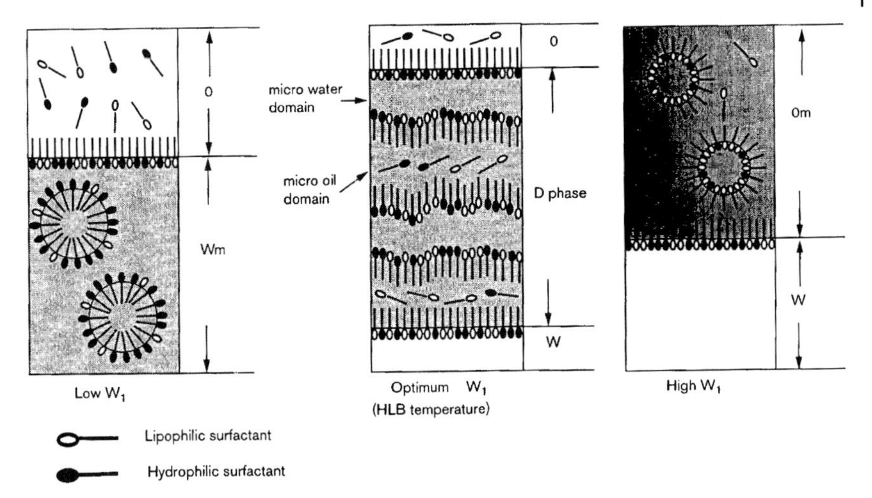

Figure 9 The schematic change in phase behavior in a mixed-surfactant system. Wm and Om are aqueous micellar and reverse micellar solution phases. W and O are excess water and oil phases. D indicates a microemulsion (surfactant) phase. W_1 is the weight fraction of lipophilic surfactant in the mixed surfactant (from Ref. [69], with permission).

Kunieda intensified the studies on phase behavior and formation of microemulsions in mixed-surfactant systems [66–76], in order to understand the relationship between maximum solubilization of microemulsions and surfactant distribution of mixed surfactants at the water/oil interface in the microemulsion phase. He developed a method to calculate the net composition of each surfactant at the interface in the bicontinuous microemulsions assuming that the monomeric solubility of each surfactant in oil is the same as in the oil microdomain of the microemulsions [69]. Using this approach, the distribution of surfactants in the different domains of bicontinuous microemulsions (Figure 9) could be quantified [70–75], even if the complete microstructure of these systems was not completely elucidated.

A very important achievement by the end of the 1990s was the understanding of the complete phase behavior of water/nonionic surfactant/oil systems. By studying the water/poly(oxyethylene) oleyl ether nonionic surfactant system as a function of polyoxyethylene chain length at constant temperature, he reported in 1997 [77] a correlation among phase behavior, packing of the surfactant alkyl chain in the self-organizing structures and the HLB value. Various self-organizing structures were found in that system (Figure 10): hexagonal (H_1, H_2), lamellar ($L\alpha$) and four kinds of isotropic (I_1, V_1, V_{2a}, V_{2b}) liquid crystals, a sponge phase (D_2) as well as aqueous and reverse micellar phases (Wm and Om). The transition from direct to reverse being produced by modulating the surfactant HLB (EO units).

Similar results were obtained in the phase behavior of the water/C_{12}EOn system as a function of EO-chain length, at constant temperature [78]. By investigating the oil-induced change in surfactant layer curvature of water/C_{12}EOn system at

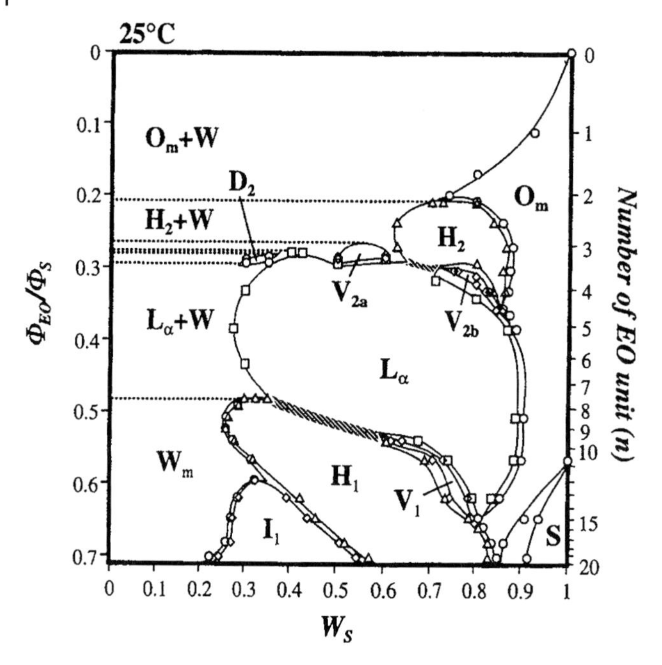

Figure 10 Phase diagram of water/poly(oxyethylene) oleyl ether(POIE) system as a function of the volume fraction of EO chain in the surfactant molecule and weight fraction of POIE at 25 °C (from Ref. [77], with permission).

constant temperature [79] and of other systems [48, 79–90] he showed that the solubilization mechanism is different depending on the type of oil: if oil has a strong tendency to dissolve in the surfactant palisade layer, the effective area per molecule, a_S, becomes large upon addition of oil, and if oil is solubilized in the deep core of aggregates a_S is almost unchanged. Thus, aromatic hydrocarbons or short-chain alkanes have a "penetration" tendency making the curvature less positive, whereas long-chain saturated hydrocarbons have a "swelling" tendency causing a change to more positive surfactant curvature, as schematically shown in Figure 11. Moreover, it was shown that the different oil-solubilization mechanism causes differences in phase transitions in liquid crystals [79, 82]. The effect of electrolytes on liquid-crystalline structures was also object of investigations in nonionic [91] and ionic [92] surfactant systems.

Kunieda investigated thoroughly the phase behavior of a wide variety of nonionic and ionic surfactants, focusing on those environmentally friendly such as sucrose esters [93–99], mono- and polyglycerol fatty acid esters [100–104], amino acid derivatives [105–107], cholesterol- and phytosterol-based polyoxyethylene alkyl ethers [108–113], conjugated esters with a glycerol residue as spacer [114–116] and gemini or dimeric [117–119]. He also investigated the phase behavior of silicone-derived amphiphiles in water and water-oil systems as well as in the molten state,

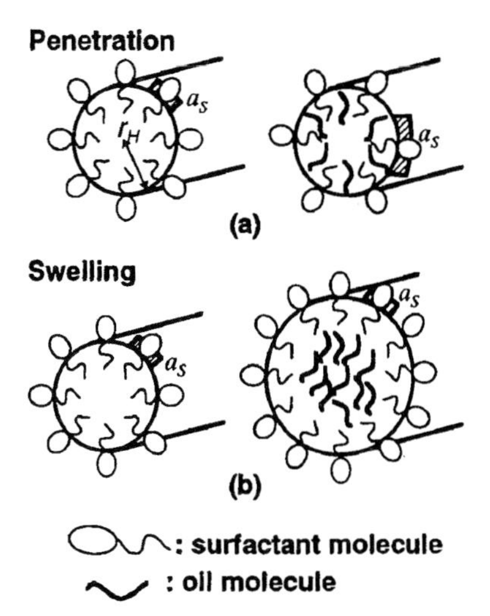

Figure 11 Schematic representation of complete penetration (a) and swelling (b) of oil. In the case of (a), r_H is unchanged while a_S is expanded upon addition of oil. In the case of (b), a_S remains constant whereas r_H increases (from Ref. [86], with permission).

making important contributions [120–129]. Another topic of his interest was the phase behavior and interactions in surfactant block-copolymer systems [130–133]. He clarified the role of the hydrophobic chain on the thermodynamics of self-aggregation and showed that surfactant is dissolved in polymer aggregates, but polymer is not dissolved in surfactant aggregates. Other surfactants investigated include fluorinated [134–136] and alkanoyl N-alkanolamides [111, 137, 138], the latter well known as foam boosters and thickening agents. He had investigated the foaming properties of surfactant solutions in earlier research [139, 140] and also more recently [119, 141–143].

A remarkable contribution in recent years was to have shown for the first time the formation of highly viscoelastic worm-like micelles (Figure 12) in mixed nonionic surfactant systems [110]. This finding allowed to clarify the relation between packing constraints of hydrophobic chains and micellar growth because the complex interactions between counterions (present in ionic surfactant systems) and headgroups had not to be taken into consideration.

He showed the formation of viscoelastic worm-like micelles in various nonionic and ionic surfactant systems and described the evolution of micellar growth namely by rheology and small-angle X ray scattering [110, 112, 118, 133, 137, 144–155]. Zero-shear viscosities 3×10^7 times that of water was reported for certain systems [118].

The deep and extensive knowledge on surfactant phase behavior inspired new research lines. In this context, the first report on the preparation of mesoporous

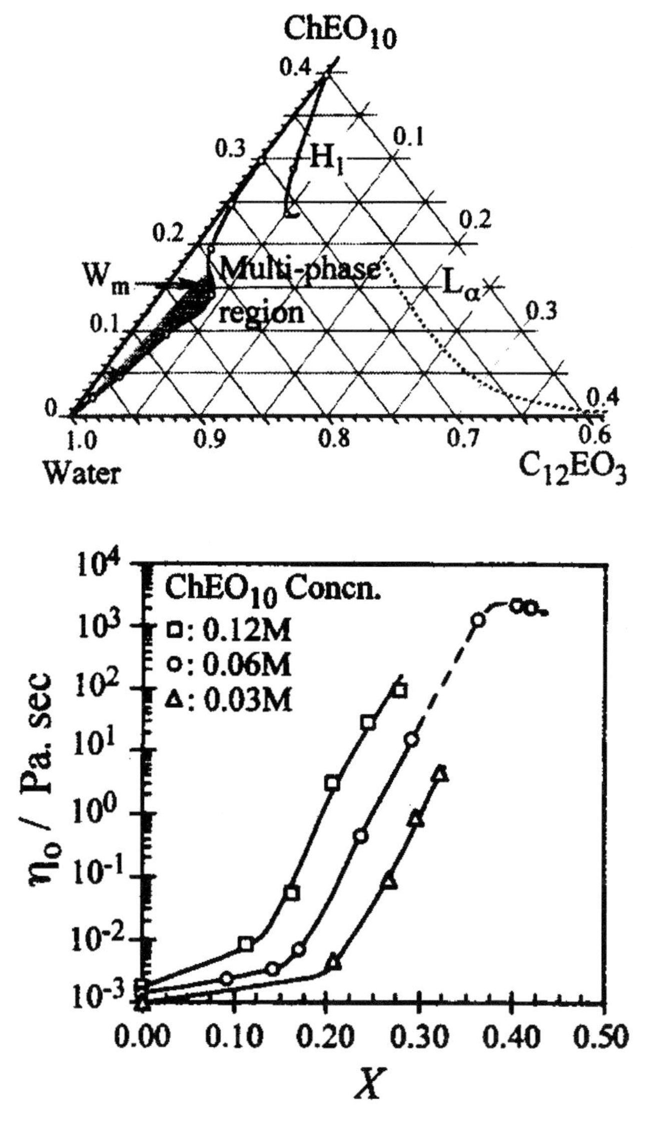

Figure 12 (a) Partial phase diagram of the water/polyoxyethylene cholesteryl ether($ChEO_{10}$)-$C_{12}EO_3$ system, in the dilute region at 25 °C, (b) Variation of zero-shear viscosity (η_0) with the mole fraction of $C_{12}EO_3$, X, at various concentrations of $ChEO_{10}$ in the system. (From Ref. [110], with permission).

silica from anionic surfactant systems appeared in 2003 [156]. He had an astonishing ability to deduce a lot of microscopic information from few macroscopic data about how surfactants behave in the presence of oil and/or water. Without doubt, his research will inspire future generations of scientists.

References

1 Kunieda, H., and Shinoda, K. (1972) Factors to increase the mutual solubility of oil and water by solubilizer. *J. Chem. Soc. Jpn.*, **11**, 2001–2006 (in Japanese).

2 Shinoda, K., and Kunieda, H. (1973) Conditions to produce so-called microemulsions: factors to increase the mutual solubility of oil and water by solubilizer. *J. Colloid Interface Sci.*, **42** (2), 381–387.

3 Kunieda, H. (1977) The mechanism of dissolution of quaternary ammonium chlorides containing two long-chain alkyl groups in water, Nikka. *J. Chem. Soc. Jpn.*, **2**, 151–156 (in Japanese).

4 Kunieda, H., and Shinoda, K. (1978) Solution behavior of dialkyldimethylammonium chloride in water: basic properties of antistatic softeners. *J. Phys. Chem.*, **82** (15), 1710–1714.

5 Kunieda, H., and Shinoda, K. (1978) Solution behavior of dialkyldimethylammonium chloride in water: basic properties of antistatic softeners. *J. Phys. Chem.*, **82** (15), 1710–1714.

6 Kunitake, T., and Okahata, Y. (1977) A totally synthetic bilayer membrane. *J. Am. Chem. Soc.*, **99**, 3860.

7 Kunieda, H., and Hyakutake, M. (1978) The effect of types of solvents and salts added on the solubilization of water in aerosol OT nonaqueous solutions. *J. Jpn. Oil Chem. Soc.*, **27** (9), 598–601 (in Japanese).

8 Kunieda, H., and Shinoda, K. (1979) Solution behavior of aerosol OT/water/oil system. *J. Colloid Interface Sci.*, **70** (3), 577–583.

9 Kunieda, H., and Sato, T. (1979) Ternary phase diagram for aerosol OT/water/oil system, a basic study on microemulsion. *J. Jpn. Oil Chem. Soc.*, **28**, 627–631 (in Japanese).

10 Kunieda, H., and Shinoda, K. (1980) Solution behavior and hydrophile–lipophile balance temperature in aerosol OT/isooctane/brine system: correlation between microemulsions and ultralow interfacial tensions. *J. Colloid Interface Sci.*, **75**, 601–606.

11 Shinoda, K., and Kunieda, H. (1987) The effect of salt concentration, temperature, and additives on the solvent property of aerosol OT solution. *J. Colloid Interface Sci.*, **118**, 586–589.

12 Kunieda, H. (1983) Phase equilibria in an ionic surfactant/brine/cosurfactant/oil system. *J. Jpn. Oil Chem. Soc.*, **32**, 393–396 (in Japanese).

13 Kunieda, H., and Friberg, S.E. (1981) Critical phenomena in a surfactant/water/oil system: basic study on the correlation between solubilization, microemulsion, and ultralow interfacial tensions. *Bull. Chem. Soc. Jpn.*, **54**, 1010–1014.

14 Shinoda, K., Hanrin, M., Kunieda, H., and Saito, H. (1981) Principles of attaining ultra-low interfacial tension: the role of hydrophile–lipophile-balance of surfactant at oil/water interface. *Colloids Surf.*, **2**, 301–314.

15 Kunieda, H., and Shinoda, K. (1982) Correlation between critical solution phenomena and ultralow interfacial tensions in a surfactant/water/oil system. *Bull. Chem. Soc. Jpn.*, **55**, 1777–1781.

16 Kunieda, H., and Shinoda, K. (1982) Phase behavior in systems of nonionic surfactant/water/oil around the hydrophile–lipophile-balance-temperature (HLB-temperature). *J. Dispersion Sci. Technol.*, **3**, 233–244.

17 Shinoda, K., Kunieda, H., Arai, T., and Saijo, H. (1984) Principles of attaining very large solubilization (microemulsion): inclusive understanding of the solubilization of oil and water in aqueous and hydrocarbon media. *J. Phys. Chem.*, **88**, 5126–5129.

18 Shinoda, K., Kunieda, H., Obi, N., and Friberg, S.E. (1981) Similarity in phase diagrams between ionic and nonionic surfactant solutions at constant temperature. *J. Colloid Interface Sci.*, **80**, 304–305.

19 Kunieda, H., and Shinoda, K. (1983) Phase behavior and tricritical phenomena in a bile salt system. *Bull. Chem. Soc. Jpn.*, **56**, 980–984.

20 Kunieda, H., and Arai, T. (1984) The tricritical point in the n-$C_8H_{17}SO_3Na$/

n-$C_4H_9OH/H_2O/n$-$C_{14}H_{30}$ system. *Bull. Chem. Soc. Jpn.*, **57**, 281–282.

21 Kunieda, H. (1987) Phase behavior and ultralow interfacial tensions around a tricritical point in a sodium taurocholate system. *J. Colloid Interface Sci.*, **116**, 224–229.

22 Kunieda, H. (1988) Tricritical phenomena in a brine/ionic surfactant/cosurfactant/oil system. *J. Colloid Interface Sci.*, **122**, 138–142.

23 Yamaguchi, Y., Aoki, R., Azemar, N., Solans, C., and Kunieda, H. (1999) Phase behavior of cationic microemulsions near the tricritical point. *Langmuir*, **15**, 7438–7445.

24 Kunieda, H., and Shinoda, K. (1985) Evaluation of the hydrophile–lipophile balance (HLB) of nonionic surfactants. I. Multisurfactant systems. *J. Colloid Interface Sci.*, **107**, 107–121.

25 Kunieda, H., and Ishikawa, N. (1985) Evaluation of the hydrophile–lipophile balance (HLB) of nonionic surfactants. II. Commercial-surfactant systems. *J. Colloid Interface Sci.*, **107**, 122–128.

26 Kunieda, H., Hanno, K., Yamaguchi, S., and Shinoda, K. (1985) The three-phase behaviour of a brine/ionic surfactant/nonionic surfactant/oil system: evaluation of the hydrophile–lipophile balance (HLB) of ionic surfactant. *J. Colloid Interface Sci.*, **107**, 129–137.

27 Kunieda, H., Asaoka, H., and Shinoda, K. (1988) Two types of surfactant phases and four coexisting liquid phases in a water/nonionic surfactant/triglyceride/ hydrocarbon system. *J. Phys. Chem.*, **92**, 185–189.

28 Yamaguchi, S., and Kunieda, H. (1988) Phase behavior of two types of isotropic surfactant phases in a brine/sodium dodecyl sulfate/hexanol/hexadecane system. *J. Jpn. Oil Chem. Soc.*, **37**, 648–653 (in Japanese).

29 Kunieda, H., Yago, K., and Shinoda, K. (1989) Two types of biosurfactant phases in a bile salt system. *J. Colloid Interface Sci.*, **128**, 363–369.

30 Kunieda, H., and Miyajima, A. (1989) Anomalous three-phase behavior in a water/ octaethyleneglycol dodecyl ether/decanol system. *J. Colloid Interface Sci.*, **129**, 554–560.

31 Kunieda, H., and Haishima, K. (1990) Overlapping of three-phase regions in a water/nonionic surfactant/triglyceride system. *J. Colloid Interface Sci.*, **140**, 383–390.

32 Kunieda, H. (1986) Correlation between the formation of a lamellar liquid-crystalline phase and the hydrophile–lipophile balance (HLB) of surfactants. *J. Colloid Interface Sci.*, **114**, 378–385.

33 Kunieda, H. (1989) Phase behaviors in water/nonionic surfactant/hydrocarbon and water/nonionic surfactant/amphiphilic oil system. *J. Colloid Interface Sci.*, **133**, 237–243.

34 Kunieda, H., and Nakamura, K. (1991) Azeotropic and critical points in a brine/ionic surfactant/long-chain alcohol system. *J. Phys. Chem.*, **95**, 1425–1430.

35 Kunieda, H., Nakamura, K., and Uemoto, A. (1992) Effect of added oil on the phase behavior in a water/ionic surfactant/alcohol system. *J. Colloid Interface Sci.*, **150**, 235–242.

36 Kunieda, H., and Miyajima, A. (1989) Anomalous Three-Phase Behavior in a Water/Octaethyleneglycol Dodecyl Ether/Decanol system. *J. Colloid Interface Sci.*, **129**, 554–560.

37 Kunieda, H., and Miyajima, A. (1989) The effect of the mixing of oils on the hydrophile–lipophile-balanced (HLB) temperature in a water/non-ionic surfactant/oil system. *J. Colloid Interface Sci.*, **128**, 605–607.

38 Kunieda, H., Nakamura, K., and Evans, F. (1991) Formation of reversed vesicles. *J. Am. Chem. Soc.*, **113**, 1051–1052.

39 Kunieda, H., Nakamura, K., Davis, H.T., and Evans, D.F. (1991) Formation of vesicles and microemulsions at HLB temperature. *Langmuir*, **7**, 1915–1919.

40 Kunieda, H., and Yamagata, M. (1992) Conditions to produce reversed vesicles. *J. Colloid Interface Sci.*, **150**, 277–280.

41 Nakamura, K., Machiyama, Y., and Kunieda, H. (1992) Formation of reversed vesicles in a mixture of ionic and nonionic amphiphiles. *J. Jpn. Oil Chem. Soc.*, **41**, 480–484 (in Japanese).

42 Kunieda, H., Akimura, M., and Nakamura, N.U.K. (1993) Reversed

vesicles: counter structure of biological membranes. *J. Colloid Interface Sci.*, **156**, 446–453.
43 Kunieda, H., Nakamura, K., Olsson, U., and Lindman, B. (1993) Spontaneous formation of reverse vesicles. *J. Phys. Chem.*, **97**, 9525–9531.
44 Kunieda, H., Makino, S., and Ushio, N. (1991) Anionic reversed vesicles. *J. Colloid Interface Sci.*, **147**, 286–288.
45 Kunieda, H., Nakamura, K., Infante, M.R., and Solans, C. (1992) Reversed vesicles by bio-compatible surfactants. *Adv. Mater.*, **4**, 291–293.
46 Ushio, N., Solans, C., Azemar, N., and Kunieda, H. (1993) Formation and stability of reverse vesicles in a sucrose alkanoate system. *J. Jpn. Oil Chem. Soc.*, **42**, 915–922 (in Japanese).
47 Kunieda, H., Kanei, N., Uemoto, A., and Tobita, I. (1994) Structure of reverse vesicles in a sucrose monoalkanoate system. *Langmuir*, **10**, 4006–4011.
48 Kunieda, H., Shigeta, K., and Suzuki, M. (1999) Phase behavior and formation of reverse vesicles in long-polyoxyethylene-chain nonionic surfactant systems. *Langmuir*, **15**, 3118–3122.
49 Nakamura, K., Uemoto, A., Imae, T., Solans, C., and Kunieda, H. (1995) Stability and size control of reverse vesicles. *J. Colloid Interface Sci.*, **170**, 367–373.
50 Olsson, U., Nakamura, K., Kunieda, H., and Strey, R. (1996) Normal and reverse vesicles with nonionic surfactant: solvent diffusion and permeability. *Langmuir*, **12**, 3045–3054.
51 Kunieda, H., and Rajagopalan, V. (1996) Formation and structure of reverse vesicles, in *Vesicles* (ed. M. Rossof), Dekker, New York, pp. 79–103.
52 Kunieda, H., Solans, C., Shida, N., and Parra, J.L. (1987) The formation of gel emulsions in a water/nonionic surfactant/oil system. *Colloids Surf.*, **24**, 225–237.
53 Kunieda, H., Yano, N., and Solans, C. (1989) The stability of gel-emulsions in a water/non-ionic surfactant/oil system. *Colloid Surf.*, **36**, 313–322.
54 Kunieda, H., Evans D.F., Solans, C., and Yoshida, M. (1990) The structure of gel-Emulsions in a water/nonionic surfactant/oil system. *Colloid Surf.*, **47**, 35–43.
55 Solans, C., Pons, R., Zhu, S., Davis, H.T., Evans, D.F., Nakamura, K., and Kunieda, H. (1993) Studies on macro-and microstructures of highly concentrated water-in-oil emulsions (gel emulsions). *Langmuir*, **9**, 1479–1482.
56 Rajagopalan, V., Solans, C., and Kunieda, H.E.S.R. (1994) Study on the stability of W/O gel emulsions. *Colloid Polym. Sci.*, **272**, 1166–1173.
57 Kunieda, H., Rajagopalan, V., Kimura, E., and Solans, C. (1994) Nonequilibrium structure of water in oil gel emulsions. *Langmuir*, **10**, 2570–2577.
58 Kunieda, H., Fukui, Y., Uchiyama, H., and Solans, C. (1996) Spontaneous formation of highly concentrated water-in-oil emulsions (gel-emulsions). *Langmuir*, **12**, 2136–2140.
59 Ozawa, K., Solans, C., and Kunieda, H. (1997) Spontaneous formation of highly concentrated oil-in-water emulsions. *J. Colloid Interface Sci.*, **188**, 275–281.
60 Kunieda, H., Ogawa, E., Kihara, K., and Tagawa, T. (1997) Formation of highly concentrated emulsions in water/sucrose dodecanoate/oil systems. *Colloid Polym. Sci.*, **105**, 239–243.
61 Rodríguez, C., Shigeta, K., and Kunieda, H. (2000) Cubic-phase-based concentrated emulsions. *J. Colloid Interface Sci.*, **223**, 197–204.
62 Kunieda, H., Ozawa, K., and Huang, K.-L. (1998) Effect of oil on the surfactant molecular curvatures in liquid crystals. *J. Phys. Chem.*, **102**, 831–838.
63 Uddin, M.H., Kanei, N., and Kunieda, H. (2000) Solubilization and emulsification of perfume in discontinuous cubic phase. *Langmuir*, **16**, 6891–6897.
64 Uddin, M.H., Rodriguez, C., Watanabe, K., Lopez-Quintela, A., Kato, T., Furukawa, H., Harashima, A., and Kunieda, H. (2001) Phase Behavior and formation of reverse-cubic-phase-based emulsion in water/poly(oxyethylene) poly (dimethylsiloxane) surfactants/silicone oil systems. *Langmuir*, **17**, 5169–5175.

65 Watanabe, K., Kanei, N., and Kunieda, H. (2002) Highly Concentrated emulsions based on the reverse-micellar-cubic phase. *J. Oleo Sci.*, **50**, 771–779.

66 Kunieda, H., and Yamagata, M. (1993) Mixing of nonionic surfactants at water-oil interfaces in microemulsions. *Langmuir*, **9**, 3345–3351.

67 Kunieda, H., and Yamagata, M. (1993) Three-phase behavior in a mixed nonionic surfactant system. *Colloid Polym. Sci.*, **271**, 997–1004.

68 Kunieda, H., Ushio, N., Nakano, A., and Miura, M. (1993) Three-phase behavior in a mixed sucrose alkanoate and polyethyleneglycol alkyl ether system. *J. Colloid Interface Sci.*, **159**, 37–44.

69 Kunieda, H., Nakano, A., and Akimaru, M. (1995) The effect of mixing of surfactants on solubilization in a microemulsion system. *J. Colloid Interface Sci.*, **170**, 78–84.

70 Kunieda, H., Nakamo, A., and Pes, M.A. (1995) Effect of oil on the solubilization in microemulsion systems including non-ionic surfactant mixtures. *Langmuir*, **11**, 3302–3306.

71 Kunieda, H., and Aoki, R. (1996) Effect of added salt on the maximum solubilisation in ionic-surfactant microemulsion. *Langmuir*, **12**, 5796–5799.

72 Pes, M.A., Aramaki, K., Nakamura, N., and Kunieda, H. (1996) Temperature-insensitive microemulsions in a sucrose monoalkanoate system. *J. Colloid Interface Sci.*, **178**, 666–672.

73 Aramaki, K., Ozawa, K., and Kunieda, H. (1997) Effect of temperature on the phase behavior of ionic-nonionic microemulsions. *J. Colloid Interface Sci.*, **196**, 74–78.

74 Kunieda, H., Ozawa, K., Aramaki, K., Nakano, A., and Solans, C. (1998) Formation of microemulsions in mixed ionic-nonionic surfactant systems. *Langmuir*, **14**, 260–263.

75 Nakamura, N., Tagawa, T., Kihara, K., Tobita, I., and Kunieda, H. (1997) Phase transition between microemulsion and lamellar liquid crystal. *Langmuir*, **13**, 2001–2006.

76 Li, X., Ueda, K., and Kunieda, H. (1999) Solubilization and phase behavior of microemulsions with mixed anionic–cationic surfactants and hexanol. *Langmuir*, **15**, 7973–7979.

77 Kunieda, H., Shigeta, K., Ozawa, K., and Suzuki, M. (1997) Self-organizing structures in poly(oxyethylene) oleyl ether-water system. *J. Phys. Chem. B*, **101**, 7952–7957.

78 Huang, L., Shigeta, K., and Kunieda, H. (1998) Phase behavior of polyoxyethylene dodecyl ether-water systems. *Prog. Colloid Polym. Sci.*, **110**, 171–174.

79 Kunieda, H., Ozawa, K., and Huang, K.L. (1998) Effect of oil on the surfactant molecular curvatures in liquid crystals. *J. Phys. Chem. B*, **102**, 831–838.

80 Aramaki, K., and Kunieda, H. (1999) Solubilization of oil in mixed cationic liquid crystal. *Colloid Polym. Sci.*, **277**, 34–40.

81 Kanei, N., Tamura, Y., and Kunieda, H. (1999) Effect of types of perfumes on the HLB temperature. *J. Colloid Interface Sci*, **218**, 13–22.

82 Kunieda, H., Umizu, G., and Aramaki, K. (2000) Effect of mixing oils on the hexagonal liquid crystalline structures. *J. Phys. Chem. B*, **104**, 2005–2011.

83 Shigeta, K., Rodríguez, C., and Kunieda, H. (2000) Solubilization of oil in discontinuous cubic liquid crystal in poly(oxyethylene) oleyl ether systems. *J. Dispersion Sci. Technol.*, **21**, 1023–1042.

84 Li, X., and Kunieda, H. (2000) Cubic-phase microemulsions with anionic and cationic surfactants at equal amounts of oil and water. *J. Colloid Interface Sci.*, **231**, 143–151.

85 Li, X., and Kunieda, H. (2000) Solubilization of micellar cubic phases and their structural relationships in the systems anionic–cationic surfactant–dodecane–water. *Langmuir*, **16**, 10092–10100.

86 Kunieda, H., Horii, M., Koyama, M., and Sakamoto, K. (2001) Solubilization of polar oils in surfactant self-organized structures. *J. Colloid Interface Sci.*, **236**, 78–84.

87 Ozawa, K., Olsson, U., and Kunieda, H. (2001) Oil-induced structural change in nonionic microemulsions. *J. Dispersion Sci. Technol.*, **22**, 119–124.

88 Kumar, A., Kunieda, H., Rodríguez, C., and López-Quintela, M.A. (2001) Studies of domain size of hexagonal liquid crystals in C12EO8/water/alcohol systems. *Langmuir*, **17**, 7245–7250.

89 Aramaki, K., Kabir, H., Nakamura, N., and Kunieda, H. (2001) Formation of oil swollen cubic phase or cubic-phase microemulsion in sucrose alkanoate systems. *Colloids Surf. A*, **183–185**, 371–379.

90 Kanei, N., Watanabe, K., and Kunieda, H. (2003) Effect of added perfume on the stability of discontinuous cubic phase. *J. Oleo Sci.*, **52** (11), 607–619.

91 Iwanaga, T., Suzuki, M., and Kunieda, H. (1998) Effect of added salts or polyols on the liquid crystalline structures of polyoxyethylene-type nonionic surfactants. *Langmuir*, **14**, 5775–5781.

92 Rodríguez, C., and Kunieda, H. (2000) Effect of electrolytes on discontinuous cubic phases. *Langmuir*, **16**, 8263–8269.

93 Aramaki, K., Kunieda, H., Ishitobi, M., and Tagawa, T. (1997) Effect of added salt on three-phase behavior in a sucrose monoalkanoate system. *Langmuir*, **13**, 2266–2270.

94 Nakamura, N., Yamaguchi, Y., Håkansson, B., Olsson, U., Tagawa, T., and Kunieda, H. (1999) Formation of microemulsion and liquid crystal in biocompatible sucrose alkanaote systems. *J. Dispersion Sci. Technol.*, **20**, 535–557.

95 Aramaki, K., Hayashi, T., Katsuragi, T., Ishitobi, M., and Kunieda, H. (2001) Effect of adding an amphiphilic solubilization-improver sucrose distearates on the solubilization capacity of nonionic microemulsions. *J. Colloid Interface Sci.*, **236**, 14–19.

96 Kanei, N., Kunieda, H. (2000) Phase behavior of water/sucrose dodecanoate/ perfume systems. *J. Jpn. Oil Chem. Soc.*, **49**, 957–966.

97 Rodríguez, C., Acharya, D.P., Hinata, S., Ishitobi, M., and Kunieda, H. (2003) Effect of ionic surfactants on the phase behavior and structure of microemulsion in sucrose fatty acid systems. *J. Colloid Interface Sci.*, **262**, 500–505.

98 Kabir, M.H., Aramaki, K., Ishitobi, M., and Kunieda, H. (2003) Cloud point and formation of microemulsions in sucrose dodecanoate systems. *Colloid Surf. A*, **216**, 65–74.

99 Rodríguez, C., Aramaki, K., Tanaka, Y., López-Quintela, M.A., Ishitobi, M., and Kunieda, H. (2005) Worm-like micelles and microemulsions in aqueous mixtures of sucrose esters and nonionic cosurfactants. *J. Colloid Interface Sci.*, **291**, 560–569.

100 Ishitobi, M., and Kunieda, H. (2000) Effect of distribution of hydrophilic chain on the phase behavior of polyglycerol fatty acid ester in water. *Colloid Polym. Sci.*, **278**, 899–904.

101 Kunieda, H., Akahane, A., and Feng, J. (2002) Ishitobi, Phase behavior of polyglycerol didodecanoates in water. *J. Colloid Interface Sci.*, **245**, 365–370.

102 Shrestha, L.K., Kaneko, M., Sato, T., Acharya, D.P., Iwanaga, T., and Kunieda, H. (2006) Phase behavior of diglycerol fatty acid esters–non polar oil systems. *Langmuir*, **22**, 1449–1454.

103 Izquierdo, P., Acharya, D.P., Hirayama, K., Asaoka, H., Ihara, K., Tsunehiro, T., Shimada, Y., Asano, Y., Kokubo, S., and Kunieda, H. (2006) Phase behavior of pentaglycerol monostearic and monooleic acid esters in water. *J. Dispersion Sci. Technol.*, **27**, 99–103.

104 Shrestha, L.K., Sato, T., Acharya, D.P., Iwanaga, T., Aramaki, K., and Kunieda, H. (2006) Phase behavior of monoglycerol fatty acid esters in nonpolar oils: reverse rodlike micelles at elevated temperatures. *J. Phys. Chem. B*, **110**, 12266–12273.

105 Yamashita, Y., Kunieda, H., Oshimura, E., and Sakamoto, K. (2003) Phase behavior of *N*-acylamino acid surfactant and *N*-acylamino acid oil in water. *Langmuir*, **19**, 4070–4078.

106 Acharya, D., Lopez-Quitela, M.A., Kunieda, H., Oshimura, E., and Sakamoto, K. (2003) Phase behavior and effect of enantiomerism on potassium *N*-dodecanoyl alaninate/water/decanol systems. *J. Oleo Sci.*, **52**, 407–420.

107 Kunieda, H., Matsuzawa, K., Makhkamov, R., Horii, M., Yamashita, Y., Yumioka, R., Koyama, M., and Sakamoto, K. (2003) Effect of amino-acid-based polar oils on the Krafft point and solubilization in ionic and nonionic surfactant solutions. *J. Dispersion Sci. Technol.*, **24** (6), 767–772.

108 Rodríguez, C., Naito, N., and Kunieda, H. (2001) Structure of vesicles in homogeneous short-chain polyoxyethylene cholesterol ether systems. *Colloids Surf. A*, **181**, 237–246.

109 Lopez-Quintela, M.A., Akahane, A., Rodriguez, C., and Kunieda, H. (2002) Thermotropic behavior of poly(oxyethylene) cholesterol ether surfactants. *J. Colloid Interface Sci.*, **247**, 186–192.

110 Acharya, D.P., and Kunieda, H. (2003) Formation of viscoelastic wormlike micellar solutions in mixed nonionic surfactant systems. *J. Phys. Chem. B*, **107**, 10168–10175.

111 Hossain, M.K., Acharya, D.P., Sakai, T., and Kunieda, H. (2004) Phase behavior of poly (oxyethylene) cholesteryl ether/novel alkanolamide/ water systems. *J. Colloid Interface Sci.*, **277**, 235–242.

112 Naito, N., Acharya, D.P., Tanimura, J., and Kunieda, H. (2004) Rheological behavior of wormlike micellar solutions in mixed nonionic systems of polyoxyethylene phytosterol–polyoxyethylene dodecyl ether. *J. Oleo Sci.*, **53**, 599–606.

113 Naito, N., Acharya, D.P., Tanimura, J., and Kunieda, H. (2005) Phase behavior of polyoxyethylene phytosteryl ether/ polyoxyethylene dodecyl ether/water systems. *J. Oleo Sci.*, **54**, 7–13.

114 Feng, J., Aramaki, K., Ogawa, A., Katsuragi, T., and Kunieda, H. (2001) Phase behavior and solution properties of sodium 3-(3-dodecanoyloxy-2-hydroxypropoxycarbonyl)- propionate in water. *Colloid Polym. Sci.*, **279**, 92–97.

115 Feng, J., Ogawa, A., Tsukahara, M., and Kunieda, H. (2002) Formation of microemulsion in NaCl aq/sodium (3-dodecanoyloxy-2-hydroxy-propyl) succinate/ glycerol mono(2-ethylhexyl) ether/oil systems. *J. Dispersion Sci. Technol.*, **23**, 29–36.

116 Kunieda, H., Masuda, N., and Tsubone, K. (2000) Comparison between phase behavior of anionic dimeric (Gemini-type) and monomeric surfactants in water and water–oil. *Langmuir*, **16**, 6438–6444.

117 Kunieda, H., Kaneko, M., Feng, J., and Tsubone, K. (2002) Formation of microemulsions with Gemini-type surfactant. *J. Oleo Sci.*, **51**, 761–769.

118 Acharya, D.P., Kunieda, H., Shiba, Y., and Aratani, K. (2004) Phase and rheological behavior of novel Gemini-type surfactant systems. *J. Phys. Chem. B.*, **108**, 1790–1797.

119 Acharya, D.P., Gutiérrez, J.M., Aramaki, K., Aratani, K., and Kunieda, H. (2005) Interfacial properties and foam stability effect of novel gemini-type surfactant in aqueous solutions. *J. Colloid Interface Sci.*, **291**, 236–243.

120 Kunieda, H., Taoka, H., Iwanaga, T., and Harashima, A. (1998) Phase behavior of polyoxyethylene trisiloxane surfactant in water and water–oil. *Langmuir*, **14**, 5113.

121 Iwanaga, T., and Kunieda, H. (2000) Effect of added salts or polyols on the cloud point and the liquid crystalline structures of polyoxyethylene modified silicone. *J. Colloid Interface Sci.*, **227**, 349–355.

122 Kunieda, H., Uddin, M.H., Horii, M., and Harashima, A. (2001) Effect of hydrophilic and hydrophobic-chain lengths on the phase behavior of A-B-type silicone surfactants in water. *J. Phys. Chem. B*, **105**, 5419–5426.

123 Uddin, M.H., Rodriguez, C., Watanabe, K., Lopez-Quintela, A., Kato, T., Furukawa, H., Harashima, A., and Kunieda, H. (2001) Phase Behavior and formation of reverse-cubic-phase-based emulsion in water/poly(oxyethylene) poly (dimethylsiloxane) surfactants/ silicone oil systems. *Langmuir*, **17**, 5169–5175.

124 Kumar, A., Uddin, M.H., Kunieda, H., Furukawa, H., and Harashima, A. (2001) Solubilization enhancing effect of A–B-type silicone surfactants in microemulsions. *J. Dispersion Sci. Technol.*, **22**, 245–253.

125 Rodriguez, C., Uddin, M.H., Watanabe, K., Furukawa, H., Harashima, A., and Kunieda, H. (2002) Self-organization, phase behavior, and microstructure of poly(oxyethylene) poly(dimethyl siloxane) surfactants in nonpolar oil. *J. Phys. Chem. B*, **106**, 22–29.
126 Kunieda, H., Uddin, M.H., Yamashita, Y., Furukawa, H., and Harashima, A. (2002) Microemulsions in poly(dimethyl siloxane)–poly(oxyethylene) copolymer (or surfactant) systems. *J. Oleo Sci.*, **51**, 113–122.
127 Kunieda, H., Uddin, M.H., Furukawa, H., and Harashima, A. (2001) Phase behavior of a mixture of poly(oxyethylene)–poly(dimethyl siloxane) copolymer and nonionic surfactant in water. *Macromolecules*, **34**, 9093–9099.
128 Uddin, M.H., Rodríguez, C., López-Quintela, A., Leisner, D., Solans, C., Esquena, J., and Kunieda, H. (2003) Phase behavior and microstructure of poly(oxyethylene)–poly(dimethylsiloxane) copolymer melt. *Macromolecules*, **36**, 1261–1271.
129 Uddin, M.H., Morales, D., and Kunieda, H. (2005) Phase polymorphism by mixing of poly(oxyethylene)–poly(dimethylsiloxane) copolymer and nonionic surfactant in water. *J. Colloid Interface Sci.*, **285**, 373–381.
130 Hossain, M.K., Hinata, S., López-Quintela, A., and Kunieda, H. (2003) Phase behavior of poly(oxyethylene)–poly(oxypropylene)–poly(oxyethylene) block copolymer in water and water–C12EO5 systems. *J. Dispersion Sci. Technol.*, **24**, 441–422.
131 Kunieda, H., Kaneko, M., López-Quintela, M.A., and Tsukahara, M. (2003) Phase behavior of a mixture of poly(isoprene)–poly(oxyethylene) diblock copolymer and poly(oxyethylene) surfactant in water. *Langmuir*, **20**, 2164–2171.
132 Aramaki, K., Hossain, M.K., Rodriguez, C., Uddin, M.H., and Kunieda, H. (2003) Miscibility of block copolymers and surfactants in lamellar liquid crystals. *Macromolecules*, **36**, 9443–9450.
133 Rodríguez-Abreu, C., Acharya, D.P., Aramaki, K., and Kunieda, H. (2005) Structure and rheology of direct and reverse liquid-crystal phases in a block copolymer/water/oil system. *Colloid Surf. A*, **269**, 59–66.
134 Kunieda, H., and Shinoda, K. (1976) Krafft points, critical micelle concentrations, surface tension, and solubilizing power of aqueous solutions of fluorinated surfactant. *J. Phys. Chem.*, **80** (22), 2468–2470.
135 Rodriguez, C., Kunieda, H., Noguchi, Y., and Nakaya, T. (2001) Surface tension properties of novel phosphocholine-based fluorinated surfactants. *J. Colloid Interface Sci.*, **242**, 255–258.
136 Sharma, S.C., Acharya, D.P., García-Roman, M., Itami, Y., and Kunieda, H. (2006) Phase behavior and surface tension of amphiphilic fluorinated random copolymer aqueous solutions. *Colloid Surf. A*, **280**, 140–145.
137 Rodríguez, C., Fujiyama, R., Sakai, T., and Kunieda, H. (2003) Phase behavior and microstructure of alkanolamide/surfactant systems. *J. Colloid Interface Sci.*, **270**, 229–235.
138 Feng, J., Kunieda, H., Izawa, T., and Sakai, T. (2004) Effect of novel alkanolamides on the phase behavior and surface properties of aqueous surfactant solutions. *J. Dispersion Sci. Technol*, **25**, 1–10.
139 Friberg, S.E., Blute, I., Kunieda, H., and Stenius, P. (1986) Stability of hydrophobic foams. *Langmuir*, **2**, 659–664.
140 Kunieda, H., and Friberg, S.E. (1986) Foams from a three-phase emulsion. *Colloids Surf.*, **21**, 17–26.
141 Kanei, N., Harigai, T., and Kunieda, H. (2005) Effect of added fragrances on the foaming properties of aqueous surfactant solutions. *J. Soc. Cosmet. Chem. Japan*, **39**, 100–108.
142 Shrestha, L.K., Aramaki, K., Kato, H., Takase, Y., and Kunieda, H. (2006) Foaming properties of monoglycerol fatty acid esters in nonpolar oil systems. *Langmuir*, **22**, 8337–8345.
143 Kunieda, H., Shrestha, L.K., Acharya, D.P., Kato, H., Takase, Y., and Gutiérrez, J.M. (2007) Super-stable nonaqueous foams in diglycerol fatty

acid esters-non polar oil systems. *J. Dispersion Sci. Technol*, **28**, 133–142.

144 Acharya, D.P., Hattori, K., Sakai, T., and Kunieda, H. (2003) Phase and rheological behavior of salt-free alkyltrimethylammonium bromide/ alkanoyl-nmethylethanolamide/water systems. *Langmuir*, **19**, 9173–9178.

145 Rodríguez, C., Hattori, K., Acharya, D.P., Sakai, T., and Kunieda, H. (2003) Phase and rheological behavior of surfactant/novel alkanolamide/water systems.. *Langmuir*, **19**, 8692–8696.

146 Acharya, D.P., Hossain, M.K., Feng, J., Sakai, T., and Kunieda, H. (2004) Phase and rheological behavior of viscoelastic wormlike micellar solutions formed in mixed nonionic surfactant systems. *Phys. Chem. Chem. Phys.*, **6**, 1627–1631; *Langmuir* **19** (2003) 8692–8696; **24** (2003) 411–422.

147 Rodríguez, C., Acharya, D.P., Maestro, A., Hattori, K., and Kunieda, H. (2004) Effect of nonionic head group size on the formation of worm-like micelles in mixed nonionic/cationic surfactant aqueous systems. *J. Chem. Eng. Soc. Jpn.*, **37**, 622–629.

148 Kunieda, H., Rodríguez, C., Tanaka, Y., and Ishitobi, M. (2004) Effects of added nonionic surfactant and inorganic salt on the rheology of sugar surfactant and CTAB aqueous solutions. *Colloid Surf. B*, **38**, 127–130.

149 Maestro, A., Acharya, D.P., Furukawa, H., Gutierrez, J.M., López-Quintela, M.A., Ishitobi, M., and Kunieda, H. (2004) Formation and disruption of viscoelastic wormlike micellar networks in the mixed-surfactant systems of sucrose alkanoate and polyoxyethylene alkylether. *J. Phys. Chem. B.*, **108**, 14009–14016.

150 Rodríguez-Abreu, C., Garcia-Roman, M., and Kunieda, H. (2004) Rheology and dynamics of micellar cubic phases and related emulsions. *Langmuir*, **22**, 5235–5240.

151 Sato, T., Hossain, M.K., Acharya, D.P., Glatter, O., Chiba, A., and Kunieda, H. (2004) Phase behavior and self-organized structures in water/ poly(oxyethylene) cholesteryl ether systems. *J. Phys. Chem. B*, **108**, 12927–12939.

152 Rodríguez-Abreu, C., Acharya, D.P., Aramaki, K., and Kunieda, H. (2005) Structure and rheology of direct and reverse liquid-crystal phases in a block copolymer/ water/oil system. *Colloid Surf. A*, **269**, 59–66.

153 Acharya, D.P., Sato, T., Kaneko, M., Singh, Y., and Kunieda, H. (2006) Effect of added poly(oxyethylene)dodecyl ether on the phase and rheological behavior of wormlike micelles in aqueous SDS solutions. *J. Phys. Chem. B*, **110**, 754–760.

154 Sato, T., Acharya, D.P., Kaneko, M., Aramaki, K., Singh, Y., Ishitobi, M., and Kunieda, H. (2006) Oil-induced structural change of wormlike micelles in sugar surfactant systems. *J. Dispersion Sci. Technol.*, **27**, 611–616.

155 Engelskirchen, S., Acharya, D.P., Garcia-Roman, M., and Kunieda, H. (2006) Effect of C12EOn mixed-surfactant systems on the formation of viscoelastic wormlike micellar solutions in sucrose alkanoate– and CTAB–water systems. *Colloid Surf. A*, **279**, 113–120.

156 Che, S., Garcia-Bennet, A.E., Yokoi, T., Sakamoto, K., Kunieda, H., Terasaki, O., and Tatsumi, T. (2003) A novel anionic surfactant templating route for synthesizing mesoporous silica with unique structure. *Nat. Mater.*, **2**, 801–805.

List of Contributors

Masahiko Abe
Tokyo University of Science
Faculty of Science and Technology
Department of Pure and Applied Chemistry
2641 Yamazaki, Noda
Chiba 278-8510
Japan
Tokyo University of Science
Institute of Colloid and Interface Science
1-3 Kagurazaka, Shinjuku
Tokyo 162-8601
Japan

Idit Amar-Yuli
The Hebrew University of Jerusalem
The Institute of Chemistry
Casali Institute of Applied Chemistry
Givat Ram Campus
Jerusalem 91904
Israel

Kenji Aramaki
Yokohama National University
Graduate School of Environment and Information Sciences
Tokiwadai 79-7, Hodogaya-ku
Yokohama 240-8501
Japan

Abraham Aserin
The Hebrew University of Jerusalem
The Institute of Chemistry
Casali Institute of Applied Chemistry
Givat Ram Campus
Jerusalem 91904
Israel

Joakim Balogh
Lund University
Division of Physical Chemistry
Center for Chemistry and Chemical Engineering
P.O. Box 124
Getingevägen 60
221 00 Lund
Sweden
University of Coimbra
Department of Chemistry
3004-535 Coimbra
Portugal

Javier Calvo-Fuentes
NANOGAP sub-nm-powder S.A.
R/da Xesta 78-A2
Parque Empresarial Novo Milladoiro
A Coruña
15895 Milladoiro – Ames
Spain

Self-Organized Surfactant Structures. Edited by Tharwat F. Tadros

ISBN: 978-3-527-31990-9

Bradley Chmelka
University of California
Department of Chemical Engineering
1210 Cheadle Hall
Santa Barbara, CA 93106
USA

Xuejun Duan
University of Santiago de Compostela
Department of Physical Chemistry
Laboratory of Magnetism and
Nanotechnology
Campus Universitario Sur
15782 Santiago de Compostela
Spain

Jordi Esquena
Institut de Química Avançada de
Catalunya (IQAC)
Consejo Superior de Investigaciones
Científicas (CSIC)
Jordi Girona 18-26
08034 Barcelona
Spain

Nissim Garti
The Hebrew University of Jerusalem
The Institute of Chemistry
Casali Institute of Applied Chemistry
Givat Ram Campus
Jerusalem 91904
Israel

Carmen González
Barcelona University
Department of Chemical Engineering
Martí i Franquès 1
08028 Barcelona
Spain

José M. Gutiérrez
Barcelona University
Department of Chemical Engineering
Martí i Franquès 1
08028 Barcelona
Spain

Masakatsu Hato
RIKEN Systems and Structural Biology
Center
1-7-22 Suehiro-cho, Tsurumi-ku,
Yokohama
Kanagawa 230-0045
Japan

Heinz Hoffmann
University of Bayreuth
BZKG
95448 Bayreuth
Germany

Toyoko Imae
Nagoya University
Research Center for Materials Science
Furo-cho, Chikusa-ku, Chikusa
Nagoya 464-8602
Japan
National Taiwan University of Science
and Technology
Graduate Institute of Engineering
43 Keelung Road, Section 4
Taipei 10607
Taiwan

Masaya Kaneko
Yokohama National University
Graduate School of Environment and
Information Science
Tokiwadai 79-7, Hodogaya-ku
Yokohama 240-8501
Japan

Helena Kaper
Christian-Albrechts-Universität Kiel
Insitut für Physikalische Chemie
Ludewig-Meyn-Str. 8
24118 Kiel
Germany

Tadashi Kato
Tokyo Metropolitan University
Department of Chemistry
1-1 Minami-Osawa, Hachioji
Tokyo 192-0397
Japan

Youhei Kawabata
Tokyo Metropolitan University
Department of Chemistry
1-1 Minami-Osawa, Hachioji
Tokyo 192-0397
Japan

Hironobu Kunieda
Yokohama National University
Graduate School of Environment and
Information Sciences
79-1 Tokiwadai, Hodogaya-ku,
Hodogaya
Yokohama 240-8501
Japan

Dietrich Leisner
Nagoya University
Research Center for Materials Science
Furo-cho, Chikusa-ku, Chikusa
Nagoya 464-8602
Japan
Metrohm Int. Headquarters
Oberdorfstr. 68
CH 9101 Herisau
Switzerland

Björn Lindman
Physical Chemistry 1
Centre for Chemistry and Chemical
Engineering
University of Lund
221 00 Lund
Sweden

M. Arturo López-Quintela
University of Santiago de Compostela
Department of Physical Chemistry
Laboratory of Magnetism and
Nanotechnology
Campus Universitario Sur
15782 Santiago de Compostela
Spain
Nagoya University
Research Center for Materials Science
Furo-cho, Chikusa-ku, Chikusa
Nagoya 464-8602
Japan
Yokohama National University
Graduate School of Environment and
Information Sciences
79-1 Tokiwadai, Hodogaya-ku,
Hodogaya
Yokohama 240-8501
Japan

Alicia Maestro
Barcelona University
Department of Chemical Engineering
Martí i Franquès 1
08028 Barcelona
Spain

Maria Miguel
University of Coimbra
Department of Chemistry
3004-535 Coimbra
Portugal

Jordi Nolla
Institut d'Investigacions Químiques i Ambientals de Barcelona (IIQAB/CSIC)
Jordi Girona 18-26
08034 Barcelona
Spain

Daisuke Nozu
Tokyo Metropolitan University
Department of Chemistry
1-1 Minami-Osawa, Hachioji
Tokyo 192-0397
Japan

Ulf Olsson
Lund University
Center for Chemistry and Chemical Engineering
Division of Physical Chemistry
P.O. Box 124
Getingevägen 60
221 00 Lund
Sweden

Skov Pedersen Jan
Aarhus University
Interdisciplinary NanoScience Center
Department of Chemistry and iNANO
8000 Aarhus C
Denmark

Carlos Rodríguez-Abreu
Institut de Química Avançada de Catalunya (IQAC)
Consejo Superior de Investigaciones Científicas (CSIC)
Jordi Girona 18-26
08034 Barcelona
Spain
International Iberian Nanotechnology Laboratory (INL)
Avda. Central No. 100
Edificio dos Congregados
4710-229, Braga
Portugal

Takaaki Sato
Waseda University
Faculty of Science & Engineering
Division of Physics and Applied Physics
Okubo 3-4-1, Shinjuku-ku
Tokyo 169-8555
Japan

Karin Schillén
Lund University
Center for Chemistry and Chemical Engineering
Division of Physical Chemistry
P.O. Box 124
Getingevägen 60
221 00 Lund
Sweden

Suraj Chandra Sharma
Tokyo University of Science
Faculty of Science and Technology
Department of Pure and Applied Chemistry
2641 Yamazaki, Noda
Chiba 278-8510
Japan

Yuwen Shen
Shandong University
Key Laboratory of Colloid and Interface Chemistry
Ministry of Education
Jinan 250100
China

Yuka Shimada
Tokyo Metropolitan University
Department of Chemistry
1-1 Minami-Osawa, Hachioji
Tokyo 192-0397
Japan

Wataru Shinoda
National Institute of Advanced
Industrial Science and Technology
(AIST)
Research Institute for Computational
Sciences (RICS) Central 2
1-1-1 Umezono, Tsukuba
Ibaraki 305-8568
Japan

Lok Kumar Shrestha
International Center for Materials
Nanoarchitectonics (MANA)
National Institute for Materials Science
(NIMS)
Namiki, Tsukuba-shi
Ibaraki 305-0044, Tsukuba
Japan

Conxita Solans
Peking University
Institut de Química Avançada de
Catalunya (IQAC)
Consejo Superior de Investigaciones
Científicas (CSIC)
Jordi Girona 18-26
08034-Barcelona
Spain

Md. Hemayet Uddin
Yokohama National University
Graduate School of Environment and
Information Sciences
79-1 Tokiwadai, Hodogaya-ku,
Hodogaya
Yokohama 240-8501
Japan

Håkan Wennerström
Lund University
Center for Chemistry and Chemical
Engineering
Division of Physical Chemistry
P.O. Box 124
Getingevägen 60
221 00 Lund
Sweden

Yun Yan
Peking University
College of Chemistry and Molecular
Engineering
State Key Laboratory for Structural
Chemistry of Unstable and Stable
Species
Beijing National Laboratory for
Molecular Science
Beijing 100871
China

Ying Zhao
Peking University
College of Chemistry and Molecular
Engineering
State Key Laboratory for Structural
Chemistry of Unstable and Stable
Species
Beijing National Laboratory for
Molecular Science
Beijing 100871
China

1
Viscoelastic Worm-Like Micelles in Nonionic Fluorinated Surfactant Systems

Suraj Chandra Sharma, Masahiko Abe, and Kenji Aramaki

1.1
Introduction

Perfluorosurfactants, like their hydrocarbon counterparts, form spherical micelles, rod-like micelles, vesicles, lamellar aggregates, and other various liquid-crystalline structures in solution, and the aggregate structures in fluorinated surfactants can be explained, like hydrocarbon surfactants, in terms of the value of critical packing parameter [1], CPP = $v/a_s l$, where v is the volume of the hydrophobic part, l its length, and a_s the average area of headgroup at the interface. Similar phase sequence is observed in fluorinated surfactant and hydrocarbon surfactant systems when surfactant concentration is changed. However, perfluorosurfactants also exhibit major important differences from hydrocarbon surfactants. They are more hydrophobic and reduce the surface tension of water to an extent, which is in general unattainable with hydrocarbon surfactants. Similarly, they show a much lower critical micelle concentration than the hydrocarbon chain surfactants of the same length do [2, 3]. Perfluorosurfactants have stiff hydrophobic chains, with their skeleton covered by a dense electron-rich environment. It is easier to pack fluorocarbon chains closely because the chains are in the all-*trans* state and less entropy is lost. As a result, these surfactants have high chemical and thermal stability. The good chemical and thermal stability of fluorocarbon surfactants is an important consideration to operate in harsh environment such as extremes of pH, high temperatures or in combination with strong oxidizing or reducing agents. Since the fluorocarbon chains are bulkier than the hydrocarbon chain, with the volume of $-CF_2$ and terminal $-CF_3$ being higher than that of the $-CH_2$ and $-CH_3$, respectively [4], a fluorinated surfactant having a very large headgroup is needed to form a spherical aggregate (CPP $\leq 1/3$), to balance the effect of bulky fluorocarbon chain [5]. Therefore, cylindrical micelles are often observed in fluorinated surfactant systems at solution conditions where spherical micelles are expected in hydrocarbon surfactant systems. These cylindrical micelles often undergo enormous one-dimensional growth and form very long and flexible aggregates, referred to as "worm-like" micelles, for which the spontaneous curvature of the end-caps is higher than the curvature along the cylindrical body. The growth

Self-Organized Surfactant Structures. Edited by Tharwat F. Tadros

ISBN: 978-3-527-31990-9

is therefore a consequence of the system to minimize the excess free energy by reducing the number of end-caps in spite of the counteracting entropy factor. When the number density of the worm-like aggregates exceeds a certain threshold value, they entangle with each other to form a transient network, similar to a solution of flexible polymers and display remarkable viscoelastic properties.

The formation and properties of viscoelastic worm-like micelles have been studied extensively, mostly in long hydrophobic chain cationic surfactant [6–15] in the presence of high concentrations of salt, which screens the electrostatic repulsions between the charged surfactant headgroups. Viscoelastic solutions of worm-like micelles have also been reported in cationic or anionic fluorinated surfactant aqueous systems, even at a relatively short fluorocarbon chain length [16–18]. The effect of the concentration of counterions and surfactant concentration on the rheological behavior and micellar growth is more or less similar in both types of surfactants [18]. There are studies on the theromoresponsive viscoelasticity in some hybrid anionic surfactants containing both fluorocarbon and hydrocarbon chain in their molecules [19–22]. Although the majority of worm-like micelles systems reported are charged, nonionic surfactants, such as ethoxylated sterols [23–25] and sucrose alkanoates [26] can also form worm-like micelles. In these systems, interfacial curvature of the aggregates can be tuned to induce sphere–rod transition and one-dimensional micellar growth via the addition of a lipophilic surfactant, such as polyoxyethylene alkyl ether or long-chain monoglyceride. A similar tendency is observed in ionic surfactants solution as well [27–31]. An aqueous solution of a surfactant with a cholesteric group as the hydrophobic part formed a viscoelastic solution at elevated temperature [32]. Knowledge about the formation and rheological properties of nonoionic systems is important not only to obtain a better understanding of the underlying basic principle of the phenomenon, but also for practical applications such as in cosmetics and toiletry products because of the absence of charged species and improved mildness to the skin.

In this chapter, a brief theoretical background on the rheological behavior of viscoelastic worm-like micelles is given. It is followed by a discussion on the temperature-induced viscosity growth in a water–surfactant binary system of a nonionic fluorinated surfactant at various concentrations. Finally, some recent results on the formation of viscoelastic worm-like micelles in mixed nonionic fluorinated surfactants in an aqueous system are presented.

1.2 Rheological Behavior of Worm-Like Micelles

The viscoelasticity of the worm-like micelles arises because of the entanglement of very long and flexible worm-like micelles to form a transient network, similar to a solution of flexible polymers. Unlike polymers, however, worm-like micelles break and re-form dynamically. When the network of worm-like micelles is deformed or the equilibrium conditions are suddenly changed, the relaxation occurs within a definite time, and the equilibrium condition is restored again. For a deformation with a time period shorter than the relaxation time, τ_R, the system exhibits an elastic property characteristic of a solid material with a Hookean constant, G_o, called the shear modulus. For a slow deformation, however, the network has sufficient time to dissipate the stress, and the viscoelastic system behaves as a viscous fluid with a zero-shear viscosity, η_o.

The rheological behavior of a viscoelastic material can be investigated by applying a small-amplitude sinusoidal deformation. The behavior can be described by a mechanical model, called the Maxwell model [33], consisting of an elastic spring with the Hookean constant, G_o, and a dashpot with the viscosity, η_o. The variation of storage modulus (G') and loss modulus (G'') with shear frequency, ω, are given by the equations

$$G'(\omega) = \frac{\omega^2 \tau_R^2}{1+\omega^2 \tau_R^2} G_o \tag{1.1}$$

$$G''(\omega) = \frac{\omega \tau_R}{1+\omega^2 \tau_R^2} G_o \tag{1.2}$$

where τ_R is the relaxation time. The parameters G' and G'' are the elastic and viscous components of the complex shear modulus, and they are related to the ability of the material to behave as an elastic or viscous material. As is evident from the Maxwell equations, in the low-frequency region, $\omega \ll \omega_c$, G' and G'' scale with ω according to $G' \approx \omega^2$ and $G'' \approx \omega$. In the high-frequency region, or more specifically, in the region of $\omega \gg \omega_c$, however, G' attains a plateau value equal to G_o, whereas G'' shows a monotonic decrease. The shear frequency corresponding to the G'–G'' crossover, ω_c, is equal to the inverse of τ_R. For worm-like micelles at particular conditions, the magnitude of τ_R is related to the average length of the worm-like micelles, whereas G_o is related to the number density of entanglement in the transient network.

Once G_o and τ_R are known, η_o can be calculated using the following relation:

$$\eta_o = G_o \tau_R \tag{1.3}$$

Alternately, the following relationship allows one to estimate η_o by extrapolating the complex viscosity values ($|\eta^*|$) to zero shear frequency:

$$|\eta^*| = \frac{(G'^2 + G''^2)^{1/2}}{\omega} = \frac{\eta_o}{\sqrt{1+\omega^2 \tau_R^2}} \tag{1.4}$$

The living-polymer model proposed by Cates and coworkers [34–36] considers the molecular-weight distribution (MWD) of the worm-like micelle to be in a thermal equilibrium, and such systems are called an equilibrium polymer or "living polymer", which is in contrast to the fixed MWD of ordinary polymer solutions. The viscoelastic behavior of the entangled worm-like micelles is described by considering two processes, reptation (i.e. reptile-like motion of the micelle along its own contour) and reversible scission of micelles, taking place at two time scales, namely, reptation time, τ_{rep}, and breaking time, τ_b. The τ_{rep} is the time required for a worm-like micelle of contour length $\bar{L}$ to pass through a hypothetical tube, and the τ_b is the average time necessary for a chain of average length $\bar{L}$ to break into two pieces. It is assumed that when a chain breaks, the two daughter chains become uncorrelated and recombine with the micellar end in random way. For fast scission kinetics ($\tau_b \ll \tau_{rep}$), a single-exponential stress decay is observed, and the viscoelastic behavior of such systems at low-frequency follows Maxwell model with a single relaxation time, τ_R given by $(\tau_b \tau_{rep})^{1/2}$ [34].

Although the Maxwell equations predict a monotonous decrease of G'' in the high- frequency region, the worm-like micelles deviate from this behavior, showing an increase of G''in the high-frequency region and a deviation from the semicircle, as well as a depression in a Cole–Cole plot of G'versus G''. This deviation is often associated with the stress relaxation by additional 'faster' processes such as Rouse modes of cylindrical micelles, analogous to polymer chain. The value of G'' at the minimum, G''_{min}, can be related to contour length, $\bar{L}$ according to the relation [35]

$$\frac{G''_{min}}{G_o} \approx \frac{l_e}{\bar{L}} \tag{1.5}$$

where l_e is the entanglement length, the contour length of the section of worm-like micelles between two entanglement points. For flexible micelles, the correlation length, ξ, which gives the mesh size of the micellar network, is related to l_e according to the relations [35]

$$l_e \approx \xi^{5/3}/l_p^{2/3} \tag{1.6}$$

The persistence length, l_p, gives an estimate of micellar flexibility. Even though the micelles are flexible, at a small length scale comparable to l_p, they behave as rigid rods. Also,

$$\xi = \left(\frac{kT}{G_o}\right)^{1/3} \tag{1.7}$$

Combining Eqs. (1.6) and (1.7) yields a relation which relates l_e to G_o [35]

$$G_o \approx \frac{kT}{l_e^{9/5}/l_p^{6/5}} \tag{1.8}$$

In the limit $\tau_b \ll \tau_{rep}$, the theoretical model predicts the following scaling laws for the viscoelastic parameters as a function of the volume fraction of surfactant (ϕ): [36]

$$G_o \approx \phi^{2.25} \tag{1.9}$$

$$\tau_R \approx \phi^{1.25} \tag{1.10}$$

$$\eta_o \approx \phi^{3.5} \tag{1.11}$$

1.3
Viscoelastic Worm-Like Micelles in Nonionic Fluorinated Surfactant System (Without Additives)

Recently, an aqueous solution of perfluoroalkyl sulfonamide ethoxylate, $C_8F_{17}SO_2N(C_3H_7)(C_2H_4O)_{10}H$, designated as $C_8F_{17}EO_{10}$ having a medium chain length (C_8)

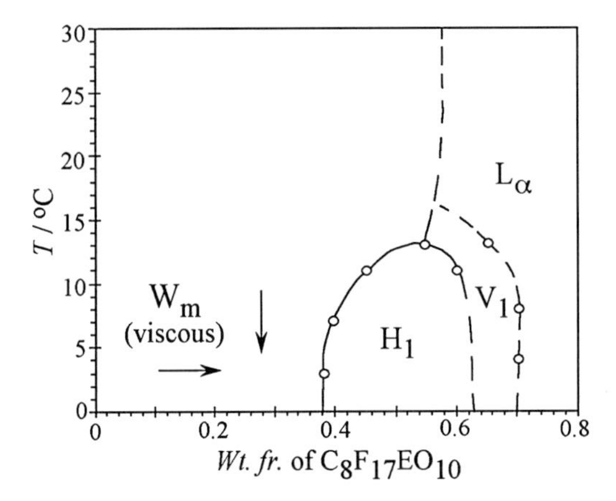

Figure 1.1 Partial phase diagram of $C_8F_{17}EO_{10}$/water binary system. W_m stands for the micellar solution, and H_1, V_1, and L_α stand for hexagonal, bicontinuous cubic, and lamellar liquid crystalline phases. The arrows in the W_m domain show the direction of viscosity increase for the systems at surfactant concentration of 15 wt% and above. At lower concentrations, viscosity increases and then decreases with increasing temperature (redrawn from Ref. [37]).

fluoroalkyl chain and relatively large headgroup (EO_{10}) has been reported to form highly viscoelastic solution of worm-like micelles at low temperatures [37]. The partial binary phase diagram of $C_8F_{17}EO_{10}$/water system is shown in Figure 1.1. This is a typical phase diagram of a nonionic surfactant, such as polyoxyethylene alkylethers (C_mEO_n) [38], with a close resemblance to the phase diagram of $C_{12}EO_6$. With increasing surfactant concentration, an aqueous micellar solution (W_m), hexagonal (H_1), bicontinuous cubic (V_1), and lamellar (L_α) liquid crystalline phases are formed successively at low temperature. Micellar solutions at low temperature become increasingly viscous with increasing surfactant concentration, and at compositions near the H_1 phase, a highly viscous or gel-like solution is formed. This gel-like solution is isotropic and does not show any sharp diffraction peak in the small-angle X-ray scattering (SAXS) spectra, which rules out possibility of the presence of the discontinuous cubic (I_1) phase. Moreover, the fact that the viscosity of the solution gradually increases with increasing concentration also indicates that the I_1 phase is not present. In fact, the formation of globular micelles is not favorable because of bulky and stiff fluorinated chain unless the headgroup is very big. With increasing temperature, the viscosity of the solution gradually decreases and finally a less viscous easily flowing isotropic solution is formed. At higher temperature, a phase separation occurs at the cloud point, which is typical of nonionic surfactant systems. Esquena *et al.* [39] have constructed the aqueous binary phase behavior of the present system over wide ranges of temperature and composition.

Figure 1.2 shows the variation of viscosity (η) as a function of shear-rate ($\dot{\gamma}$) for 1 wt% surfactant solution at different temperatures. At low temperature (5 °C), η

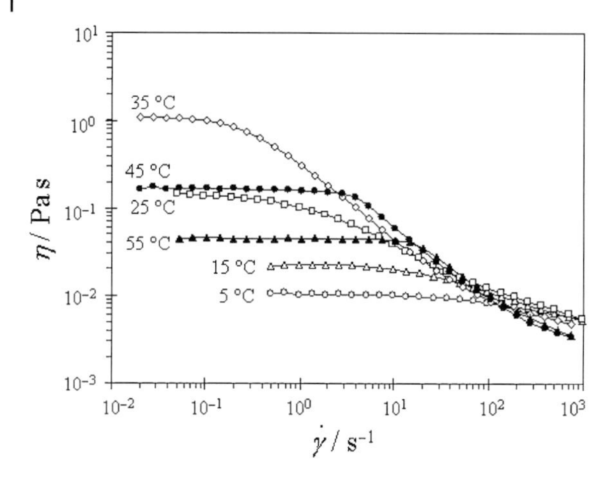

Figure 1.2 Steady shear-rate ($\dot{\gamma}$) vs. viscosity (η) curves of 1 wt% surfactant system at various temperatures (from Ref. [37]).

is independent of $\dot{\gamma}$, that is, Newtonian flow behavior is observed up to $\dot{\gamma}$ ~100 s^{-1}. At higher values of $\dot{\gamma}$ (>100 s^{-1}), the viscosity decreases with increasing $\dot{\gamma}$, and this "shear thinning" behavior is typical of worm-like micelles. It should be that the viscosity of the surfactant solution is significantly higher than that of water or micellar solutions generally observed in surfactants solutions at such a low concentration of 1 wt%. Therefore, cylindrical micelles are expected to be present even at low temperatures. Upon an increase of temperature up to 35 °C, the critical $\dot{\gamma}$ for shear thinning shifts gradually to lower value and also the viscosity in the plateau region (low $\dot{\gamma}$) increases. This suggests that the system is becoming more "structured" with increasing temperature. When the temperature is increased further to 45 °C, the viscosity decreases and the Newtonian region appears up to higher $\dot{\gamma}$, which corresponds to the structural modification in the system. A comparison of the steady shear-rate curves at 25 and 45 °C shows that, although the viscosity at plateau region (low $\dot{\gamma}$) is nearly the same, the shear thinning behavior at higher shear-rate is not similar, which might be the result of a difference in the structure of the system at the respective temperatures. Rheological measurements were carried out only up to 55 °C because phase separation occurs at higher temperature.

The increase of viscosity with temperature in the nonionic surfactant can be understood in terms of the decrease in the interfacial curvature of the aggregates due to progressive dehydration of the EO chain. This would induce a sphere–rod transition in the aggregate shape or induce one-dimensional growth if the rod-like aggregates are already formed. Formation of end-caps in the cylindrical aggregates becomes unfavorable with increasing temperature because of the high free-energy cost of the formation of hemispherical ends and consequently one-dimensional growth is favored. The trend observed up to 35 °C is consistent with that view. However, above 35 °C, a turning point occurs and viscosity begins to decrease as

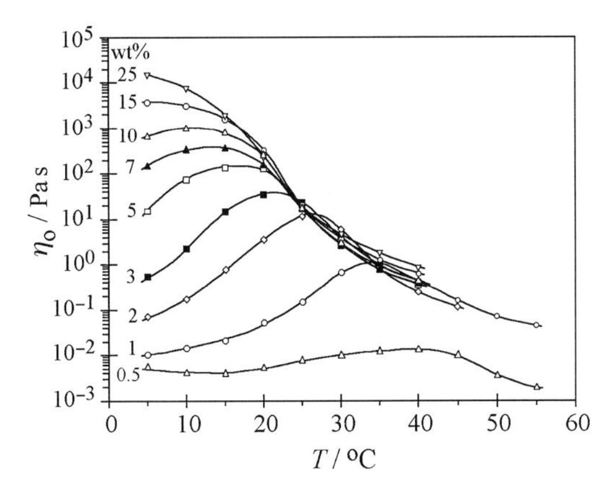

Figure 1.3 Variation of zero-shear viscosity (η_o) as a function of temperature at different concentrations (in wt%) of surfactant in water (from Ref. [37]).

if the micellar length changes in the opposite direction, possibly due to an increase in the extent of micellar scission kinetics. The explanation for this change in the rheological behavior is that with increasing temperature, the energy cost of the free ends of the micelles become higher, and therefore the free ends would fuse at the cylindrical part of its own or other micelles, thus forming micellar joints, or branching in the network structure. Such joints can slip along the cylindrical body, thereby allowing a faster and easier way of stress relaxation [40, 41]. Branching points also restricts the alignment of micelles under shear [42], causing an increase in critical $\dot{\gamma}$. In a number of surfactant systems, such micellar connections or branching points have been detected by cryogenic transmission electron microscopy [43–46].

Figure 1.3 shows the variation of zero-shear viscosity (η_o) as a function of temperature at different surfactant compositions. At 0.5 wt% surfactant concentration, the viscosity of the solution is low, and a gradual but small increase in viscosity is observed with increasing temperature up to 40 °C; then it decreases with a further increase in temperature and ultimately a phase separation takes place. For 1 wt% surfactant concentration, the increase in viscosity with an increase in temperature is clearly visible, with an increase of about 2 orders of magnitude in viscosity upon increasing temperature from 5 to 35 °C, and finally, the viscosity decreases with a further increase in temperature. The trend of the viscosity–temperature curve is essentially similar with increasing surfactant concentration, but the temperature for the viscosity maximum ($T_{\eta\text{-max}}$) gradually shifts to a lower value, and there is a significant increase in the viscosity at temperatures below $T_{\eta\text{-max}}$. Highly viscoelastic solutions are formed at low temperatures and high surfactant concentration. For 25 wt% surfactant concentration, the viscosity decreases with increasing temperature from 5 °C, and judging from the trends of curves, $T_{\eta\text{-max}}$ is well below 5 °C, but the behavior is essentially the same as that observed at lower concentrations. The viscosity of the micellar solution is very sensitive to the concentration at low

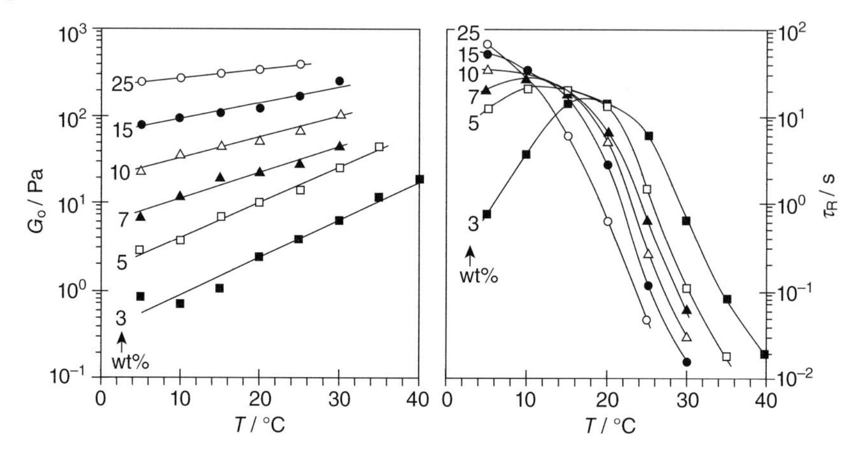

Figure 1.4 Variation of shear modulus (G_o) and stress relaxation time (τ_R) as a function of temperature at different concentrations (in wt%) of surfactant in water (from Ref. [37]).

temperatures, but at and above 25 °C, the viscosity is nearly the same over a wide range of concentrations (2–25 wt%).

With increasing surfactant concentration, the degree of hydration of the ethylene oxide chain decreases. The decrease in the hydration would decrease the spontaneous curvature of the aggregate and hence the energy required for the formation of hemispherical end-caps of the cylindrical micelles increases. Consequently, one-dimensional micellar growth is favored and viscosity increases. Therefore, the effect of increasing surfactant concentration is similar to that of increasing temperature. Formation of micellar joints is expected to take place at lower temperatures with increasing surfactant concentration and, hence the viscosity maximum shifts gradually toward the lower temperature.

The estimated values for G_o and τ_R are plotted as a function of temperature at different surfactant concentrations as shown in Figure 1.4. It can be seen that G_o increases monotonically with increasing temperature, which may be taken as evidence of an increase in the network density of the worm-like micelles at all concentrations. However, the extent of the increase of G_o with temperature decreases with increasing surfactant concentration. At 3 wt% surfactant concentration, G_o increases by 1 order of magnitude when temperature changes by 20 °C, whereas at 25 wt% the change is significantly small.

On the other hand, the τ_R of the viscoelastic solutions at low concentrations, for example, in a 3 wt% surfactant system, first increases promptly with increasing temperature, reaches the maximum value at ~20 °C, and then decreases. When the surfactant concentration increases from 3 to 5 wt% and above, the trend remains essentially the same, but the position of the τ_R maximum slowly moves to lower temperature. At higher concentrations of surfactant (e.g., at 10 and 15 wt%), the τ_R maxima are around 5 °C. No maximum is seen at 25 wt% surfactant concentration, but τ_R starts decreasing from 5 °C. However, judging from the trends of the τ_R–T curves, the τ_R maxima for these concentrations is expected to fall below 5 °C and the τ_R approaches a very large value (e.g., ~67 s) for a 25 wt% system at 5 °C, which indicates the presence of very long micelles.

The observed trend in τ_R clearly suggests that with increasing temperature some structural modifications occur in the network that allow the system to release the stress quickly. The shortening of micelles is not possible because the continuous growth of G_o clearly indicates that network density increases with temperature. These results are consistent with the formation of micellar joints or branching because sliding of the branching point along the micellar length can provide a fast stress relaxation mechanism.

The variation of microstructural changes in shape and size of the micellar aggregates of 1 wt% surfactant solution at different temperatures (namely, 15, 35, 45, and 55 °C) can be obtained by SAXS technique. These temperatures correspond to the different regions of viscosity temperature curves, namely, the viscosity-growth region (see Figure 1.2), the viscosity-maximum region, the viscosity-decline region after the maximum, and near the phase boundary, respectively. The SAXS data are analysed by the generalized indirect Fourier transformation (GIFT) method [47–49]. Figure 1.5a shows the scattering intensity $I(q)$–q curves at different temperatures. The absence of a peak or maximum in the low-q regime indicates a negligible interparticle interaction. The decay of $I(q)$ at $\sim q^{-1}$ in the low-q region suggests that the aggregates have rod-like local structure at 15, 35 and 45 °C. However, at 55 °C, a very small but noticeable increase in the slope in the $q < 0.7\,\text{nm}^{-1}$ region was observed that suggests that the aggregate shape is mainly cylindrical but a lamellar-like structure is gradually developing. Lamellar-like aggregates are characterized by $I(q)$ decaying at $\sim q^{-2}$ in the low-q region. The structural change with increasing temperature can be seen from the corresponding pair-distance distribution function (PDDF), $p(r)$, curves shown in Figure 1.5b. The PDDF patterns at 15, 35 and 45 °C exhibit a typical feature of the rod-like particles as illustrated by a pronounced peak in the low-r regime and an extended tail in the high-r side. The distance in the low-r region at which $p(r)$ decreases sharply after the maximum gives a cross-sectional diameter of the aggregates,

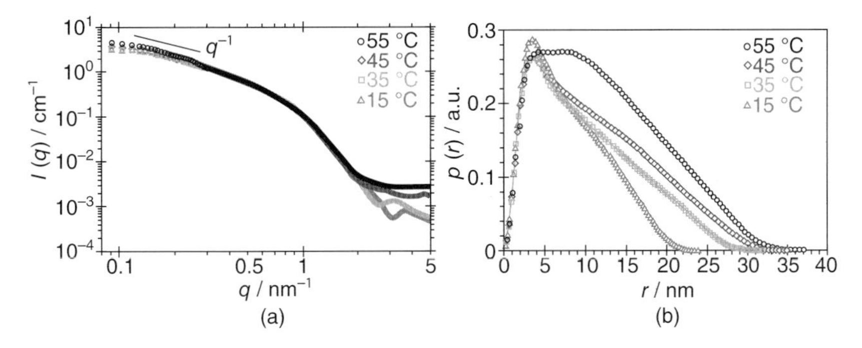

Figure 1.5 (a) Scattered intensity, $I(q)$, from SAXS measurement of the 1wt% surfactant solution at different temperatures and (b) the corresponding pair-distance distribution functions (PDDF), $p(r)$ (from Ref. [37]).

whereas the value of r at which $p(r)$ becomes zero gives the maximum dimension of the aggregate (D_{max}), that is, the length of the cylindrical micelles. The PDDFs of the aggregates show that the D_{max} increases with increasing temperature, but the cylindrical cross-sectional diameter of the rod-like aggregates is almost constant (~5.8 nm). The D_{max} depends on the maximum of resolution (or q_{min}) of the measurement, and it may not provide the actual length of the micelles, which are believed to have their contour lengths in the several hundred nanometers or even micrometer range. Nevertheless, the observed trends in the variation of D_{max} values can be taken as evidence of micellar growth. With a further increase in temperature to 55 °C, the D_{max} continues to increase and a bulge in the middle-r region of $p(r)$ function appears. Such a change in the $p(r)$ function suggests a gradual evolution of the local structure toward a lamellar pattern [50].

Temperature-induced micellar growth and evolution of low-curvature structures at higher temperature as obtained by SAXS measurements provide a direct evidence of the formation of micellar joints or branching. Addition of lipophilic nonionic surfactant such as trioxyethylene dodecyl ether to the dilute solutions of several ionic as well as hydrophilic nonionic surfactant induces one-dimensional micellar growth and causes the viscosity to sharply increase to a maximum and then decrease [23, 25, 26, 30, 31]. Finally, a two-phase region consisting of aggregates with low curvature (L_α phase or vesicles) and a micellar solution appears. The mechanism of transformation of worm-like micelles to structures with low curvature is still not clearly understood. Most probably, formation of micellar joints and the growth of the joints to a bilayer structure is the mechanism. It is reasonable to expect a similar type of structural evolution in the present system because increasing temperature and the addition of a lipophilic surfactant both decrease the interfacial curvature of the aggregates.

1.4 Viscoelastic Worm-Like Micelles in Mixed Nonionic Fluorinated Surfactant Systems

Mixtures of perfluoroalkyl sulfonamide ethoxylates, $C_8F_{17}SO_2N(C_3H_7)(C_2H_4O)_nH$ (abbreviated as $C_8F_{17}EO_n$, $n = 20$, 1, and 3) are known to form viscoelastic worm-like micelles in water [51, 52]. Incorporation of a cosurfactant in the palisade layer of micellar aggregates reduces the effective area per molecule a_s, which results in a decrease in the interfacial curvature of the aggregate, thus leading to micellar growth. Partial ternary phase diagrams of water/$C_8F_{17}EO_{20}$/$C_8F_{17}EO_1$ and water/$C_8F_{17}EO_{20}$/$C_8F_{17}EO_3$ systems at 25 °C are shown in Figure 1.6.

In water/$C_8F_{17}EO_{20}$ binary system, a micellar solution (W_m) is formed over a wide range of surfactant concentration at 25 °C. Because of the very big headgroup, $C_8F_{17}EO_{20}$ forms globular types of micelles above the critical micelle concentration (CMC), which has been found to be 0.023 mM at 25 °C despite its bulky and stiff fluorocarbon chain [53]. The detailed aqueous binary phase behavior and the micellar structure of the $C_8F_{17}EO_{20}$ in water are described elsewhere [54]. The micellar solution of $C_8F_{17}EO_{20}$ can solubilize a significant amount of $C_8F_{17}EO_1$ or

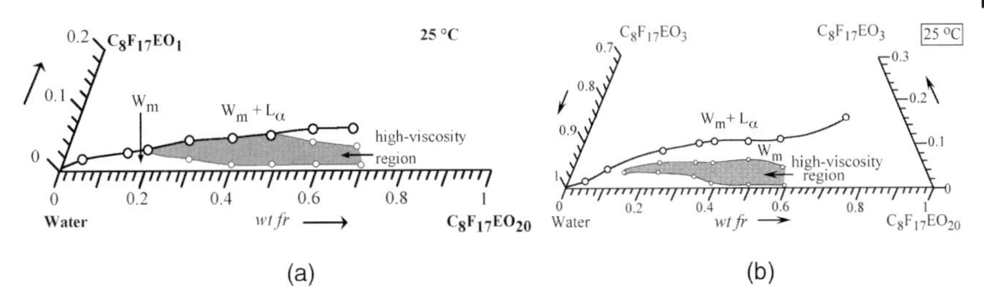

Figure 1.6 Partial phase diagrams of (a) water/$C_8F_{17}EO_{20}$/$C_8F_{17}EO_1$ and (b) water/$C_8F_{17}EO_{20}$/$C_8F_{17}EO_3$ ternary systems. W_m stands for the isotropic micellar solution and L_α is the lamellar liquid-crystalline phase (redrawn from Ref. [51, 52]).

$C_8F_{17}EO_3$, which is evident from the height of the W_m domain in the ternary phase diagram. Due to the bulky and stiff hydrophobic tail and small hydrophilic group, $C_8F_{17}EO_1$ or $C_8F_{17}EO_3$ itself cannot form discrete aggregates in water. Due to the same reason, incorporation of $C_8F_{17}EO_1$ or $C_8F_{17}EO_3$ in the aggregates of $C_8F_{17}EO_{20}$ reduces the average headgroup area at the interface or, in other words, reduces the interfacial curvature, and beyond the solubilization limit of the W_m phase, the L_α phase separates out from the isotropic solution. Upon successive addition of $C_8F_{17}EO_1$ or $C_8F_{17}EO_3$ to the micellar solution of $C_8F_{17}EO_{20}$, no significant change in viscosity occurs in the dilute solution of the surfactant, but at higher concentration (above 15 or 20 wt% of $C_8F_{17}EO_{20}$), viscosity increases gradually at first, then promptly and a viscous solution is observed. The shaded area in the phase diagrams shows the approximate region of viscous solution inside the W_m domain. The samples inside the high-viscosity zone show flow-birefringence, which is seen when the samples are viewed under crossed polarizers while being shaken. With further addition of $C_8F_{17}EO_1$ or $C_8F_{17}EO_3$ viscosity decreases and ultimately a phase separation occurs. It should be noted that the range of surfactant concentrations in which viscoelastic worm-like micellar solutions are formed is much higher than the range reported in the literature.

Figure 1.7 shows the variation of zero-shear viscosity (η_o) as a function of the weight fraction of $C_8F_{17}EO_1$ in total surfactant, W at two different $C_8F_{17}EO_{20}$ concentrations (25 and 35 wt%). It can be seen that above a certain value of W, the viscosity increases steeply and attains the maximum, followed by a decline. When $C_8F_{17}EO_{20}$ concentration is increased from 25 to 35 wt%, a rapid viscosity growth occurs at lower concentration of $C_8F_{17}EO_1$, or in other words, the η_o–W curve shifts toward lower W values, which can be attributed to the decrease in the effective cross-sectional area per surfactant molecule (a_s) in the aggregate with increasing concentration of $C_8F_{17}EO_{20}$. The trend is essentially similar to that observed in the worm-like micellar solutions formed in mixed systems of hydrocarbon surfactants. At compositions around the viscosity maxima the systems exhibit viscoelasticity.

Figure 1.8 shows the change in G_o and τ_R with W for 35 wt% $C_8F_{17}EO_{20}$ + $C_8F_{17}EO_1$ and 25 wt% $C_8F_{17}EO_{20}$ + $C_8F_{17}EO_1$ systems. These parameters were estimated by fitting of the experimental data from frequency-sweep measurements, especially

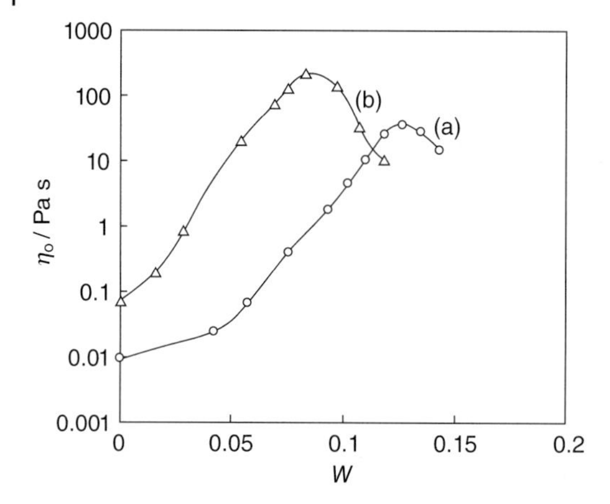

Figure 1.7 Variation of zero-shear viscosity (η_o) as a function of the weight fraction of $C_8F_{17}EO_1$ in total surfactant, *W* for the (a) 25 wt% $C_8F_{17}EO_{20}$ + $C_8F_{17}EO_1$ and (b) 35 wt% $C_8F_{17}EO_{20}$ + $C_8F_{17}EO_1$ systems at 25 °C (from Ref. [51]).

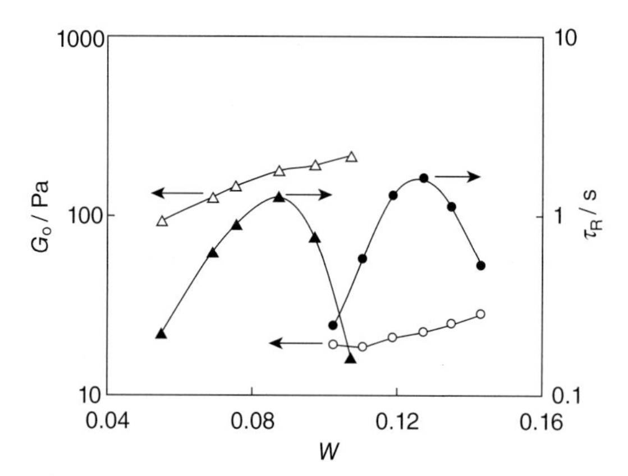

Figure 1.8 Variation of shear modulus (G_o) and relaxation time (τ_R) as a function of the mixing fraction of $C_8F_{17}EO_1$ in total surfactant (*W*) in the (a) 25 wt% $C_8F_{17}EO_{20}$ + $C_8F_{17}EO_1$ and (b) 35 wt% $C_8F_{17}EO_{20}$ + $C_8F_{17}EO_1$ systems at 25 °C. Solid lines are given for visual guide only (from Ref. [51]).

the data in the low-frequency region, to the Maxwell equations. The shift of the η_o and τ_R curves toward the lower *W* values in the η_o–*W* (Figure 1.7) and τ_R–*W* (Figure 1.8) plots upon increasing the $C_8F_{17}EO_{20}$ concentration in the mixed system corresponds to the higher extent of micellar growth. This is also obvious from the increase in G_o or network density upon increasing surfactant concentration. The lower value of τ_R at the maximum in the 35 wt% $C_8F_{17}EO_{20}$ + $C_8F_{17}EO_1$ system in comparison to that in the 25 wt% $C_8F_{17}EO_{20}$ + $C_8F_{17}EO_1$ system should not be considered as a lower extent of micellar growth in the former system. Instead, it might have arisen from the fact that with increasing surfactant concentration the

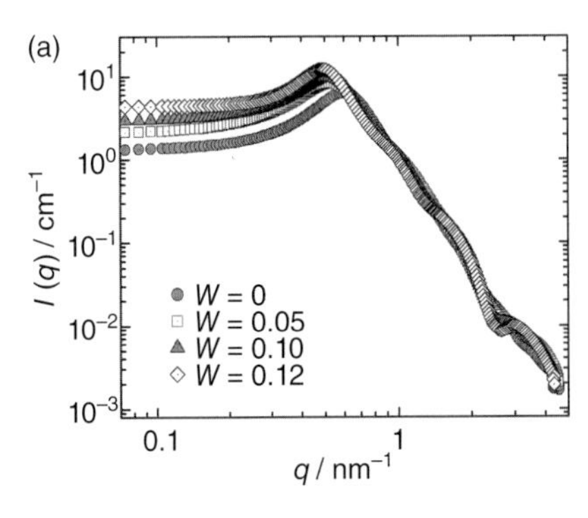

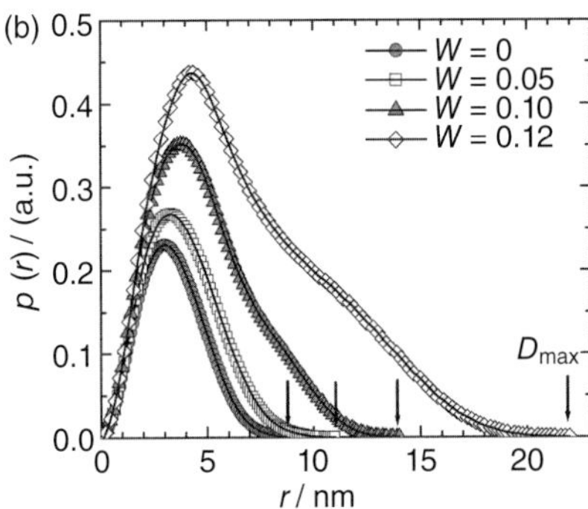

Figure 1.9 (a) Normalized X-ray scattered intensities, $I(q)$ of the 25 wt% $C_8F_{17}EO_{20}$ + $C_8F_{17}EO_1$ system at different weight fraction of $C_8F_{17}EO_1$ in total surfactant, W at 25 °C and (b) the corresponding PDDF, $p(r)$, functions. The arrows in panel "b" represent the maximum dimension of the micelles (from Ref. [54]).

spontaneous curvature decreases and the system favors the micellar branching at lower values of W in order to minimize the energy cost of the formation of end-caps. Continuous growth of G_o in the given composition range where η_o and τ_R decrease shows that after branching the network density grows until the phase separation occurs. There are indications that the local structure at branching points evolve toward bilayer structure [37] and ultimately separates out. The viscosity growth, and the change of G_o and τ_R as a function of the mixing fraction of $C_8F_{17}EO_3$ in total surfactant (W) for the 25 wt% $C_8F_{17}EO_{20}$ + $C_8F_{17}EO_3$ and the 35 wt% $C_8F_{17}EO_{20}$ + $C_8F_{17}EO_3$ systems at 25 °C show a similar trend to that obtained for the water/$C_8F_{17}EO_{20}$/$C_8F_{17}EO_1$ system.

A supportive structural evidence for the rheological data of worm-like micelles can be obtained from SAXS technique for the water/$C_8F_{17}EO_{20}$/$C_8F_{17}EO_1$ ternary system [54]. Figure 1.9 shows scattering functions, $I(q)$, and the corresponding pair-distance distribution functions (PDDF), $p(r)$, deduced with the GIFT (generalized indirect Fourier transformation) analysis of the SAXS data for 25 wt% $C_8F_{17}EO_{20}$ + $C_8F_{17}EO_1$ systems at different W at 25 °C.

The local maximum in the scattering functions at ~0.5 nm^{-1} is a clear signature of strong intermicellar interactions, which is quite unavoidable as the system is concentrated (25 wt% $C_8F_{17}EO_{20}$). On increasing the value of W from $W = 0.05$ to 0.12, the interaction peaks shift towards forward direction (towards low q) and the forward scattering intensity increases. Such behavior in the scattering function can be taken as evidence of micellar growth. The micellar growth induced by the fluorocarbon cosurfactant $C_8F_{17}EO_1$ can be clearly seen in the $p(r)$ curves shown in Figure 1.9b. A symmetrical, nearly bell-shaped $p(r)$ curve of the 25 wt% surfactant corresponds to the globular or ellipsoid prolate type of micellar structure with nearly homogeneous electron density distribution inside the particle. With increasing W, an asymmetry in the shape of the $p(r)$ curve is developed and grows parallel to the W, indicating the micellar growth. With successive increase of W, D_{max} as indicated by arrows increases, and at $W = 0.12$ (the composition at which

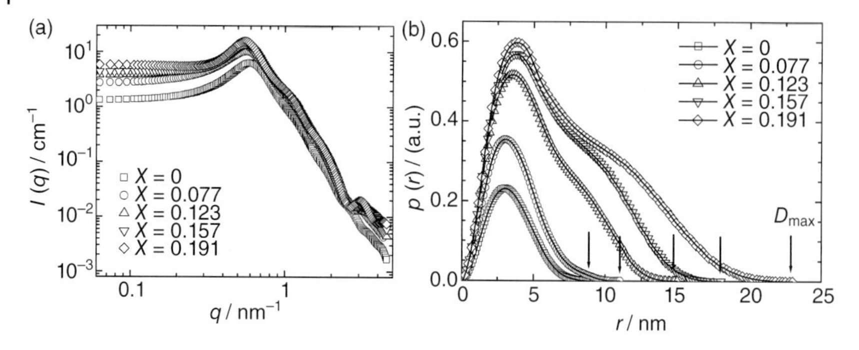

Figure 1.10 (a) Normalized X-ray scattered intensities, $I(q)$ of the 25 wt% $C_8F_{17}EO_{20}$ + $C_8F_{17}EO_3$ system at different weight fraction of $C_8F_{17}EO_3$ in total surfactant, X at 25 °C and (b) the corresponding PDDF, $p(r)$, functions. The arrows in panel "b" represent the maximum dimension of the micelles (from Ref. [52]).

the maximum viscosity is obtained, see Figure 1.7), D_{max} increases greatly. A pronounced peak in the low-r side with an extended tail in the higher-r side of the $p(r)$ curve is a clear signature of cylindrical micelles. The similar results of scattering functions, $I(q)$, and the corresponding pair-distance distribution functions (PDDF), $p(r)$, of the SAXS data for the 25 wt% $C_8F_{17}EO_{20}$ + $C_8F_{17}EO_3$ systems at different weight fractions of $C_8F_{17}EO_3$ in total surfactants, X at 25 °C are shown in Figure 1.10. Here also, as shown in Figure 1.10a, the forward scattering intensity increases and the interaction peak shifts toward the forward direction (low-q side) with increasing X from $X = 0.077$ to 0.191 indicating micellar growth. As can be seen from Figure 1.10b, when $C_8F_{17}EO_3$ is added, the D_{max} shifts to the higher-r side, and PDDF curves become asymmetric, indicating elongated aggregates.

1.5 Summary

Fluorinated surfactants tend to form aggregates with low curvature, such as cylindrical aggregates, under solution conditions where hydrocarbon surfactants would form spherical aggregates. The packing constraints of the perfluoroalkyl chain is considered to be reason for the formation of flexible cylindrical aggregates in water even at low surfactant concentrations and low temperature for the perfluoroalkyl sulfonamide ethoxylate system, and with an increase surfactant concentrations, these cylindrical micelles undergo one-dimensional growth and form very long and flexible worm-like micelles that entangle and form a network making a viscoelastic solution. The increase in viscosity with increasing temperature for the nonionic surfactant system is mainly attributed to the decrease in the spontaneous curvature of the aggregates and increased energy cost for the formation of hemispherical end-caps and consequently favors one-dimensional growth. Above a certain temperature, viscosity begins to decrease, not because of the shrinking of micelles but because the system tends to eliminate the free ends by forming micellar joints in the network and such changes in the microstructure result in a decrease in the viscosity and stress relaxation time but the network structure is retained. A viscoelastic solution of worm-like micelles is formed in an aqueous solution of highly hydrophilic nonionic fluorinated surfactant, perfluoroalkyl sulfonamide ethoxylate at relatively high surfactant concentration when a hydrophobic amphiphile (cosurfactant) is added. Addition of cosurfactant reduces the interfacial curvature of the aggregates and induces one-dimensional micellar growth. With successive addition of cosurfactant, the viscosity increases rapidly to form viscoelastic solutions, then decreases after the maximum, and ultimately a phase separation occurs. Increasing surfactant or cosurfactant concentration in the mixed nonionic system increases the extent of micellar growth that is mainly attributed to the decrease in the spontaneous curvature of the aggregates and consequently a progressive increase in the energy cost for the formation of the hemispherical end-caps of the aggregates.

References

1 Israelachvili, J.N., Mitchell, D.J., and Ninham, B.W. (1976) *J. Chem. Soc., Faraday Trans. 2*, **72**, 1525.

2 Shinoda, K., Hato, M., and Hayashi, T. (1972) *J. Phys. Chem.*, **76**, 909.

3 Kunieda, H., and Shinoda, K. (1976) *J. Phys. Chem.*, **80**, 2468.

4 Tanford, C. (1980) *The Hydrophobic Effect: Formation of Micelles and Biological Membranes*, 2nd edn, John Wiley & Sons, Inc., New York.

5 El Moujahid, C., Ravey, J.C., Schmitt, V., and Stébé, M.J. (1998) *Colloids Surf. A*, **136**, 289.

6 Kern, F., Lemarechal, P., Candau, S.J., and Cates, M.E. (1992) *Langmuir*, **8**, 437.

7 Khatory, A., Lequeux, F., Kern, F., and Candau, S.J. (1993) *Langmuir*, **9**, 1456.

8 Kim, W.-J., Yang, S.-M., and Kim, M. (1997) *J. Colloid Interface Sci.*, **194**, 108.

9 Kim, W.-J., and Yang, S.-M. (2000) *J. Colloid Interface Sci.*, **232**, 225.

10 Imai, S., and Shikata, T. (2001) *J. Colloid Interface Sci.*, **244**, 399.

11 Vethamuthu, M.S., Almgren, M., Brown, W., and Mukhtar, E. (1995) *J. Colloid Interface Sci.*, **174**, 461.

12 Hartmann, V., and Cressely, R. (1997) *Colloids Surf. A*, **121**, 151.

13 Rehage, H., and Hoffmann, H. (1988) *J. Phys. Chem.*, **92**, 4712.

14 Montalvo, G., Rodenas, E., and Valiente, M. (2000) *J. Colloid Interface Sci.*, **227**, 171.

15 Ponton, A., Schott, C., and Quemada, D. (1998) *Colloids Surf. A*, **145**, 37.

16 Wang, K., Karlsson, G., Almgren, M., and Asakawa, T. (1999) *J. Phys. Chem. B*, **103**, 9237.

17 Knoblich, A., Matsumoto, M., Murata, K., and Fujiyoshi, Y. (1995) *Langmuir*, **11**, 2361.

18 Hoffmann, H., and Würtz, J. (1997) *J. Mol. Liq.*, **72**, 191.

19 Abe, M., Tobita, K., Sakai, H., Kondo, Y., Yoshino, N., Kasahara, Y., Matsuzawa, H., Iwahashi, M., Momozawa, N., and Nishiyama, K. (1997) *Langmuir*, **13**, 2932.

20 Tobita, K., Sakai, H., Kondo, Y., Yoshino, N., Iwahashi, M., Momozawa, N., and Abe, M. (1997) *Langmuir*, **13**, 5054.

21 Tobita, K., Sakai, H., Kondo, Y., Yoshino, N., Kamogawa, K., Momozawa, N., and Abe, M. (1998) *Langmuir*, **14**, 4753.

22 Danino, D., Weihs, D., Zana, R., Orädd, G., Lindblom, G., Abe, M., and Talmon, Y. (2003) *J. Colloid Interface Sci.*, **259**, 382.

23 Acharya, D.P., and Kunieda, H. (2003) *J. Phys. Chem. B*, **107**, 10168.

24 Acharya, D.P., Hossain, Md.K., Feng, J., Sakai, T., and Kunieda, H. (2004) *Phys. Chem. Chem. Phys.*, **6**, 1627.

25 Naito, N., Acharya, D.P., Tanimura, K., and Kunieda, H. (2004) *J. Oleo Sci.*, **53**, 599.
26 Maestro, A., Acharya, D.P., Furukawa, H., Gutiérrez, J.M., López-Quintela, M.A., Ishitobi, M., and Kunieda, H. (2004) *J. Phys. Chem. B*, **108**, 14009.
27 Herb, C.A., Chen, L.B., and Sun, W.M. (1994) *Structure and Flow in Surfactant Solutions*, C.A. Herb and R.K. Prud'homme (eds), ACS Symposium Series 578, American Chemical Society, Washington, DC, pp. 153–166.
28 Rodriguez, C., Acharya, D.P., Hattori, K., Sakai, T., and Kunieda, H. (2003) *Langmuir*, **19**, 8692.
29 Acharya, D.P., Hattori, K., Sakai, T., and Kunieda, H. (2003) *Langmuir*, **19**, 9173.
30 Acharya, D.P., Sato, T., Kaneko, M., Singh, Y., and Kunieda, H. (2006) *J. Phys. Chem B*, **110**, 754.
31 Acharya, D.P., Shiba, Y., Aratani, K., and Kunieda, H. (2004) *J. Phys. Chem. B.*, **108**, 1790.
32 Sato, T., Hossain, Md.K., Acharya, D.P., Glatter, O., Chiba, A., and Kunieda, H. (2004) *J. Phys. Chem. B*, **108**, 12927.
33 Larson, R.G. (1999) *The Structure and Rheology of Complex Fluids*, Oxford University Press, New York.
34 Cates, M.E., and Candau, S.J. (1990) *J. Phys.:Condens. Matter*, **2**, 6869.
35 Granek, R., and Cates, M.E. (1992) *J. Chem. Phys.*, **96**, 4758.
36 Cates, M.E. (1988) *J. Phys. Fr.*, **49**, 1593.
37 Acharya, D.P., Sharma, S.C., Rodríguez -Abreu, C., and Aramaki, K. (2006) *J. Phys. Chem. B*, **110**, 20224.
38 Mitchell, D.J., Tiddy, G.J.T., Waring, L., Bostock, T., and MacDonald, M.P. (1983) *J. Chem. Soc., Faraday Trans. 1*, **79**, 975.
39 Esquena, J., Rodríguez, C., Solans, C., and Kunieda, H. (2006) *Micropor. Mesopor. Mater.*, **92**, 212.
40 Candau, S.J., and Oda, R. (2001) *Colloids Surf. A*, 183–185.
41 Khatory, A., Kern, F., Lequeux, F., Appell, J., Porte, G., Morie, N., Otta, A., and Urbach, W. (1993) *Langmuir*, **9**, 933.
42 Croce, V., Cosgrove, T., Dreiss, C.A., King, S., Maitland, G., and Hughes, T. (2005) *Langmuir*, **21**, 6762.
43 Lin, Z. (1996) *Langmuir*, **12**, 1729.
44 Danino, D., Talmon, Y., Levy, H., Beinert, G., and Zana, R. (1995) *Science*, **269**, 1420.
45 In, M., Aguerre-Chariol, O., and Zana, R. (1999) *J. Phys. Chem. B*, **103**, 7747.
46 Zana, R. (2002) *Adv. Colloid Interface Sci.*, **97**, 205.
47 Brunner-Popela, J., and Glatter, O. (1997) *J. Appl. Crystallogr.*, **30**, 431.
48 Weyerich, B., Brunner-Popela, J., and Glatter, O. (1999) *J. Appl. Crystallogr.*, **32**, 197.
49 Brunner-Popela, J., Mittelbach, R., Strey, R., Schubert, K.-V., Kaler, E.W., and Glatter, O. (1999) *J. Chem. Phys.*, **21**, 10623.
50 Moitzi, C., Freiberger, N., and Glatter, O. (2005) *J. Phys. Chem. B*, **109**, 16161.
51 Sharma, S.C., Acharya, D.P., and Aramaki, K. (2007) *Langmuir*, **23**, 5324.
52 Sharma, S.C., Shrestha, R.G., Shrestha, L.K., and Aramaki, K. (2009) *J. Phys. Chem. B*, **113**, 1615.
53 Sharma, S.C., Shrestha, L.K., and Aramaki, K. (2007) *J. Dispersion Sci. Technol.*, **28**, 577.
54 Shrestha, R.G., Shrestha, L.K., Sharma, S.C., and Aramaki, K. (2008) *J. Phys. Chem. B*, **112**, 10520.

2
Structure of Nonionic Surfactant Micelles in Organic Solvents: A SAXS Study

Lok Kumar Shrestha and Kenji Aramaki

2.1
Introduction

2.1.1
Reverse Micelles

Reverse micelles have an inverted structure in comparison to the conventional normal micelles in aqueous systems. Therefore, they are often known as inverse or inverted micelles. In reverse micelles, the micellar cores consist of a hydrophilic polar component and the shells consist of lipophilic nonpolar part of the surfactant molecules. The dipole–dipole interaction between the hydrophilic headgroups acts as one of the driving forces for the formation of reverse micelles in organic solvents. Reverse micelles are mostly observed in the ternary mixtures of surfactant/ water/oil, mostly in oil-rich regions [1–3]. Furthermore, reverse micelles have also been observed in aqueous systems of lipophilic surfactant in surfactant-rich regions [4, 5]. In most of the studies carried out in the past, water was regarded as an essential component in the formulation of reverse micelles. Only a few reports exist in the literature of surfactant science that describe the formation of reverse micelles in organic solvents without water addition [6–10].

Studies on reverse micelles have attracted significant interest over the years because of their wide range of applications. Their utility has been demonstrated in various chemical reactions. Reverse micelles have been shown to stabilize reactive species that are insoluble in nonpolar solvents, and have also been used as a size-controlling microreactor for different aqueous chemical reactions [11, 12]. They facilitate reactions by stabilizing reactants such as radicals in nonpolar media and by increasing the local concentration of reactants prior to reaction [13]. Furthermore, they have been found to be an excellent model for biological membranes [14]. One of the most interesting features of reverse micelles is their efficiency to solubilize water or other polar solvents. Water is often solubilized in the core of the reverse micelles, and it can swell the micelles to many times their empty size [15]. Reverse micelles have been used as template for the synthesis of nanomaterials for a long time [16–22]. It has been found that the structure of the

Self-Organized Surfactant Structures. Edited by Tharwat F. Tadros

ISBN: 978-3-527-31990-9

nanomaterials largely depends on the structure (size and shape) of the template micelles [23].

In most of the studies of reverse micelles, ternary mixtures of surfactant/water/oil systems such as ternary mixtures of water/aerosol OT (AOT)/oils or water/lecithin/oils were mostly considered [24–30]. Some authors have also formulated reverse micelles in mixed-solvent systems [31, 32]. Similarly; reverse micelles have also been formulated in polymer/oil systems [33]. The formation of reverse micelles in nonionic surfactant/oil binary systems without the addition of water is still a matter of discussion. Fewer amphiphiles are reported to form reverse micelles in surfactant/oil binary systems. The hydrophilic moiety of polyglycerol fatty acid esters is found to be more solvophobic compared to the conventional ethylene oxide (EO) -based nonionic surfactants and, hence, tend to form micellar aggregates in nonpolar solvents [34, 35]. Glycerol-based nonionic surfactants are drawing considerable interest from food or cosmetic scientists due to their biocompatibility and biodegradability. Polyglycerol fatty acid esters are edible nonionic surfactants, which have a polyglycerol chain as a hydrophilic moiety and mostly used in food products, cosmetics and pharmacy. They can be used as emulsifiers, dispersants, solubilizers, rheology modifiers, flavor and fragrance carriers, *etc.* Their biocompatibility and biodegradability, together with the capability of forming α-gel phases make them useful in food industry. The α-gel phase has been found to be efficient in emulsion, foam and dispersion stabilization [36, 37]. Recently, the diglycerol fatty acid esters have been found to be effective foam stabilizers both in aqueous and nonaqueous systems as it forms lamellar liquid crystal or α-solid dispersion in the diluted regions [38–41].

In the first part of this chapter, the nonaqueous phase behavior of mono- and diglycerol fatty acid esters with different alkyl chain length in a variety of organic solvents will be described and in the second part the structure of glycerol-based nonionic surfactant micelles confirmed by small-angle X-ray scattering (SAXS) measurements in different organic solvents depending on the composition, temperature, nature of surfactant and added water or glycerol will be discussed.

2.1.2
Theoretical Background on SAXS

The scattering intensity $I(q)$ depends on the scattering angle or scattering vector, q, and is the complex square of the scattering amplitude, $F(q)$, which is the Fourier transform of the scattering length density difference $\Delta\rho(r)$, describing the scattering particle in real space. In solution scattering, say micellar solutions, or microemulsions or colloidal dispersions, one measure the spatial average of these functions and finally will have an expression as described by Equation 2.1.

$$I(q) = 4\pi \int_0^\infty p(r) \frac{\sin qr}{qr} \mathrm{d}r \tag{2.1}$$

where $p(r)$ is the pair-distance distribution function (abbreviated as PDDF) and contains information about the particle size, shape, and internal core-shell structure of the particles. In general, the intraparticle scattering contributions are related to the form factor, $P(q)$, theoretically corresponding to the Fourier transformation of $p(r)$ and the interparticle interference scattering is connected to the structure factor $S(q)$. The $S(q)$ is given as the Fourier transformation of the total correlation function, $h(r) = g(r) - 1$, as

$$S(q) = 1 + 4\pi n \int_0^{\infty} [g(r) - 1] r^2 \frac{\sin qr}{qr} \mathrm{d}r \tag{2.2}$$

where n is the particle number density and $g(r)$ is the pair-correlation function.

For monodisperse spherical particle systems, the total scattering intensity, $I(q)$, is simply be given as

$$I(q) = nP(q)S(q) \tag{2.3}$$

For dilute systems, the interparticle interaction is minimum and can be neglected so that the $S(q)$ is identical to unity and $I(q)$ is simply given by $P(q)$. Under this condition Equation 2.1 modifies to

$$P(q) = 4\pi \int_0^{\infty} p(r) \frac{\sin qr}{qr} \mathrm{d}r \tag{2.4}$$

For a particle of an arbitrary shape having a scattering density difference of $\Delta\rho(r)$, the pair-distance distribution function (PDDF) is given by

$$p(r) = r^2 \Delta\tilde{\rho}^2(r) \tag{2.5}$$

where $\Delta\tilde{\rho}^2(r)$ is the convolution square of $\Delta\rho(r)$ averaged over all directions in space. For spherical particles, averaging is not necessary because scattering density difference, or electron density difference, $\Delta\rho_s(r)$, is only a function of the radial position and deconvolution [42] of PDDF gives the radial contrast profile.

The functional form of the $I(q)$, and $p(r)$ can be used to determine the geometry of the aggregates in terms of shape and size [43] and also the internal structure of the scattering objects as for example micelles in the surfactant solution systems. The SAXS data of the reverse micelles are fitted (smoothing of statistical noise), desmeared (the geometrical broadening effect), and finally Fourier transformed to the real space. The Fourier transformation gives the $p(r)$-function, that is, the structure of the reverse micelles in the real space. The shape of the $p(r)$-function determines the basic geometry of the aggregates (spherical, cylindrical, or planar) both in homogeneous and inhomogeneous systems. It also clearly indicates the transformation from spherical to cylindrical micelles. However, it should be remembered that the technique based on indirect Fourier transformation (IFT) is acceptable only for very dilute solutions with amphiphile concentrations less than 1%, that is, when there is no interparticle interaction or $S(q)$ is unity. For the semidilute systems say with amphiphile concentrations exceeding 1%, the interparticle interaction comes into play and one cannot neglect the contribution of $S(q)$ to the total scattering intensity and one must have to consider it during the IFT process. This can simply be done by evaluating the SAXS data by updated version of IFT so called generalized indirect Fourier transformation (GIFT) method [44–46], which allows determination of the form factor and the structure factor simultaneously from the scattering data without assuming any form factor model. Nevertheless, one needs an appropriate assumption for the interparticle interaction potential and the closure relation for the structure factor. In the limited cases of well-defined colloidal systems, $P(q)$, may be fixed from a measurement at very low concentration or theoretical calculation, and the experimental structure factor, $S(q)^{\text{exp}}$, can be deduced by dividing the normalized scattered intensity, $I(q)/c$, by the defined $P(q)$. However, for self-assembled systems like micellar solutions and microemulsion droplets, $P(q)$ is a function of concentration and temperature, and has to be determined from experiments.

In the calculation, a model of the averaged structure factor for a hard-sphere (HS) interaction potential, $S(q)^{\text{av}}$ is used [47, 48], which considers the Gaussian distribution of the interaction radius σ for individual monodisperse systems for polydispersity m, and a Percus–Yevick (PY) closure relation to solve Ornstein–Zernike (OZ) equation. The detailed theoretical description on the method has been reported elsewhere [49–51].

The cross-sectional diameter of cylindrical micelles can be determined by the deconvolution of the SAXS data, if $\Delta\rho_c(r)$ is only a function of the radial position within the cross section. If the axial length of the cylindrical particle is at least three times longer than the cross-sectional diameter, it is possible to obtain the radial profile, $\Delta\rho_c(r)$, which is related to the PDDF of cross section $p_c(r)$ as [52]

$$p_c(r) = r\Delta\tilde{\rho}_c^{\,2}(r) \tag{2.6}$$

The cross-sectional PDDF can be calculated from the scattered intensity via

$$I(q)q = \pi L I_c(q) = 2\pi^2 L \int_0^{\infty} p_c(r) J_0(qr)\,\mathrm{d}r \tag{2.7}$$

where $J_0(qr)$ is the zeroth-order Bessel function. The indirect Fourier transformation of Equation 2.7 yields $p_c(r)$, which is then used to calculate $\Delta\rho_c(r)$ by the deconvolution technique [42, 53].

2.2 Phase Behavior

The knowledge of the phase behavior of surfactants or polymers in water or in oils is a basic understanding of the properties of these systems, and it is very important for surfactant industrial applications. It has been found that depending on the temperature or composition, the surfactants form a variety of self-assembled structures in water and oils or in both [54–63]. In comparison to the aqueous systems, the surfactant/oil systems offer less variety of self-assembled structures [9, 64–68]. In the following sections, the phase behavior of mono- and diglycerol fatty acid esters in a variety of organic oils will be described.

2.2.1 Phase Behavior of Monoglycerol Fatty Acid Ester/Oil Systems

The phase behavior of monoglycerol fatty acid esters with different alkyl chain length in liquid paraffin oil, squalane, and squalene is described first, then the phase behavior in the dilute regions of the monoglycerol fatty acid esters in linear chain alkanes is discussed. This section is based on Ref. [64], and [69], respectively.

2.2.1.1 Phase Behavior in Liquid Paraffin, Squalane, and Squalene

Binary phase diagrams of monoglycerol fatty acid esters (C_mG_1 where $m = 8$, 10, and 12 are the number of carbon in alkyl chain of the surfactant): glycerol α-monooctanoate (C_8G_1), glycerol α-monodecanoate ($C_{10}G_1$), and glycerol α-monododecanoate ($C_{12}G_1$) in different nonpolar oils liquid paraffin, squalene and squalane, in the whole composition range at atmospheric pressure are shown in Figure 2.1.

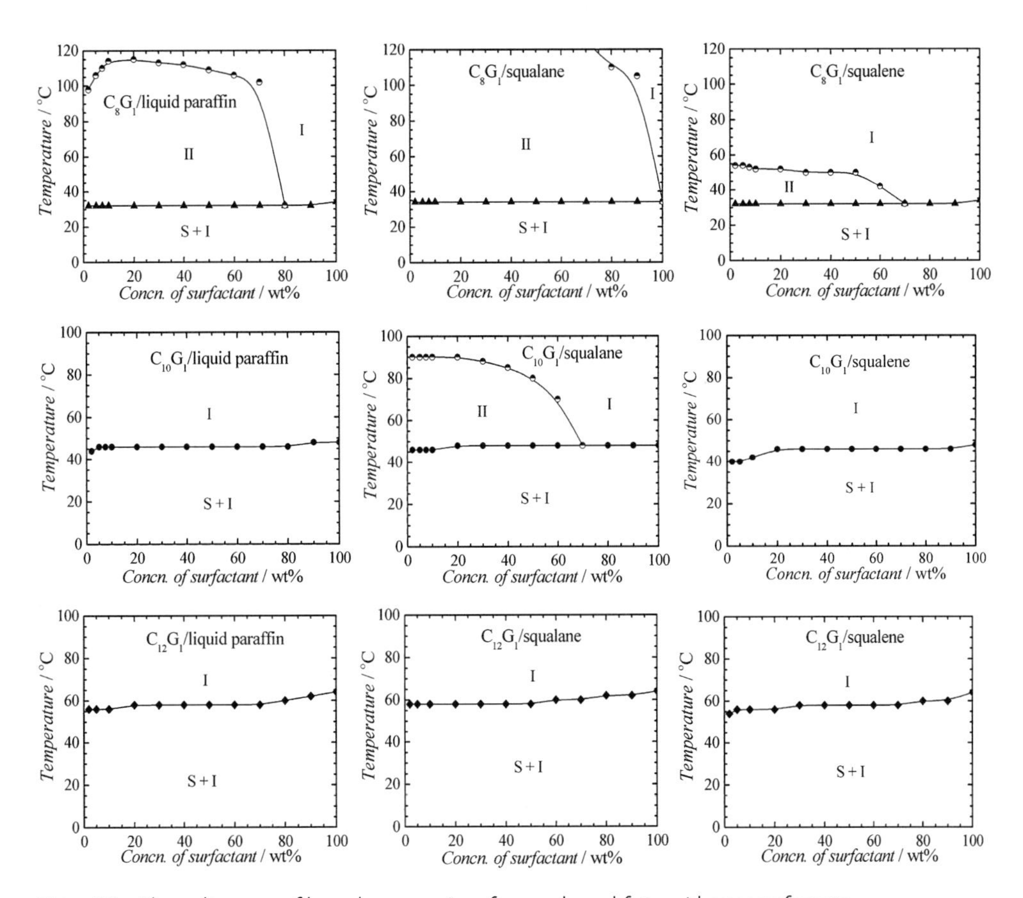

Figure 2.1 Phase diagrams of homologous series of monoglycerol fatty acid ester surfactants in different nonpolar oils liquid paraffin, squalane and squalene (S = solid, I = isotropic single liquid phase, and II = isotropic two-liquid phases).

At lower temperatures surfactant solid and oil phases are in equilibrium and with increasing temperature, the surfactant solid melts at temperatures depending mainly on the lipophilic chain length of surfactant. With increasing hydrocarbon chain length of the surfactant, the solid to liquid transition temperature increases monotonically, but the melting temperature of the solid phase is almost constant at all compositions. It is obvious from Figure 2.1 that the C_mG_1, which have only one glycerol unit as a hydrophilic moiety in their molecules, do not form any liquid-crystalline phases in the nonpolar oils. Instead, the solid surfactant melts to form an isotropic single- or two-liquid phase system depending on the systems and compositions.

The turbid solutions are obtained in the II liquid phases region in all the C_8G_1/oil systems and also in the $C_{10}G_1$/squalane system above the solid melting temperatures. The turbid solutions separate into two transparent liquid phases after a few minutes. Judging from the solubility curve, one phase is an oil-rich phase with less surfactant and the other phase is the surfactant rich phase containing a considerable amount of oil. Figure 2.1 shows that the size of the two-liquid phases domain depends on hydrocarbon chain length of the surfactant and also on the nature of oils. The width of the two liquid phases domain in the phase diagrams determines the miscibility of surfactants in oils. In the case of the same surfactant C_8G_1, it is most miscible with squalene (the narrowest two liquid phase domain) and least miscible with squalane (the widest two-liquid phase domain). With increasing alkyl chain length the surfactants become increasingly lipophilic and, hence, the mutual solubility of surfactant and oil increases (see Figure 2.1, there is no two-liquid phase domain in $C_{12}G_1$/oil systems).

The two-liquid phase transforms to single-phase isotropic solution upon further increasing temperature. The two-to-single phase transition temperature is the highest for the least miscible surfactant-oil system, i.e. the C_8G_1/squalene system. The upper critical solution temperature of the C_8G_1/squalane system could be more than 120 °C and is not included in the phase diagram. The boundary between the single- and two-phase regions correspond to the cloud-point curve for a water–poly(oxyethylene) type of surfactant systems [70–74], in which the surfactant becomes less hydrophilic with increasing temperature and ultimately surfactant-rich phase separates out from the solvent-rich phase. In the surfactant/oil systems, however, the penetration of oil in the surfactant chain increases and the miscibility of oil and surfactant increase with temperature. Consequently, the phenomenon opposite to clouding occurs, that is, an isotropic solution of reversed aggregate is formed.

2.2.1.2 Phase Behavior in n-Alkanes

In the preceding section, it was seen that the monoglycerol fatty acid esters could not form any liquid-crystalline phases in liquid paraffin, squalane, and squalene. Phase-behavior studies have shown that the surfactants of this class do not form any liquid-crystalline phases in *n*-alkanes also. At normal room temperature, there is a solid phase in equilibrium with an excess oil phase in glycerol α-monolaurate

($C_{12}G_1$)/alkane, and glycerol α-monomyristate ($C_{14}G_1$)/alkane systems [69, 75]. The dilute systems of the $C_{12}G_1$ and $C_{12}G_1$ in alkanes (octane to hexadecane) are essentially a dispersions of solid, which transformed into an isotropic reverse micellar solution phase upon heating. This indicates that the dissolution tendency of the $C_{12}G_1$ and $C_{14}G_1$ in alkanes increases with increasing temperature. This might be due to enhanced interpenetrating of oils to the hydrocarbon chain of the surfactant. It has been found that the melting temperature of the solid phase is largely dependent to the alkyl chain length of the oils, but it is practically the same with the change of surfactant concentration. The melting temperature increases linearly with the total carbon number in the hydrocarbon skeleton of alkanes [69, 74]. Increasing the carbon number of oils, say from octane to hexadecane, is expected to hinder the penetration of oils to the surfactant chain and, hence, the melting temperature increases. In the fixed oil system, the melting temperature increases with increasing lipophilic chain length of the surfactant, that is, changing surfactant from $C_{12}G_1$ to $C_{14}G_1$.

2.2.2
Phase Behavior of Diglycerol Fatty Acid Ester/Oil Systems

In this section, the nonaqueous phase behavior of homologous series of diglycerol fatty acid esters in a variety of nonpolar oils will be discussed. The binary phase behavior of these surfactants in liquid paraffin, squalane, and squalene will be described first and then the phase behavior in *n*-alkanes and aromatic oils. The data presented in this section are rendered from Ref. [9] and [65], respectively.

2.2.2.1 Phase Behavior in Liquid Paraffin, Squalane, and Squalene

The phase behavior of diglycerol fatty acid esters (C_mG_2, $m = 12$, 14, and 16) in liquid paraffin oil, squalane, and squalene is discussed. This section compares the nonaqueous phase behavior of mono- and diglycerol-based nonionic surfactants in different oils. Phase diagrams of surfactant/oil binary systems for the C_mG_2 ($m = 12$–16) surfactants in a wide range of temperature and concentration at atmospheric pressure are shown in Figure 2.2. Contrary to the C_mG_1 surfactants, the C_mG_2 surfactants form a variety of self-assembled structures in nonpolar oils liquid paraffin oil, squalane, and squalene.

One can see a surfactant solid phase at lower temperatures, which does not swell with oil and its melting temperature is practically constant over a wide range of compositions. As can be seen in Figure 2.2 the melting temperature increases with increasing hydrocarbon chain length of the surfactant. In each surfactant system, the melting temperature is practically the same upon addition of oil. The melting of the pre-surfactant is observed in WAXS patterns. As shown in Figure 2.3, there appears one sharp peak in a wide-angle regime at lower temperatures whose position (q-value) corresponds to the characteristic distance ~0.42 nm in real space. This high-q peak is the signature of solid state of surfactant that arises from the crystalline structure of the hydrophobic tail. It diminishes at a temperature between

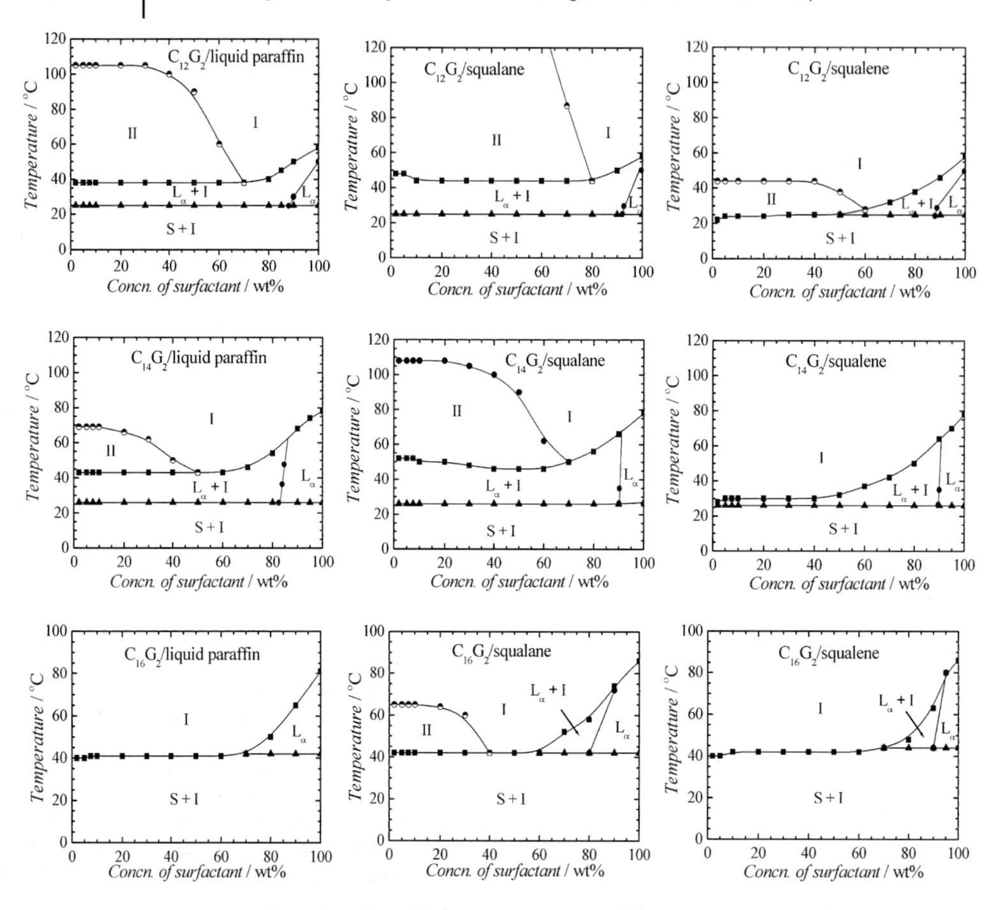

Figure 2.2 Phase diagrams of C_mG_2 (m = 12–16) in different nonaqueous solvent systems over a wide range of temperatures and concentrations at atmospheric pressure (S = solid, L_α = Lamellar liquid crystal, I = isotropic solution, and II = isotropic two-liquid phase).

25 and 35 °C. The solid shows a lamellar structure in the SAXS region, with a strong single peak in the WAXS region. Hence, this crystal is considered to be α-solid or L_β phase [76–79]. After disappearance of the WAXS peak, we still observe equidistant sharp reflections in the small-angle region ($q < 2$), indicating that the solid surfactant is melted and the L_α phase is formed.

After the solid surfactant is melted, a liquid crystal is formed, whose structure is identified to a lamellar liquid crystal confirmed by SAXS. The lamellar liquid crystal (L_α) phase can solubilize some amount of oil and get swollen, but with successive addition of oil, a two-phase region consisting of excess oil and the L_α phase, in which the interlayer spacing is unchanged with the compositions, is obtained see Figure 2.4. In all the systems, a lamellar-liquid-crystal (L_α) phase is formed in a concentrated region after melting of the solid phase and in the dilute regions reverse vesicles are formed at temperature between the solid melting

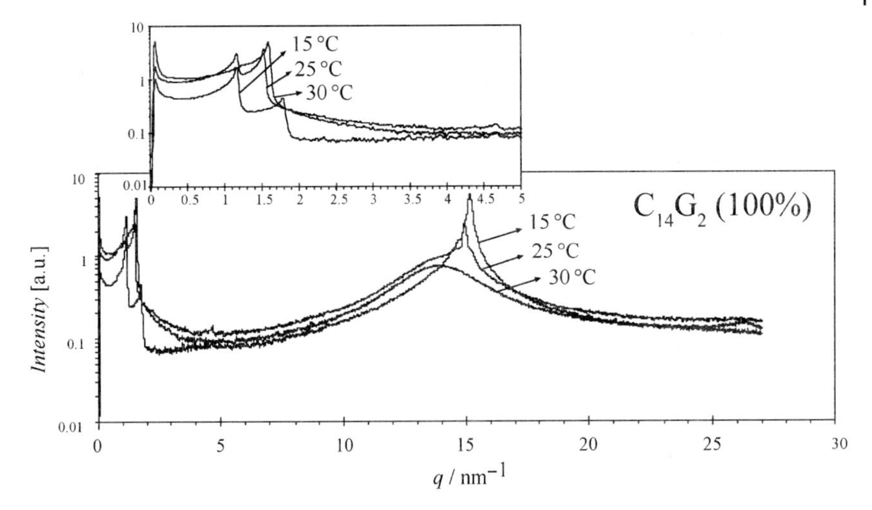

Figure 2.3 Small-and wide-angle X-ray scattering patterns of pure $C_{14}G_2$ at different temperatures.

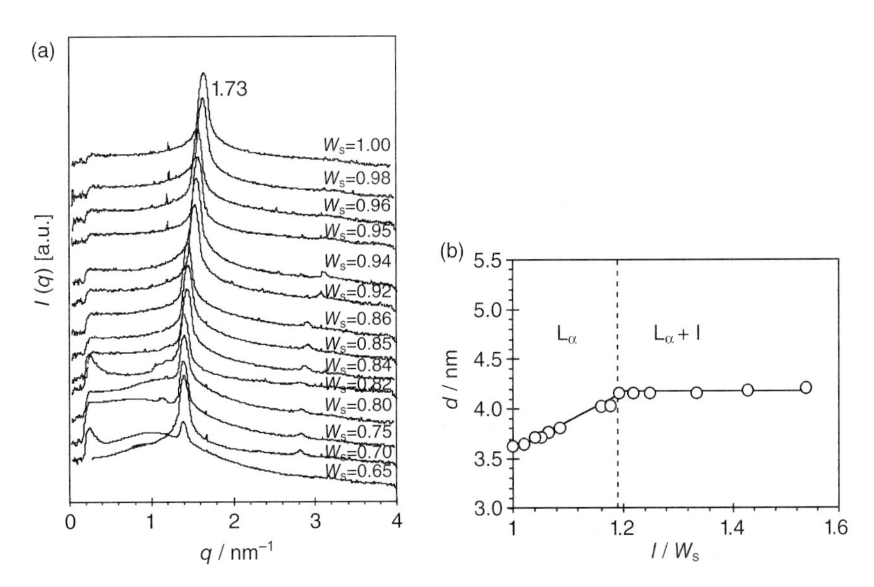

Figure 2.4 (a) SAXS patterns of $C_{14}G_2$/liquid paraffin systems at 35 °C at various surfactant weight fractions, W_S, and (b) the corresponding variation of the interlayer spacing, d, as a function of I/W_s.

temperature and isotropic two- or single-phase regions. At higher temperatures, above the L_α + I regions, there are two liquid-phase regions in all the $C_{12}G_2$/oil systems and are absent from the $C_{14}G_2$/squalene system. This two-liquid phase region is present only in squalane with $C_{16}G_2$ surfactant and it appears soon after the solid phase melts. As mentioned earlier, this two-phase boundary corresponds to the cloud-point curve of nonionic surfactant aqueous solutions. However, instead of being less soluble in water at high temperature for the cloud point, the surfactant becomes more soluble in the organic solvents at high temperature. Namely, the effect of temperature on the solubility is opposite to the clouding phenomenon. When the hydrocarbon chain of diglycerol surfactant decreases, the two-phase region becomes wider. In the case of a fixed surfactant, the surfactant is most miscible with squalene (narrowest two-phase regions) and the order of dissolutions tendency is squalene > liquid paraffin > squalane. These results show that hydrophilic moiety (diglycerol group) is more insoluble in oils compared to the conventional poly(oxyethylene) type nonionic surfactants.

Similar to the C_mG_1/oil systems, the turbid solution in the C_mG_2/oil systems separates into two transparent liquid phases after a few minutes. As mentioned earlier, judging from the solubility curve, one phase is an oil-rich phase with less surfactant and the other phase is the surfactant phase containing a considerable amount of oil. With increasing temperature, the penetration of oil in the surfactant chain increases and flat or nearly flat aggregates transform to cylindrical or spherical aggregates, depending on the lipophilic chain length of the amphihile. The present result confirms the penetrating theory of Ninham and coworkers [80]. When water is added, three coexisting phases (water, surfactant, and oil phases) would appear. In fact, in the water/C_mG_2/oil systems, a three-phase region is observed. Judging from the width of two-phase regions, $C_{14}G_2$/squalene is most soluble compared with $C_{14}G_2$/liquid paraffin and $C_{14}G_2$/squalane. This tendency is the same for other surfactant systems. Note that squalene has 6 double bonds in the molecule and is more polar than the other two. The liquid paraffin is a blend of hydrocarbons but the molecular weight of squalane (saturated hydrocarbon) is larger.

2.2.2.2 Phase Behavior in Alkanes and Aromatic Oils

From the preceding section, it is very clear that the diglycerol-based nonionic surfactants have the potential to form a variety of self-assembled structures in long-chain branched oils squalane and squalene and in blends of hydrocarbon oils like liquid paraffin. Then, the question arose how do these surfactants behave in aromatic oils and in *n*-alkanes? To answer this question, the nonaqueous phase behavior of a particular surfactant diglycerol monomyristate ($C_{14}G_2$) in *n*-alkanes (octane to hexadecane) and in two different aromatic oils was constructed. Figure 2.5 shows the binary phase diagrams of $C_{14}G_2$/alkane systems and for the sake of comparison a phase diagram with a cycloalkane (cyclohexane) is also presented. The Figure 2.5 also includes the phase diagrams of $C_{14}G_2$ in phenyl octane (an aromatic oil).

We note that the phase behavior of $C_{14}G_2$ in alkanes is pretty much similar to the phase behavior in liquid paraffin or squalane. One can see a solid, lamellar liquid crystal, and an isotropic reverse micellar phase with successively increasing temperature in the surfactant axis. The solid surfactant at lower temperatures can

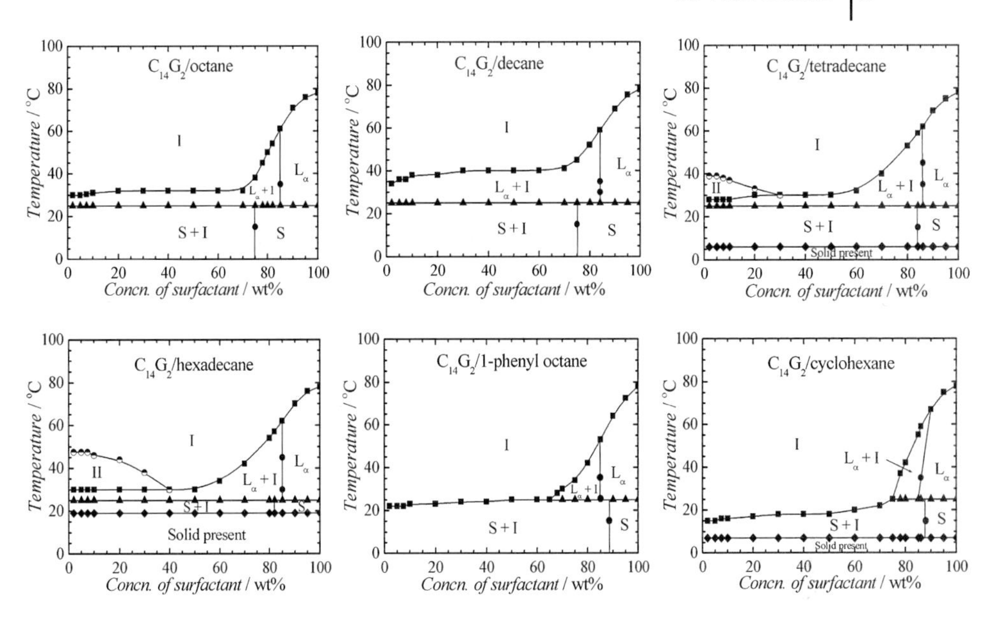

Figure 2.5 Phase diagrams of $C_{14}G_2$ in alkanes (from octane to hexadecane) and two different aromatic oils phenyl octane and ethyl benzene (S = solid, L_α = Lamellar liquid crystal, I = isotropic solution, and II = isotropic two-liquid phases).

solubilize some amount of oil and the extent of oil solubilization depends on the nature of the oils, as confirmed by the plot of interlayer spacing distance, *d*, as a function of reciprocal of weight fraction of the surfactant. The value of *d* increases with oil concentration and after a certain concentration of oil attains a plateau value [65]. The constant *d* value represents the saturation point of oil solubilization and no more oil can be incorporated in the solid phase. Further addition of oil results in dispersion of solid phase in oil. The solid phase transforms into a lamellar liquid-crystalline phase upon heating. The L_α phase potentially solubilizes some amount of oil and gets swollen, but with successive addition of oil, a two-phases region consisting of an excess oil phase and L_α phase emerges. In the dilute regions of long-chain alkanes, reverse vesicles are observed [65]. Contrary to the *n*-alkane systems, an isotropic reverse micellar solution phase exists in the dilute region of $C_{14}G_2$/aromatic oil system at normal room temperature. At higher temperatures, the L_α phase melts to either isotropic single-liquid phase or turbid two-phase solutions, as shown in Figure 2.5. The turbid solution region is mainly observed in the case of thc longer-chain alkanes such as in tetradecane and hexadecane, and it is wider in the latter system. The turbid solution separates into two transparent liquid phases after a few minutes and transforms to an isotropic single-phase solution consisting of reverse micellar aggregates upon heating, Figure 2.5.

2.3
Structure of Reverse Micelles

A number of important interfacial phenomena, such as solubilization or detergency depend on the micellization behavior of the amphiphilic molecules. Considering the recent upward trend of utilizing micellar solutions as a template for the preparation of nanomaterials, such as nanocrystals or mesoporous materials, controlling freely the size and shape of micellar aggregates is of central importance from the application viewpoint because the geometry of micellar aggregates largely affects the structure of the products [11, 12, 17, 18]. The studies on the self-aggregation of amphiphilic molecules have a long history. However, much less is known about the formulations and structures of the reverse micellar aggregates in nonpolar organic solvents if compared to aqueous systems.

It is readily known that when surfactant molecules are added in nonpolar organic solvents, there is a tendency to minimize contact between the solvent and the hydrophilic headgroup of the amphiphile and thus reverse micelles are formed with an opposite structure compared to the conventional normal micelles in aqueous media. In contrast to the aqueous systems, generally, the structure of the nonpolar oils is largely unaltered by the presence of the amphiphile molecules. The interaction of the surfactant hydrophobic tail with the solvent molecules is just as favored as with other surfactant tails. In such a case, there would be no strong tendency for the formation of large micellar aggregates. However, the dipole–dipole interaction between the hydrophilic headgroups may act as one of the driving forces of the formation of reverse micelles in nonpolar media. In addition, the transfer free energy of a hydrophilic headgroup from a hydrophilic environment to oil may provide insights into the mechanism of the reverse micellar formation. The following sections will describe the structures of reverse micelles based on mono- and diglycerol-based nonionic surfactants in different organic solvents and the tunable parameters that control the geometry of the reverse micelles.

2.3.1
Monoglycerol Fatty Acid Ester-Based Reverse Micelles

2.3.1.1 Structure of Micelles in Liquid Paraffin, Squalane, and Squalene

In this section, the structures of monoglycerol based reverse micelles in a long-chain branched and a blend of hydrocarbon oils will be described and are based on Ref. [64]. As shown in Figure 2.1, the monoglycerol fatty acid ester nonionic surfactants (C_mG_1) form an isotropic solution at higher temperatures after melting of the solid phase. The structures of the reverse micelles formed in these single-phase regions are characterized by the small-angle X-ray scattering (SAXS) technique. In the following subsections, how temperature, lipophilic chain length of the surfactant, composition, solvent, and added water affect the structure of reverse micelles will be discussed.

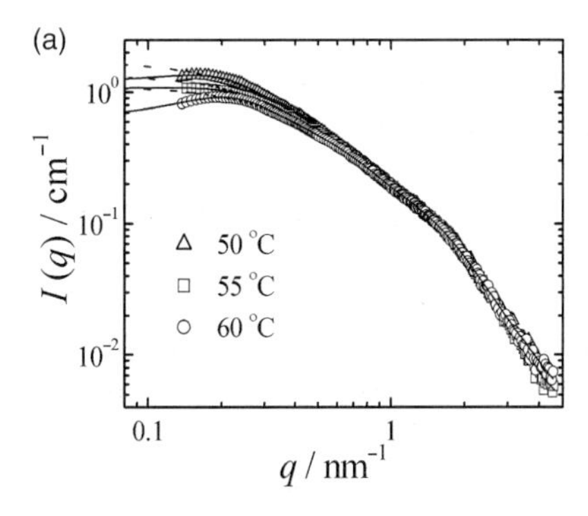

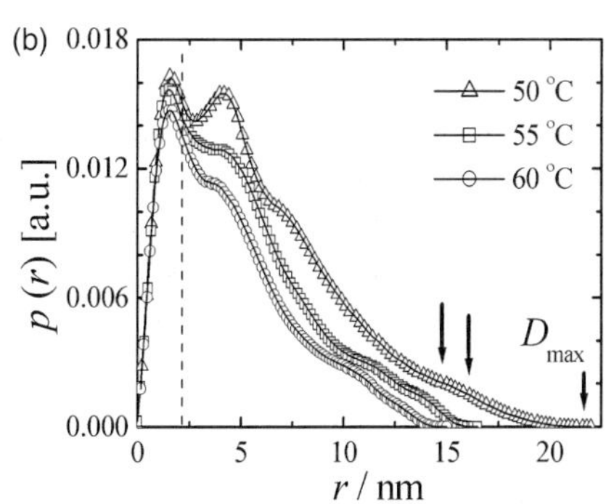

Figure 2.6 (a) The X-ray scattered intensities, $I(q)$, of the 5 wt% $C_{10}G_1$/liquid paraffin in absolute unit at different temperatures (50, 55, and 60 °C), and (b) the pair-distance distribution functions (PDDFs), $p(r)$, extracted from these scattering curves with the GIFT method. Solid and broken lines in (a) represent GIFT fit and the calculated (total) form factor for n particles existing in unit volume, $nP(q)$, respectively. The same representation will continuously be used for all successive Figures displaying scattering functions. The arrows in (b) indicate the maximum dimension of micelles, D_{max}, and the broken line on the inflection point of $p(r)$ after the maximum highlights the core diameter.

Effect of Temperature Temperature-induced microstructure transition of the reverse micelles observed in $C_{10}G_1$/liquid paraffin system as obtained by SAXS is presented in Figure 2.6, which shows the scattering functions, $I(q)$, and the resulting $p(r)$-functions for the 5 wt% system at different temperatures (50, 55, and 60 °C).

The strong dependence of scattering intensity on the scattering vector in the $I(q)$ *vs.* q curves gives a clear picture on the formation of aggregate structure in the 5 wt% $C_{10}G_1$/ liquid paraffin systems. Minute observation of Figure 2.6a reveals that the scattering intensity in the forward direction ($I(q=0)$) monotonically decreases with increasing temperature without changing the scattering behavior in the high-q regions. Alternatively, one can say that the low-q slope of the $I(q)$ increases with decreasing temperature. The decreasing low-q slope of the $I(q)$ and the resulting decrease in the maximum dimension, D_{max}, with the rise of temperature provide direct evidence for deceasing micellar size with increasing temperature. As shown in Figure 2.6b, all the $p(r)$-curves exhibit a sharp peak in the low-r regime and an extended tail to the high-r side, which is the typical feature of an elongated rod-like particle with nearly homogeneous electron density distribution [45]. The inflection point that appears on the slightly higher r-side of the maximum in the $p(r)$-curves semiquantitatively gives the cross-sectional diameter of the hydrophilic core. This cross-sectional diameter is almost unchanged at all temperatures studied. However, the maximum dimension of the elongated particles, D_{max}, decreases from ~22 to ~15 nm upon increasing temperature from 50 to 60 °C.

Unlike the aqueous surfactant systems, the miscibility of solvent and surfactant increases in surfactant–oil systems with the rise of temperature because the surfactant molecules become increasingly lipophilic. The increased solubilization may also be due to increased thermal agitation, which would increase the space available for the solubilization in the micelle. With increasing temperature, the penetration of oil in the surfactant chain increases in consistence with the penetration model of Ninham and coworkers [80, 81]. Therefore, the length of the cylindrical aggregates decreases, which is essentially a rod-to-sphere type transition in the micellar structure. Besides, increasing temperature increases the surfactant's hydrophobic character and the van der Walls interaction between the hydrocarbon chain of surfactant and oil increases and this eventually increases the oil-solubilization capacity of the surfactant. This is in good agreement with the phase behavior; more and more oils are solubilized at higher temperature, as shown in Figure 2.1. Moreover, it has been shown that increasing temperature decreases the aggregation number of aggregates in nonpolar media [82] and less elongated or spherical-type particles are favorable at elevated temperatures.

Effect of Alkyl Chain Length of Surfactant Figure 2.7 shows the scattering functions, $I(q)$, and the corresponding $p(r)$-curves, for the 5 wt% $C_{10}G_1$ and $C_{12}G_1$ in squalene at 60 °C. The data provide direct evidence of micellar elongation (an increase in D_{max} with an extended tail in $p(r)$-curves) with decreasing hydrocarbon chain length of the monoglycerol-based nonionic surfactant. This result is in agreement with the phase diagrams. Namely, the short hydrocarbon chain surfactant tends to separate from an oil phase, as shown in Figure 2.1.

As can be seen in Figure 2.7a the scattering intensities of the short-chain surfactant $C_{10}G_1$ is much higher over a wide q range compared to the long-chain surfactant $C_{12}G_1$ indicating the formation of bigger particles in the former system. The difference in the micellar structure depending on the alkyl chain of the surfactant can better be seen in the $p(r)$-curves of Figure 2.7b, which shows that the

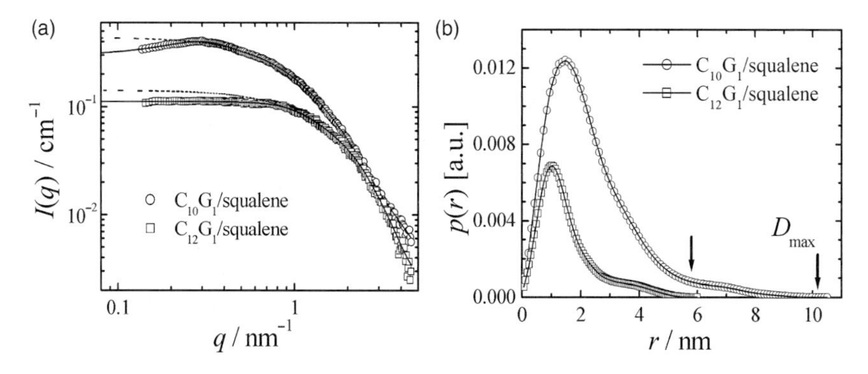

Figure 2.7 (a) The X-ray scattered intensities $I(q)$, and (b) the corresponding $p(r)$-functions for the 5 wt% $C_{10}G_1$/squalene and $C_{12}G_1$/squalene systems at 60 °C.

ellipsoidal prolate-type micelles with maximum diameter ~6 nm observed in the $C_{10}G_1$/squalene system changes to more elongated particles with maximum diameter ~10 nm in the $C_{12}G_1$/squalene system.

Compared to the aqueous systems, much less is known regarding the aggregation numbers of reverse micelles in nonpolar media. However, from the data available, it is obvious that the average aggregation number of reverse micelles in nonpolar media increases with an increase in dipole–dipole attraction or intermolecular bonding between the polar headgroups, which decreases with increase in the alkyl chain length of the surfactant, number of the carbon chain per surfactant molecule, the steric requirements of the chain close to polar headgroups, and the temperature. The present result supports the theory of Ruckenstein and coworkers [82] that predicts a decrease in the aggregation number with increasing hydrocarbon chain of surfactants.

Effect of Surfactant Concentration Monoglycerol-based nonionic surfactant micelles in organic solvent show micellar growth with concentration. Figure 2.8 shows the scattering functions, $I(q)$, and the corresponding $p(r)$-curves for the $C_{12}G_1$/squalene as a function of surfactant concentration at 60 °C. As expected, increasing surfactant concentration increases the total scattering intensity due to the increased number density of the scattering objects, indicating a pronounced elongation in the micellar shape or an increase in the aggregation number.

A monotonous growth in the maximum size of the micelles with surfactant concentration can be seen in the $p(r)$-curves. As can be seen in Figure 2.8b, all the $p(r)$-curves exhibit the typical feature of an elongated rod-like particle in view of a pronounced peak in the low-r regime and an extended tail to the high-r side. The absence of a local maximum and minimum on the lower-r side indicates nearly homogeneous electron density distribution (no core-shell structure) in scattering particles. This behavior is similar to the surfactant/water systems, where

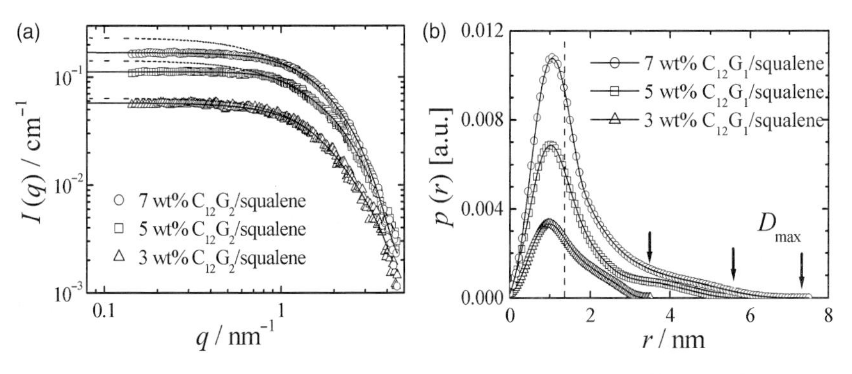

Figure 2.8 (a) The scattering curves of the $C_{12}G_1$/squalene systems at different surfactant concentrations obtained on absolute scale at 60 °C, and (b) the resulting $p(r)$-functions.

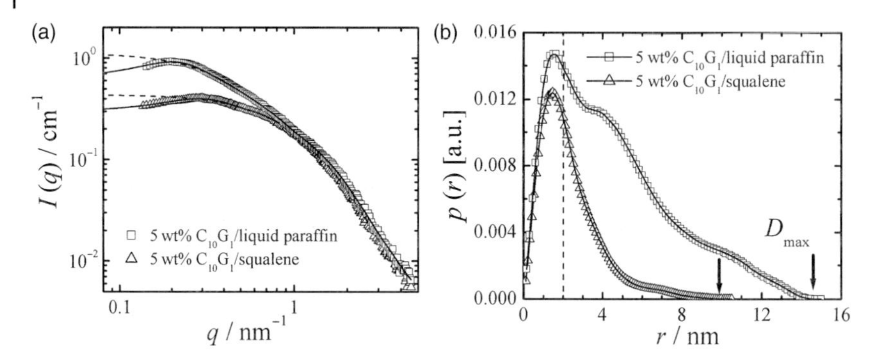

Figure 2.9 (a) The scattered intensities of 5 wt% $C_{10}G_1$/squalene, 5 wt% $C_{10}G_1$/liquid paraffin obtained in absolute unit at 60 °C, and (b) the corresponding $p(r)$-curves.

the micellar growth with increasing surfactant concentration is caused due to entropy effects. Thus, once can say that the micellar growth (sphere-to-rod transition) is expected to occur in conventional as well as in the reverse micellar system.

Effect of Oil and Added Water The SAXS data show that the micellar growth is favored upon changing the oil from squalene to liquid paraffin. The scattering functions, $I(q)$, and the resulting $p(r)$-curves for the $C_{10}G_1$/squalene and $C_{10}G_1$/liquid paraffin systems at 60 °C are shown in Figure 2.9. One can clearly see in the scattering functions that the $I(q)$ for the $C_{10}G_1$/liquid paraffin shows a higher low-q slope than the $C_{10}G_1$/squalene, corresponding to an enhanced one dimensional micellar growth in the former system, which is also evident from the corresponding $p(r)$-curves shown in Figure 2.9b. The drop that appears at $q <$ ~0.2–0.3 nm^{-1} in the $C_{10}G_1$/liquid paraffin system shows that the system has an interparticle interference effect. A small bump that appears at $r \sim 5$ nm in the $p(r)$-curves of the $C_{10}G_1$/liquid paraffin system is come from the interparticle interference effect. In the squalene system, where smaller ellipsoidal prolate-type aggregates are formed, the mutual dissolution of surfactant and oil is the highest among three oils, see phase behavior in Figure 2.1. The result shows that when the mutual miscibility of surfactant/oil increases, the aggregates become shorter. This fact, supported by the effect of temperature and lipophilic chain length on the micellar size discussed above, may be explained in terms of the increased penetration of the miscible oil in the lipophilic layer of the aggregate.

It is well known that a small amount of water induces micellization of the poly(oxyethylene)-type nonionic surfactants in nonpolar organic solvents. Even when surfactant aggregation does not occur, or the aggregation number is very small in a particular solvent in the absence of other materials, the addition of an insoluble solvent like water, may give rise to aggregation with consequent solubilization of the additives [83]. Addition of water is expected to decrease the spontaneous curvature as a result of increase in the headgroup repulsion. Besides, the added water solubilized in the interior of the micelle core in the nonpolar medium

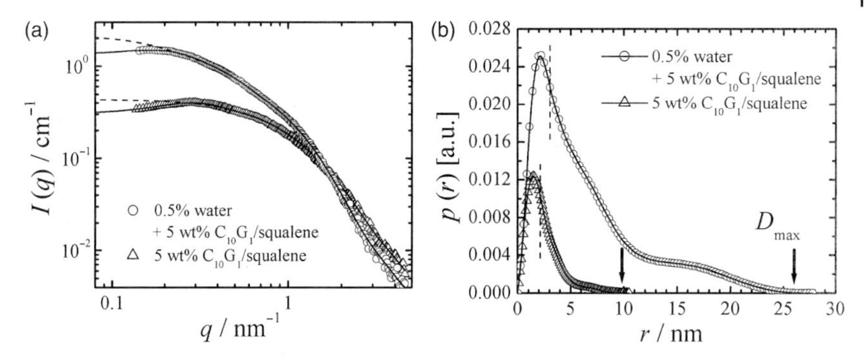

Figure 2.10 (a) The scattered intensities, $I(q)$, of the 5 wt% $C_{10}G_1$/squalene system without and with water in absolute unit at 60 °C, and (b) the corresponding $p(r)$-curves.

has been shown to cause an increase in the aggregation number [84]. Figure 2.10 shows the $I(q)$ and corresponding $p(r)$-functions for the 5 wt% $C_{10}G_1$/squalene system with and without water. Upon addition of 0.5 wt% of water to 5 wt% $C_{10}G_1$/squalene the low-q scattering intensity is enhanced strongly indicating a dramatic elongation in the reverse micellar structure, which is further proven by Figure 2.10b. A shift in the position of the inflection point after the maximum of the $p(r)$-curves to the higher r-side indicates an increase in cross-sectional diameter of the hydrophilic core. Hence, the data provide direct structural evidence, showing that added water induces a drastic elongation and simultaneously a slight expansion of the core diameter of the reverse micelles.

2.3.1.2 Structure of Micelles in n-Alkanes

This section is based on Ref. [69] and will describe the structure of monoglycerol-based nonionic surfactant micelles in *n*-alkanes. The effects of solvent properties, temperature, the alkyl chain length of surfactant, surfactant concentration, and added water or glycerol will be mainly discussed. The glycerol α-monomyristate ($C_{14}G_1$) is taken as the main surfactant.

Effect of Solvent Properties As mentioned in Section 2.2.1.2, the $C_{14}G_1$ could not form any self-assembled structures in *n*-alkanes at room temperature and tends to separate from the oil phase. In dilute regions this surfactant forms a solid dispersion, which upon heating transforms into an isotropic reverse micellar solution. Assuming that the solvent properties have a significant impact on the inverse micellar structure in the surfactant/oil binary systems, the SAXS measurements are performed on the 5 wt% $C_{14}G_1$ system in different *n*-alkanes (*n*-octane, *n*-decane, *n*-tetradecane, and *n*-hexadecane) at 70 °C. This is the temperature at which all the systems exist in an isotropic solution. Interestingly, the $C_{14}G_1$ has a potential to form the inverse micelles with different structures depending on the chain length of alkanes without external water addition. The scattering functions, $I(q)$, and the corresponding $p(r)$-functions, extracted from the GIFT method for the

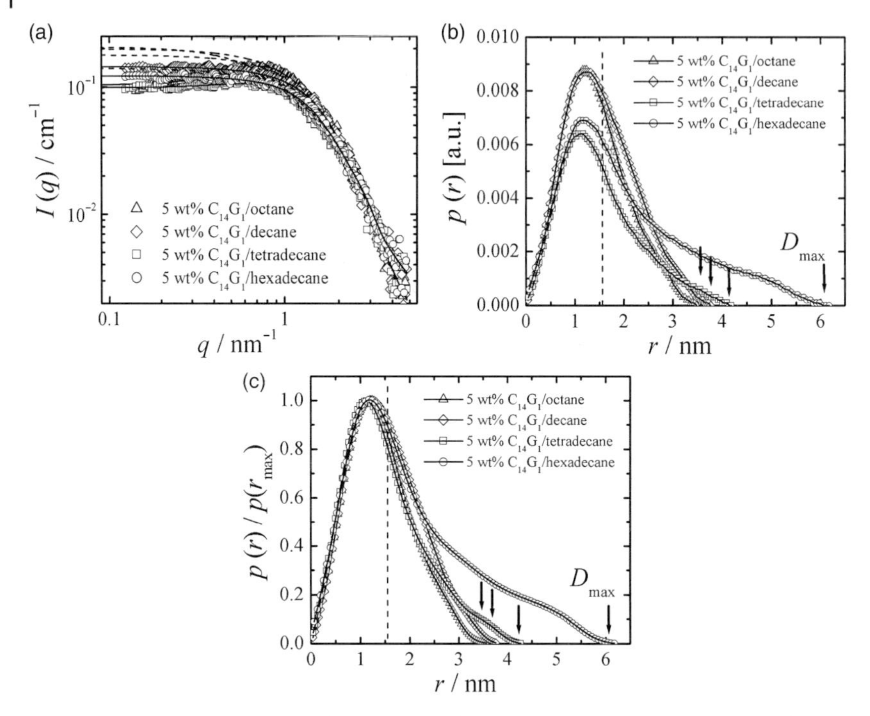

Figure 2.11 (a) The X-ray scattered intensities, $I(q)$, of 5 wt% $C_{14}G_1$/oil systems in absolute units at 70 °C, (b) The $p(r)$-functions (PDDFs) extracted with GIFT analysis, and (c) the normalized $p(r)$-functions ($p(r)/ p(r_{max})$). The arrows and broken lines in panel b and c indicate the maximum length of micelles, D_{max}, and the cross-sectional diameter of the reverse micellar core, respectively.

$C_{14}G_1$/oil systems displayed in Figure 2.11 compare the effects of chain length of alkanes on the aggregation behavior of the $C_{14}G_1$.

The $I(q)$ changes mainly in the low-q region keeping high-q scattering functions virtually identical upon changing oil from octane to hexadecane (Figure 2.11a). Minute observation reveals that the low-q slope of the scattering function, which is proportional to the higher-r side of the $p(r)$ curves increases with increasing chain length of the alkanes. Thus, without going into the depth, merely judging from the scattering behavior one can say that changing oil from octane to hexadecane favors the micellar growth. The structure of the scattering particles depending on the solvent properties can be read out from the $p(r)$-functions shown in Figure 2.11b. A slightly asymmetric shape without a local maximum and minimum in the $p(r)$-curve of the 5 wt% $C_{14}G_1$/octane system confirms slightly elongated ellipsoidal prolate-type particles with nearly homogeneous electron density distribution inside the micellar core. The maximum dimension, D_{max} of the reverse micelle goes on increasing with the chain length of oil. A pronounced peak in the low-q region with an extended tail in the high-q regime of the $p(r)$ in the $C_{14}G_1$/

hexadecane system indicates the rod-like structure of the scattering particles. The inflection point observed on the slightly higher-r side of the maximum in the $p(r)$ curves (as indicated by a dotted line at r ~2 nm in Figures 2.11b and c) qualitatively highlights the cross-sectional diameter of the observing particles. The position of the inflection point is practically unchanged upon changing oil from octane to hexadecane indicating that the alkyl chain length of the oils does not induce significant changes in the internal structure of the reverse micelles. Note that as expected from the identical contrast of oils and hydrophobic part of the surfactant, the SAXS could efficiently detect only the hydrophilic core of the reverse micelles. Figure 2.11c presents normalized $p(r)$-curves, $p(r)/p(r_{max})$, for better visibility of the micellar growth.

The formation of smaller prolate-type particles in octane and bigger rod-like particles in hexadecane system can be attributed to the penetration tendency of octane to the lipophilic chain of the surfactant. The long-chain oil cannot penetrate the lipophilic chain of the surfactant and thus the critical packing decreases and, hence, rod-like particles are more likely to form. Besides, it is possible that the transfer free energy of the glycerol moiety from a hydrophilic environment to the hydrophobic oily environment differs depending on the chain length of the oils and determines the aggregation behavior of surfactants in oils.

Effect of Temperature Figure 2.12 shows the $I(q)$ and the resulting $p(r)$ for the 5 wt% $C_{14}G_1$/tetradecane system at different temperatures (60 and 70 °C). The decreasing trend of the low-q intensity with temperature without affecting the scattering behavior in the high-q regions can be seen in the $I(q)$ curves. Such feature in the scattering functions is the signature of micelle shortening. Increasing temperature from 60 to 70 °C, leads to a decrease of the micellar size from ~4.9 nm to 4.1 nm, see Figure 2.12b, that is, increase in temperature by 10 °C decreases the micellar size by ~16%. Thus, the present SAXS data quantitatively show that the

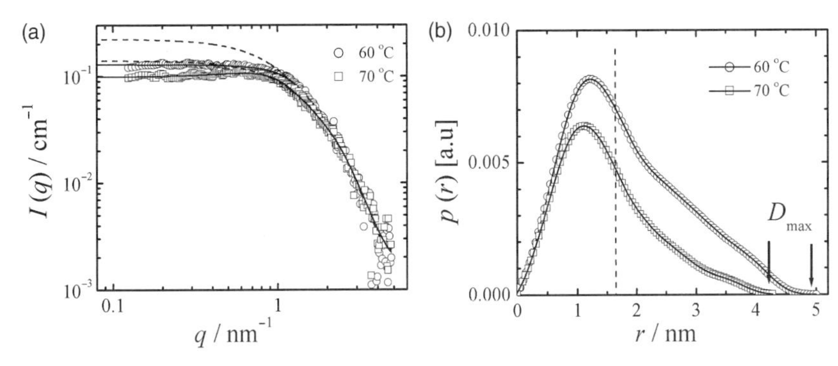

Figure 2.12 (a) The normalized X-ray scattering intensities, $I(q)$, of the 5 wt% $C_{14}G_1$/tetradecane system at 60 and 70 °C, and (b) the corresponding $p(r)$-curves. Arrows and broken line in panel b highlight the maximum dimension, D_{max}, and core diameter of the micelles, respectively.

length of the aggregates decreases with increasing temperature, which is essentially a rod-to-sphere-type transition in the micellar structure.

Micelle shortening with temperature in nonaqueous system can be regarded as a common phenomenon. As proven by the phase-behavior study, contrary to the aqueous systems of poly(oxyethylene)-type nonionic surfactant, the miscibility of oil and surfactant in nonaqueous systems increases with temperature. Thermal agitation increases with temperature, and so does the space available for the solubilization in the micelle. Furthermore, the hydrophobicity of the surfactant increases at higher temperatures so that the van der Walls interaction between the hydrocarbon chain of the surfactant and the oils increases, which leads to an increase in the oil-solubilization capacity of the surfactant. Furthermore, increasing temperature increases the oil penetration tendency to the surfactant chain. As a result, the packing parameter increases and hence the length of the reverse micelles decreases. Note that increasing the temperature induces similar effects to those from a decrease in the alkyl chain length of the oils. In both cases, the interpenetration of oil and surfactant chain increases and micelles shrink, that is, a rod-to-sphere transition is favorable.

Effect of Surfactant Concentration Figure 2.13 shows the $I(q)$ and the corresponding $p(r)$ curves for the $C_{14}G_1$/octane system at different concentrations at 60 °C. As generally expected, with increasing volume fraction of the surfactant, the total scattered intensity increases due to increased scattering volume. As can be seen from Figure 2.13a, increasing surfactant concentration (expressed in wt%) increases the total scattering intensities, and even if after normalization of the intensities by concentration the low-q intensity becomes higher for higher concentration. The higher forward scattering intensity at higher concentration is the signature of larger scattering particles, which can be clearly seen from the real space functions shown in Figure 2.13b.

Figure 2.13b clearly shows that all the pair-distance distribution functions curve to exhibit the typical feature of elongated particle, as confirmed by the pronounced peak in the low-r regime and an extended tail to the high-r side. Increasing surfactant concentration increases the maximum length of the particle, D_{max}, as highlighted by arrows in Figure 2.13b. As was mentioned before, the position of the inflection point seen on the higher-r side of the maximum in the $p(r)$-curves semi-quantitatively indicates the internal structure that is, the cross section of the micelles. Since the position of the inflection point seems to be unaffected by the volume fraction of the surfactant, as indicated by the broken line, we would expect a similar internal structure (cross-sectional diameter) of the elongated inverse micellar aggregates at all surfactant concentrations studied. Growth to the maximum length of the micelles keeping the cross section constant is the signature of the one-dimensional micellar growth. The $p(r)$ functions show that the maximum length of the particles changes from ~3.2 to ~6 nm upon increasing surfactant concentration from 3 to 15 wt%, while the hydrophilic core diameter estimated from the position of the inflection point stays virtually constant at

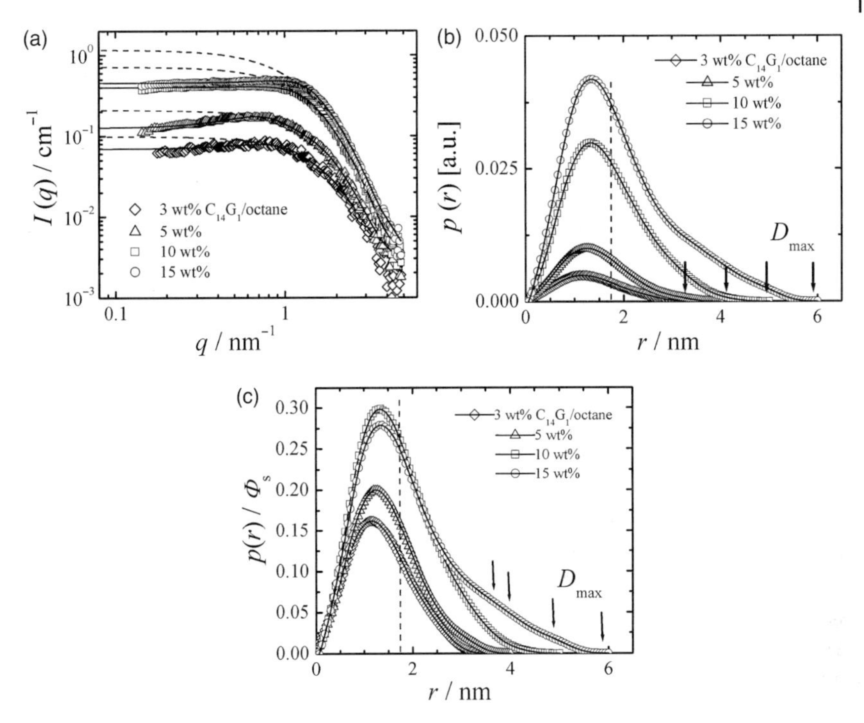

Figure 2.13 (a) The scattering curves $I(q)$ of the $C_{14}G_1$/octane systems at different surfactant concentrations obtained on absolute scale at 60 °C, and (b) the corresponding $p(r)$-functions. The growth of the micelles can be seen best in panel (c) where the curves are divided by the surfactant volume fraction Φ_s. Without growth all curves would be on top of each other.

~2 nm. The growth of the micelles can be seen best in Figure 2.13c where the curves are divided by the surfactant volume fraction Φ_s. Without growth all curves would be on top of each other. Thus, the present SAXS data clearly highlight the onset of one-dimensional micellar growth with surfactant concentration. With increasing surfactant concentration, the volume fraction of the hydrophilic part increases and causes thermodynamic instability in the system due to increased free energy. As a result, the micelles tend to grow to achieve thermodynamic stability at the cost of excess free energy. Besides, considering the general postulation of spontaneous curvature in a reverse system, increasing surfactant concentration decreases the packing parameters, and hence elongated micelles are expected to form at higher surfactant concentrations.

Effect of Added Water or Glycerol The formation of reverse micelles and, hence, the existence of critical micelle concentration (*cmc*) in surfactant/oil binary systems is still a matter of open discussion. Surfactants usually do not form reverse micelles in organic solvents in absolute dry conditions unless a small amount of water or other polar additives are added [85–89]. However, traces of water can always be present as impurities. Added water or other polar additives tend to solubilize in the interior of the reverse micellar core and cause an increase in the aggregation number [84]. In the studies of the effect of ionic liquid on the size of the reverse micelles in supercritical CO_2, Li *et al.* [90], found that size of the reverse micelles increases considerably on increasing the amount of solubilized ionic liquid.

The effects of added water and glycerol on the geometry of reverse micelles formed in $C_{14}G_1$/oil systems at fixed temperature (60 °C) will be discussed in the following. First, the water effect is discussed. Figure 2.14 shows the $I(q)$ and the corresponding $p(r)$-curves for the $C_{14}G_1$/octane system at different concentrations of water at 60 °C. The forward scattering intensity $I(q = 0)$ is strongly enhanced and the scattering curve of the high-q region (~2 nm^{-1}) shifts towards the forward direction upon addition of water (Figure 2.14a). This indicates that the added water causes simultaneous changes in the maximum dimension and the cross-sectional structure of reverse micellar aggregates that is, two-dimensional micellar growth.

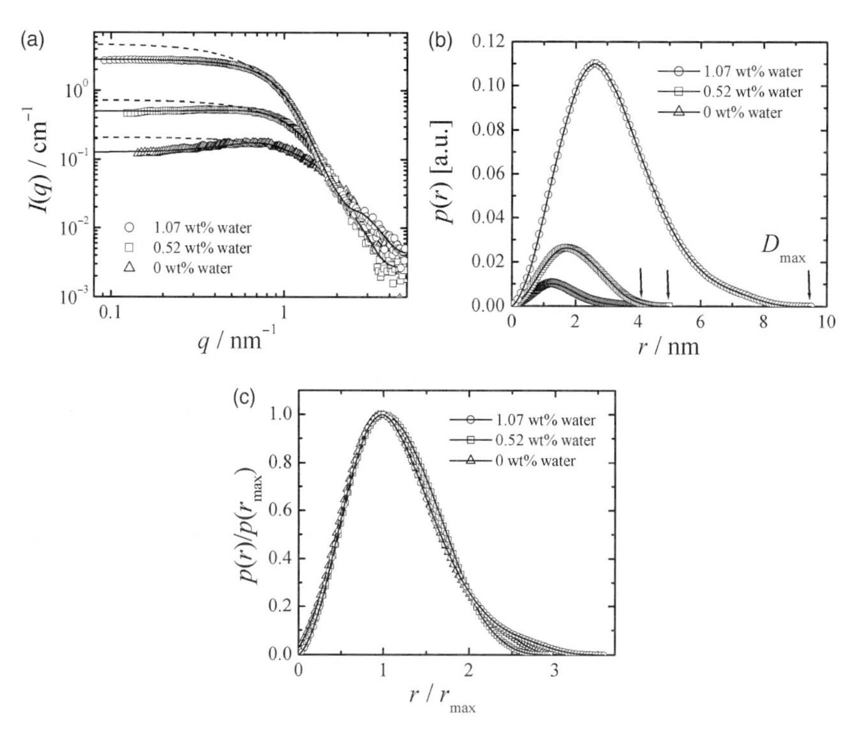

Figure 2.14 (a) The X-ray scattering intensities $I(q)$ of 5 wt% $C_{14}G_1$/octane with different concentration of added water obtained in absolute unit at 60 °C, (b) the corresponding $p(r)$-functions, and (c) a "master curve" plot of normalized $p(r)$-function, $p(r)/p(r_{max})$ *vs.* r/r_{max}.

The position of r_{max} of the maximum of $p(r)$ and maximum length of the micelles (D_{max}) increases with increasing water concentration, as can be seen in Figure 2.14b. This indicates the formation of swollen micelles with a water pool at the micellar core. All $p(r)$ functions fall on a "master curve" when plotted on relative axes r/r_{max} and $p(r)/p(r_{max})$ as shown in Figure 2.14c. This confirms that the shape does not change, only the size is increasing with the amount of water. The present results are in good agreement with the results described in Section 2.3.1.1. Such changes are possibly due to the fact that the water molecules have a tendency to make a water pool at the micellar core. Besides, water may go very close to the hydrophilic headgroup and form strong hydrogen bonding with the glycerol molecule so that the overall hydrophilic size of the surfactant increases, which favors micellar growth due to a lower critical packing parameter. Note that around 1.35 wt% water can be solubilized in the micellar core of the 5 wt% $C_{14}G_1$/octane system at 60 °C.

The effect of the mixing fraction of glycerol to water on the reverse micellar structure is discussed here. Only 0.65 wt% glycerol was soluble in the 5 wt% $C_{14}G_1$/octane system at 60 °C. Figure 2.15 shows the SAXS results on the glycerol-incorporated samples. For the sake of comparison, SAXS data of glycerol- and water-free samples is also presented. Minute observation of $I(q)$ *vs.* q curves reveal that the forward scattering intensity is decreasing with increasing mixing fraction of glycerol. This indicates a reduction of aggregation number and size of the micelles with increasing mixing fraction of glycerol in the system. As can be seen from the $p(r)$-curves in Figure 2.15b, the maximum dimension of the particles, D_{max} ca. 9.5 nm in the 5 wt% $C_{14}G_1$/octane + 1% gly:water (0 : 100) system, reduces to D_{max} ~ 6 nm in the 5 wt% $C_{14}G_1$/octane + 1% gly:water (50 : 50) system, that is, micellar size is reduced by ~38%. In order to have a better visibility on the effect of mixing fraction of glycerol to the micellar cross section, the plot of normalized $p(r)$-function *vs.* r is shown in Figure 2.15c, which shows that the position of r_{max} shifts towards the lower-r side on increasing the glycerol content in the system indicating the decrease in the cross-sectional diameter of the micelles. Figure 2.15d shows the "master plot", that is, the plot of normalized $p(r)$-function *vs.* r/r_{max}, which shows hardly any shape change, the size is reduced with increasing glycerol content and the micelles become a bit more globular. This change can easily be understood as a reduction of headgroup hydration by the presence of glycerol.

Increasing the temperature has a comparable effect on headgroup hydration as replacement of water by glycerol. The effect of temperature on the structure of micelles in the water-containing system is shown in Figure 2.16. Increasing the temperature from 60 to 90 °C leads to decreased scattering intensities in the forward direction see Figure 2.16a, indicating again a decrease in micellar size. As can be seen in Figure 2.16b, the D_{max} decreases from ~9.5 nm to 5.3 nm, that is, micellar size decreases by ~45% upon increasing the temperature from 60 to 90 °C. Judging from the shape of the normalized $p(r)$-functions *vs.* r/r_{max} shown in Figure 2.16c, one can say that temperature does only moderately modify the shape of the particles to be more globular, but the size shrinks essentially with increasing temperature.

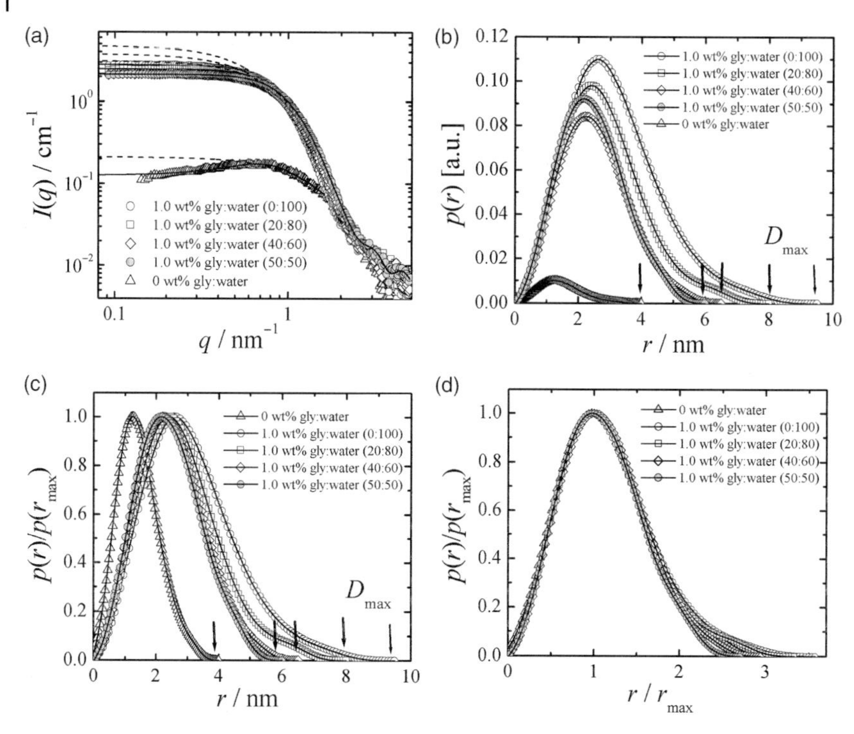

Figure 2.15 Effect of mixing fraction of glycerol on the reverse micellar structure: (a) The X-ray scattering intensities, $I(q)$, of the 5 wt% $C_{14}G_1$/octane + 1.0% gly:water systems at different mixing fractions of glycerol and water at 60 °C, (b) the corresponding pair-distance distribution functions (PDDFs), $p(r)$, (c) the normalized $p(r)$-function *vs.* r, and (d) the normalized $p(r)$-function *vs.* r/r_{max}.

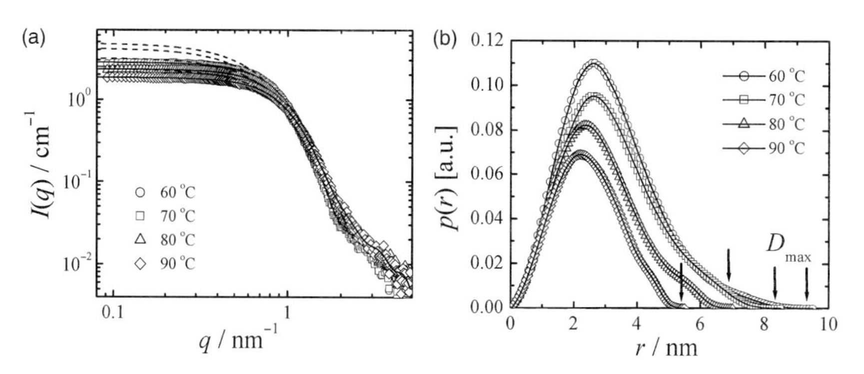

Figure 2.16 (a) The X-ray scattering intensities $I(q)$ of 5 wt% $C_{14}G_1$/octane + 1.0% water at different temperatures (60 °C to 90 °C), and (b) the corresponding $p(r)$-functions. The arrows in panel b indicate the maximum dimension of the scattering particles.

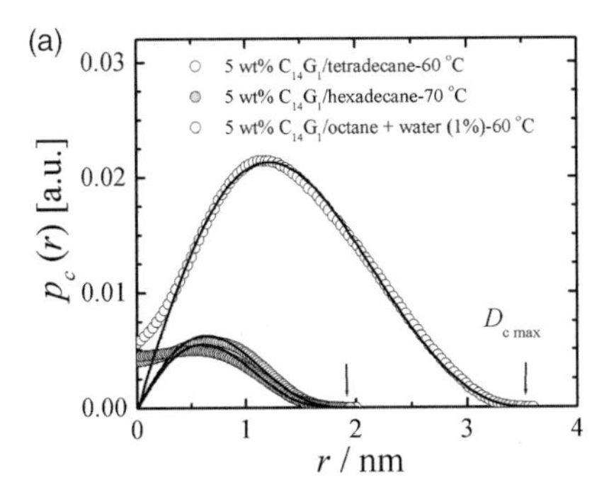

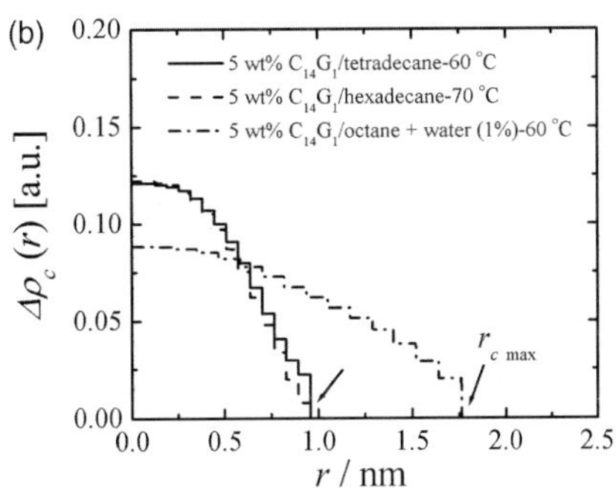

Figure 2.17 Direct cross-sectional (core) structure analysis: (a) The cross-sectional PDDF, $p_c(r)$, for the $C_{14}G_1$/oil systems with and without water, and (b) The corresponding electron density profile, $\Delta\rho_c(r)$, calculated via deconvolution from $p_c(r)$. Solid lines in panel (a) represent DECON fit.

Judging from the inflection point of the $p(r)$-curves located on the higher-r side of the maximum (Figures 2.11–2.13), the core diameter for all the $C_{14}G_1$/alkane systems is calculated in the range of 1.8–2.0 nm. To complement this estimation, the direct analysis for the cross-sectional structure is carried out, that is, deconvolution of the cross-section PDDF $p_c(r)$ into the radial difference electron density distribution profile $\Delta\rho_c(r)$ [42, 53]. The results are presented in Figure 2.17. For the sake of comparison the cross-sectional $p_c(r)$ functions for 5 wt% $C_{14}G_1$/octane + 1% water is also included in Figure 2.17.

Quantitative estimation of the cross-sectional diameters ~1.9 nm judged from the $D_{c\ max}$ in $p_c(r)$ of the $C_{14}G_1$ in tetradecane and hexadecane systems are very close to the value estimated from the inflection point in the total $p(r)$-function and almost identical in all linear-chain oils (data not shown). Addition of water significantly increases the cross-sectional diameter of the micelles indicating the formation of water pool at the micellar core.

2.3.2 Structure of Reverse Micelles Based on Diglycerol Fatty Acid Esters

In the following sections, the structure of reverse micelles in diglycerol-based nonionic surfactants will be described. First, the micellar structure in a long-chain branched oil and mixed hydrocarbon oil will be discussed. Then, the structure of micelles in diglycerol monomyristate ($C_{14}G_2$) in n-alkanes and aromatic oils will be discussed. The data presented in this section are rendered from Ref. [9], and [67], respectively.

2.3.2.1 Structure of Reverse Micelles in Liquid Paraffin and Squalane

As was described in Section 2.2.2.1, the diglycerol fatty acid esters (C_mG_2, m = 12–16) efficiently form a variety of self-assembled structures in nonpolar oils; liquid

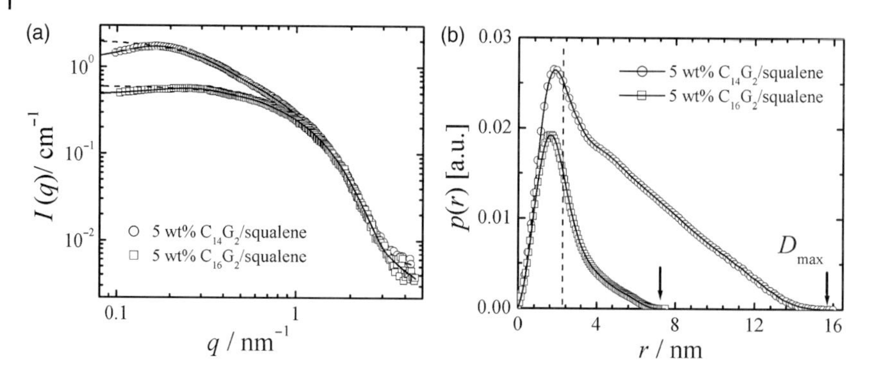

Figure 2.18 (a) The scattering functions of the 5 wt% $C_{14}G_2$/squalene and $C_{16}G_2$/squalene systems in absolute units at 50 °C, and (b) the $p(r)$-functions. The arrows and a broken line in panel b indicate the maximum length and cross-sectional diameter of reverse micelles, respectively.

paraffin, squalane and squalene. The phase behavior (Figure 2.2) shows that the surfactants of this class form α-solid phase near the surfactant axis at lower temperatures, which transform into L_α phase upon heating. The L_α phase swells with oil and there is a vesicular dispersion in the dilute regions. The L_α phase transforms into an isotropic single or two-phase solutions phase upon further heating. The SAXS measurements are performed in the isotropic single-phase solution to see the structure of aggregates. In the following section, the effect of a surfactant chain, oil and added water will be described.

Effect of Surfactant Figure 2.18 shows the SAXS results showing the effect of hydrocarbon chain length of diglycerol surfactant on the structure of reverse micelles formed in a long-chain branched oil squalene at 50 °C. The surfactant dependence of reverse micellar structure can be clearly seen in Figure 2.18a. The low-q scattering intensity is strongly enhanced upon decreasing hydrocarbon chain length of the surfactant without affecting the scattering intensities in the cross-sectional regions in the high-q region. Such a feature in the scattering function is a clear signature of sphere-to-rod-type transition in the aggregate structure.

Coming to the $p(r)$-functions in Figure 2.18b, the $p(r)$-curve of the $C_{14}G_2$/squalene system exhibits the typical feature of a long cylinder-like particle with maximum length ca. ~16 nm judging from a pronounced peak in the low-r regime and an extended tail to the high-r side. The absence of local maximum and minimum on the lower-r side of the peak position indicates a homogeneous electron density distribution (no core-shell structure) in the observing particles. With increasing hydrocarbon chain length of the surfactant, the maximum length of the particles decreases and a feature of short rod-like particles with maximum length ca. ~7 nm are seen in the $p(r)$-curve. This result is in agreement with the phase diagrams and also similar to the monoglycerol based surfactant systems, Figure 2.7. Namely,

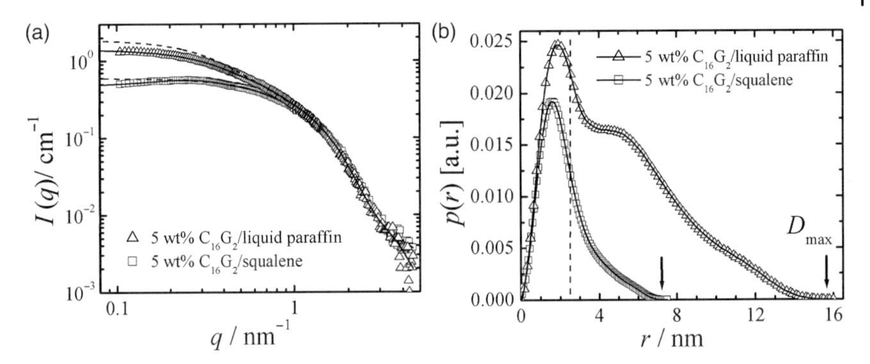

Figure 2.19 (a) The scattering functions of 5 wt% $C_{16}G_2$ in liquid paraffin and squalene in absolute unit at 50 °C, and (b) the $p(r)$-functions. The arrows and a broken line in panel b indicate the maximum length and cross-sectional diameter of reverse micelles.

the short-hydrocarbon chain surfactant tends to separate from oil, as shown in the Figure 2.2, resulting in the reduction of surface area per surfactant on the micellar hydrophobic/hydrophilic interface. Note that the effect of alkyl chain length of the surfactant is more crucial in the diglycerol surfactant system. It is obvious from Figure 2.18b that the cross-sectional diameter of the hydrophilic core is almost the same in both systems.

Effect of Oils Depending on the nature of solvent oils the structure of diglycerol-based nonionic surfactant micelles differ significantly, as shown in Figure 2.19 in which the scattering function and the corresponding $p(r)$-function for the $C_{16}G_2$ in liquid paraffin and squalene at 50 °C are presented. The difference in the scattering behavior, mainly in the low-q region depending on the oil can be clearly seen in Figure 2.19a. The forward scattering intensity of $C_{16}G_2$/liquid paraffin system is much higher than the $C_{16}G_2$/squalene system and low-q intensity decay with q following a $\sim q^{-1}$ behavior, indicating the formation of cylindrical-type particles in the former system [45].

As can be seen in the $p(r)$-functions, the reverse micelle in $C_{16}G_2$/squalene is much shorter than that of the $C_{16}G_2$/liquid paraffin system. This tendency is again in good agreement with the anticipated tendency from the phase diagram. In the squalene system, the mutual dissolution of surfactant and oil is the highest among the three oils.

Effect of Water Water brings similar changes in the structure of diglycerol surfactant micelles as was observed in the monoglycerol surfactant systems. The only difference is that more water can be solubilized in the micellar core in the diglycerol surfactant system. Upon addition of 0.14 wt% of water to the 5 wt% $C_{14}G_2$/squalene system, the dramatic elongation of reverse micelles takes place, see Figure 2.20. A slight increase of the cross section of the hydrophilic core is observed in terms of the slight shift of the inflection point after the maximum in

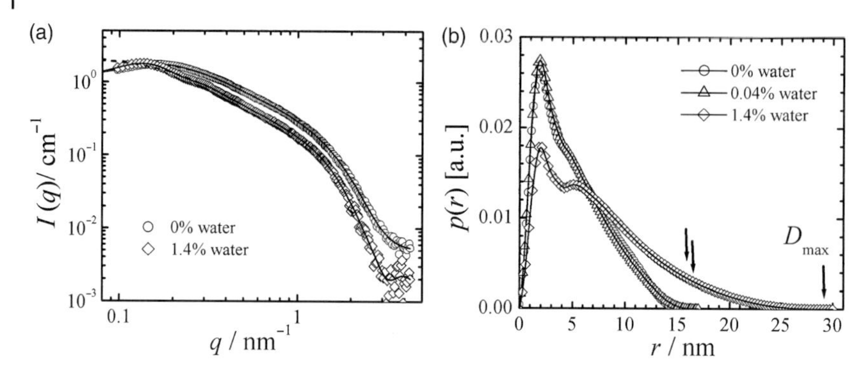

Figure 2.20 (a) The scattering functions of 5 wt% $C_{14}G_2$/squalene with and without water in absolute units at 50 °C, and (b) the $p(r)$-functions.

the $p(r)$-functions. Hence, water induces mainly the elongation of micelles. Note that surfactant samples themselves contain a small amount of water. It is practically difficult to remove water completely and to measure SAXS data in an absolutely dry state. For example, the present $C_{14}G_2$ contains 0.66% of water [38], which means that in the 5 wt% $C_{14}G_2$/squalene solution, at least 0.03 wt% water is already involved before adding water. Hence, upon addition of water, the water content in the system is roughly changed from 0.03 wt% to 0.17 wt%. Generally, it is difficult to characterize the reverse micellar structure under a perfectly dry condition but the data presented in this chapter unambiguously provides the direct structural evidence that water induces drastic elongation of reverse micelles.

2.3.2.2 Structure of Reverse Micelles Alkanes and Aromatic Oils

In the following section, structure of diglycerol fatty acid ester nonionic surfactant micelles in *n*-alkanes (straight-chain hydrocarbon oils from octane to hexadecane), and in two aromatic oils (phenyl octane and ethyl benzene) will be described.

Solvent Dependence of Reverse Micellar Structure Diglycerol monomyristate ($C_{14}G_2$) nonionic surfactant spontaneously forms a variety of reverse micellar aggregates depending on the nature of the organic solvents. The scattering functions, $I(q)$, and the corresponding $p(r)$-functions for the $C_{14}G_2$/oil systems are displayed in Figure 2.21, which compares the effects of aromatic oils, and straight-chain hydrocarbon oils on the aggregation behavior of the $C_{14}G_2$ surfactant.

As can be seen from Figure 2.21b, the $p(r)$-curves of 5 wt% $C_{14}G_2$ in ethyl benzene exhibits a symmetric bell shape, assigned to a typical feature of a nearly spherical micelle, where the core radius is estimated to be ~1.5 nm from the maximum dimension D_{max} ~ 3 nm. A slight elongation of the micellar structure is attained with the cross-sectional diameter virtually unchanged, when the 1-phenyl octane replaces ethyl benzene. Ethyl benzene and 1-phenyl octane are aromatic hydrocarbon oils having some degree of polarity. Hence, it is expected that the aromatic ring go all the way to the hydrophilic/hydrophobic interface of micelles globular type of micelles are more likely to form. A slight elongation in the case of 1-phenyl octane compared to the ethyl benzene may possibly be caused due to the fact that the former oil cannot reach as deep a center of micelle as the latter does due to a less polar character with the longer alkyl chain. Cyclohexane is less polar than these aromatic oils so that more elongated aggregate formation may be expected. However, the structural behavior of the $C_{14}G_2$/cyclohexane system is not as straightforward as can successively be explained as a simple counterpart of the aromatic oil-based systems. Judging from the slightly asymmetric shape of the $p(r)$-curve and the lack of a local maximum and minimum, the $C_{14}G_2$/cyclohexane system seems to have a slightly elongated, ellipsoid prolate-like shape with a nearly homogenous scattering length density distribution inside the core. The D_{max} is considerably increased compared to those of the aromatic oil-based systems, accompanied by an increase of the cross-sectional diameter as can be seen in Figure 2.21b. Surprisingly, $D_{max} \sim 6.5$ nm far exceeds twice the extended hydrophilic

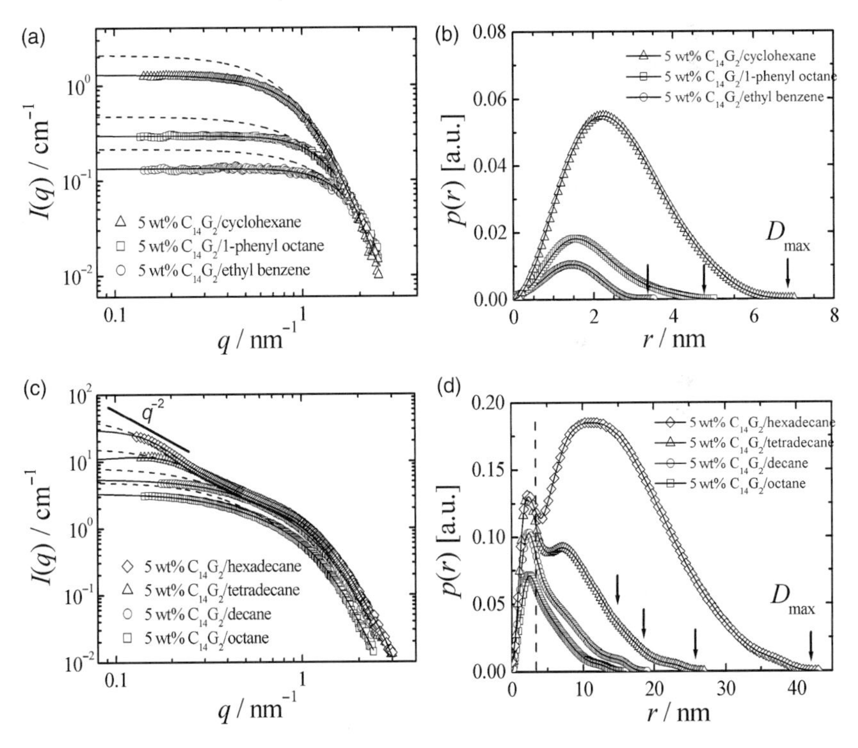

Figure 2.21 The X-ray scattered intensities, $I(q)$, of 5 wt% $C_{14}G_2$/oil systems at higher temperature 50 °C in absolute unit and the $p(r)$-functions: (a) the $I(q)$ of the $C_{14}G_2$/oil systems in aromatic oils and cyclohexane, and (b) the corresponding $p(r)$-functions. (c) The $I(q)$ of the $C_{14}G_2$ in n-alkanes, and (d) the corresponding $p(r)$-functions. The arrows in panel b and d indicate the maximum dimension of micelles, D_{max}, and the broken line on the inflation point of $p(r)$ on the high-r side of the maximum highlights the core diameter.

chain length of the $C_{14}G_2$ surfactant. To explain all these features, one has to consider that cyclohexane is dissolved in the hydrophilic core and nearly randomly distributed, despite its nonpolar nature.

When straight-chain hydrocarbon oils are used as solvents, the scheme of micellar growth turns out to be fairly different from aromatic oils; it becomes highly sensitive to the chain length of the oil, leading to the formation of long cylindrical to planar-like aggregates. As shown in Figure 2.21, upon changing oils from octane to hexadecane, we observed the successively increasing scattered intensity together with an increasing slope in the low-q region, which clearly represents the signature of micellar growth with increasing chain length of oils [43]. The resulting $p(r)$-functions in octane and decane systems exhibit typical signature of ellipsoid prolate particles having a nearly homogenous electron density distribution. With tetradecane, besides a rapid increase of D_{max}, a bump appears in the $p(r)$-curve on the higher-r side of its maximum around $r \sim 8$ nm. When the solvent is replaced with the longer-chain hexadecane, the low-q slope of $I(q)$ becomes markedly steeper and reaches $\sim q^{-2}$. The small bump seen in $p(r)$ of the $C_{14}G_2$/tetradecane system grows into a pronounced broad maximum for the $C_{14}G_2$/hexadecane, accompanying a drastic increase of D_{max}. All these features confirm a planar-like aggregate formation in the $C_{14}G_2$/hexadecane system. Note that tetradecane and 1-phenyl octane have almost equivalent molecular weights. However, SAXS experiments have shown that very long cylindrical particles are produced in the former, whereas in the latter, only slightly elongated particles are present, which highlights the markedly different effects of linear-chain hydrocarbon oils and aromatic oils. In the aromatic oil-based systems, the micellar growth is almost surely governed by the polarity of the aromatic ring, which is proven by the fact that the chain length has no major effect on the micellar growth despite a large difference of hydrocarbon chain length. In contrast, there appears a drastic effect of the hydrocarbon chain length on the micellar growth when linear-chain hydrocarbon oils are used as solvent, which may be explained in terms of the free-energy landscape or their penetration to the lipophilic part of the surfactant.

The recent development in the free-energy approach [91–93] has given a convincing explanation to the exponential decrease in the critical micellar concentration (*cmc*) with the hydrocarbon chain length and its nonmonotonous temperature dependence for normal micellar systems like poly(oxyethylne) alkylethers in aqueous media. These arise essentially from the liner and nonmonotonic trends of the oil-water transfer free energy respectively as a function of hydrophobic chain length and temperature. Although it is difficult to calculate the transfer free energy from the SAXS data, it can be speculated that the transfer free energy of the hydrophilic part of the surfactant from the hydrophilic environment to oils with different hydrocarbon chain length may be of central importance for the determination of the self-assembly of the amphiphile. It is naturally expected that the oil-to-oil transfer free energy of the hydrophobic moiety of $C_{14}G_2$ will be negligible for different chain-length hydrocarbon oils. Considering the hydrophilic nature of the diglycerol group, its transfer free energy from a hydrophilic environment to oil will considerably differ depending on the chain length of the hydrocarbon oils,

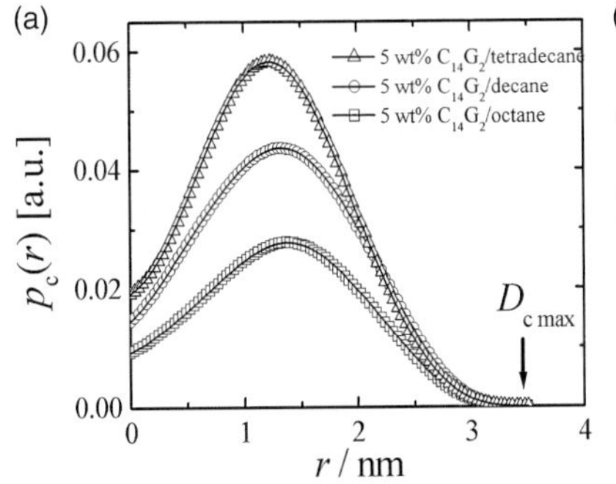

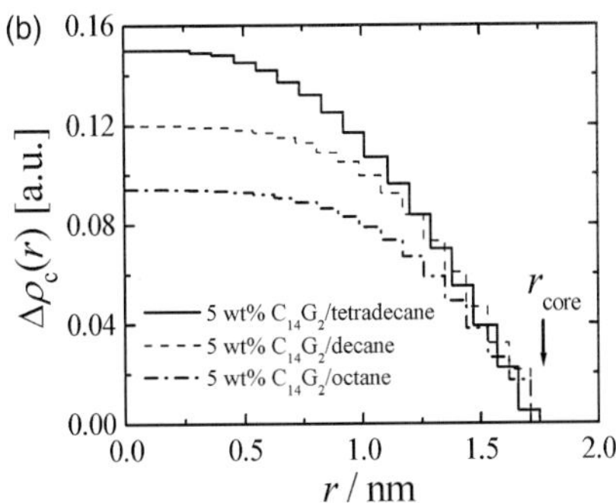

Figure 2.22 The model-free cross-sectional (core) structure analysis. (a) The cross-sectional pair-distance distribution functions, $p_c(r)$, for the $C_{14}G_2$/oil systems with octane, dectane, and tetradecane, and (b) the cross-sectional radial electron density profile $\Delta\rho_c(r)$ obtained from the deconvolution of the $p_c(r)$.

as oil-to-water transfer free energies of different chain-length hydrocarbons sensitively do.

For elongated ellipsoidal prolate, rod-like or cylindrical scattering objects as observed in the $C_{14}G_2$/oil systems, for example, with octane, decane and tetradecane, a model-free analysis of the micellar cross-sectional structure is available, under the assumption that the scattering length density profile of the cross section is simply a function of the radial position. The core diameter for all $C_{14}G_2$/oil systems with linear-chain hydrocarbon oils is approximately in the range of 3.0–3.5 nm when using an inflection point of $p(r)$ located on the higher-r side of the sharp maximum as its measure. To confirm this estimation, the model-free approach to the cross-sectional structure is used.

If judged from $D_{c\ max}$ in $p_c(r)$ shown in Figure 2.22, the $C_{14}G_2$/oil systems with octane, decane, and tetradecane, give the nearly identical (maximum) core diameters of ~3.5 nm, which are quantitatively complementary to that estimated from the inflection point in $p(r)$.

Concentration-Induced 1D Micellar Growth The $C_{14}G_2$ surfactant favors micellar growth with increasing surfactant concentration. In Figure 2.23 the scattering functions, $I(q)$, and the resulting $p(r)$-functions of the $C_{14}G_2$/octane system at different surfactant concentrations at 50 °C are presented.

As can be seen from Figure 2.23a, increasing surfactant concentration increases the total scattering intensity, which is more than expected simply from the increasing scattering volume according to surfactant concentration. In concentration < 5 wt%, the increasing low-q slope of $I(q)$ versus q in Figure 2.23a is due to one-dimensional micellar growth, which can be clearly seen from the real space functions in Figure 2.23b. The D_{max} increases with concentration and is clearer in the normalized $p(r)$-curves, $p(r)/p(r_{max})$. Theoretically, the formation of a long cylindrical (homogeneous) particle leads to a q^{-1} behavior in the forward intensity. The

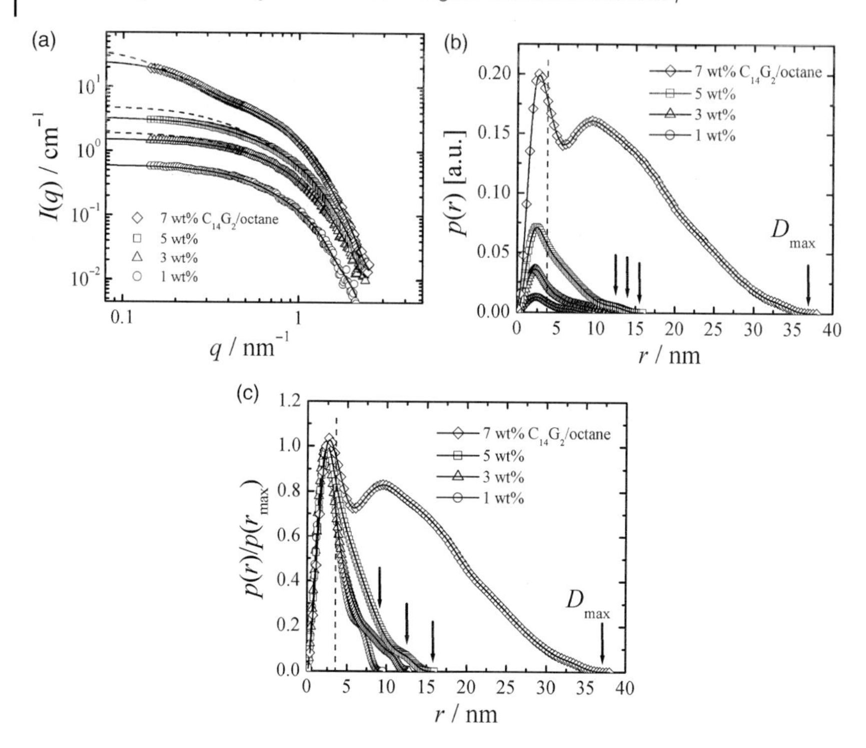

Figure 2.23 (a) The scattering curves $I(q)$ of $C_{14}G_2$/octane systems at different surfactant concentrations obtained on absolute scale at 50 °C, (b) the corresponding $p(r)$-functions, and (c) the normalized $p(r)$-functions, $p(r)/p(r_{max})$.

internal density fluctuation, as expected for the actual systems, also affects the low-q slope. At lower surfactant concentration < 5 wt%, the low-q slope of $I(q)$ is a little smaller than q^{-1}, although the micellar structure is already elongated. At the highest concentration (7 wt%), the slope of the scattering function $I(q)$, in the low-q region, gets markedly steeper, which exceeds q^{-1}. Simultaneously, the $p(r)$-curve exhibits an additional feature of a pronounced bump on the higher-r side of the peak (around $r \sim 10$ nm). These observations indicate the onset of two-dimensional growth of aggregates from cylindrical to planar shape [94, 95]. Thus, the presence of flat particles is highly expected at higher surfactant concentrations.

Temperature-Induced Rod-to-Sphere Transition Figure 2.24 shows the scattering functions, $I(q)$, and the resulting $p(r)$-functions, for a 5 wt% $C_{14}G_2$/octane system at different temperatures (40, 50, 60, and 70 °C). The low-q slope of $I(q)$ monotonically decreases with increasing temperature without changing the high-q intensities in q >1.0 nm^{-1} related to the local structure of aggregates see inset of Figure 2.24a. The D_{max} in the $p(r)$-curves successively decreases from ~17 to ~10 nm on increasing temperature from 40 to 70°C, while the position of the characteristic inflection located on the higher-r side of the sharp maximum is virtually unchanged, see Figure 2.24b.

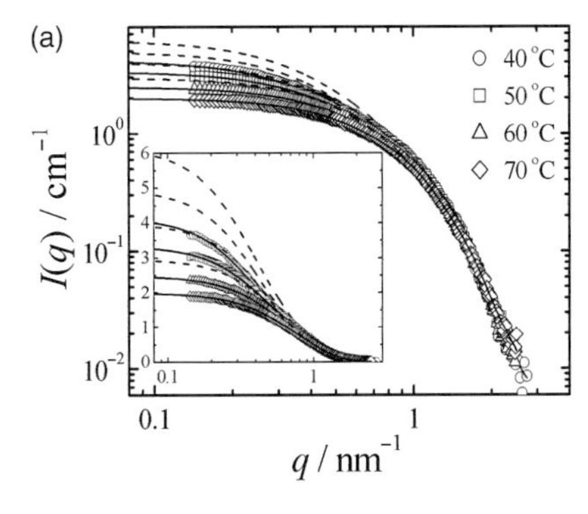

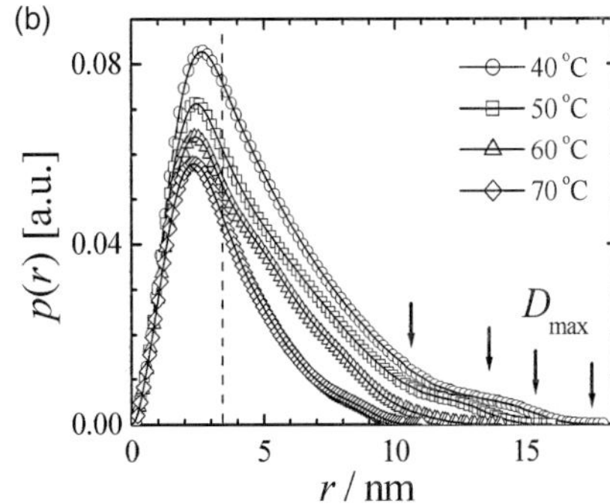

Figure 2.24 (a) The X-ray scattered intensities $I(q)$ of the 5 wt% $C_{14}G_2$/octane system in absolute unit at different temperatures, and (b) the corresponding $p(r)$-functions. The inset in panel a highlights the $I(q)$ in the low-q region. Arrows and a broken line in panel b highlight D_{max} and the core diameter, respectively.

The phase diagrams clearly showed that the miscibility of oil and surfactant increases with temperature. As mentioned earlier, increasing temperature increases the interpenetration of oil to the lipophilic chain of the surfactant, which tends to increase the critical packing parameter and the increasing critical parameter in reverse systems favors the formation of aggregates with more negative curvature. Nevertheless, one should keep in mind that the $C_{14}G_2$ contains a small amount of water (0.66%) as an impurity, that is, the headgroup is hydrated and with increasing temperature, similar to poly(oxyethylene) type nonionic surfactant systems, the dehydration of the headgoup takes place and thus favors the rod-to-sphere-type transition in nonaqueous system.

Water-Induced 2D Micellar Growth The addition of water tends to increase the spontaneous curvature pursuant to headgroup repulsion and favors micellar growth in the nonpolar oil systems. This may also be related to the problem of the transfer free energy briefly discussed in the previous section. Moreover, water-induced micellar growth and an increase in the aggregation number of reverse micellar aggregates has been known for quite some time [84]. The effects of added water on the structure of reverse micelles based on $C_{14}G_2$ obtained from SAXS measurements are presented in Figure 2.25. Two major changes can be seen in the scattering behavior of the 5 wt% $C_{14}G_2$/cyclohexane system upon addition of water. First, the forward scattering intensity in the low-q region and hence the low-q slope are strongly enhanced with increasing water concentration, indicating a significant growth to the maximum dimension of the micelles. Next, the scattering intensity in the high-q cross-sectional region ($q > 1.0\,\text{nm}^{-1}$) shifts towards the forward direction, indicating growth to the cross-sectional diameter of the micelles. That is, the added water causes 2D micellar growth; simultaneous changes in the maximum dimension and the cross-sectional structure of reverse micellar aggregates. As can be seen in Figure 2.25b, upon addition of 0.6 wt% of water to the 5 wt% $C_{14}G_2$/cyclohexane system, a slight increase of the cross section

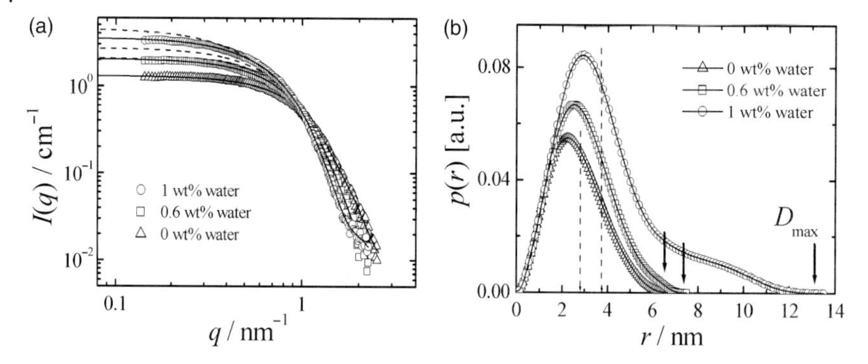

Figure 2.25 (a) The X-ray scattering intensities $I(q)$ of 5 wt% $C_{14}G_2$/cyclohexane at different concentration of water obtained in absolute unit at 50 °C, and (b) the corresponding $p(r)$-functions. Solid and broken lines in panel a represent GIFT fit and the calculated form factor for n particles existing in unit volume, $nP(q)$, respectively.

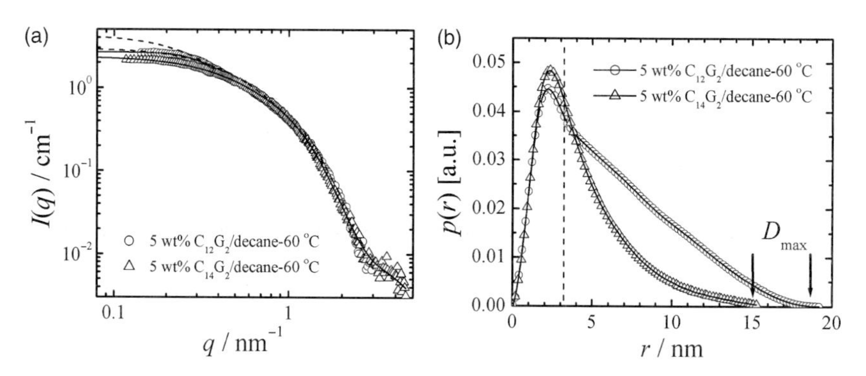

Figure 2.26 (a) The X-ray scattering intensities $I(q)$ of 5 wt% $C_{12}G_2$ and $C_{14}G_2$ in decane in absolute unit at 60 °C, and (b) the corresponding $p(r)$-functions.

of the hydrophilic core is observed in terms of the slightly shifted inflection point seen on the higher-r side of the maximum in $p(r)$.

With further addition of water say at 1 wt% of water a drastic elongation in the micellar structure can be seen in Figure 2.25b. The D_{max} of the micelle nearly doubles over that without water. One can expect phase separation at higher water concentration when the excess water cannot be incorporated in the micellar core. Interestingly, the 5 wt% $C_{14}G_2$/cyclohexane system can solubilize ~4.9% water in the micellar core and further addition of water results in a turbid solution.

Surfactant Chain-Length Dependence of Micellar Structure Figure 2.26 shows the SAXS results of surfactant chain-length dependence of micellar structure in diglycerol surfactants. The oil, composition, temperature, and headgroup size of the surfactant are fixed. Only the surfactant chain length is varied from C_{14} to C_{12}. A notable difference in the scattering behavior, mainly in the low-q region can be seen in Figure 2.26a. Note that the scattering intensities in the higher-q cross-section region are identical in both systems. This makes sense because SAXS estimates the micellar core and in both the systems the core consists of the same size diglycerol moiety.

The effect of lipophilic chain length of the surfactant can better be seen in $p(r)$-functions, Figure 2.26b. Unlike in monoglycerol surfactant systems, where only the size of the particles change with the changes in the alkyl chain length of the surfactant, in diglycerol surfactant systems, both the shape and size of the micelles change with a small change in the lipophilic chain of the surfactant. A pronounced peak in the low-r side with a linear decay of $p(r)$-functions in the higher-r side can be taken as evidence of the long cylinder-type micelles in the short-chain surfactant $C_{12}G_2$/decane system. A pronounced peak in the low-r side with a downward convex shape of the $p(r)$-functions in the intermediate-r side can be taken as the formation of elongated ellipsoidal prolate-type micelles observed in the $C_{14}G_2$/decane system. The maximum length of the micelles in the $C_{12}G_2$/decane system is nearly 25% bigger than the $C_{14}G_2$/decane system, see downward arrows in Figure 2.26b. Although there is a significant change in the shape and size, the micellar cross section, as indicated by a broken line after the maximum in the $p(r)$-functions, is practically the same in both the systems. This result shows that the surfactant chain length can also be a major parameter in the structure control of reverse micelles in diglycerol-based nonionic surfactant systems.

Growth Control by Headgroup Size of the Surfactant The structure of nonionic surfactant micelles depending on the composition, temperature, solvent nature, water addition, and surfactant's chain length have been discussed so far. In the following section, the growth control of glycerol-based nonionic surfactant by headgroup size will be discussed. In order to study this effect two surfactants $C_{14}G_1$ and $C_{14}G_2$ are considered. Note that both the surfactants have the same lipophilic chain length and the only difference is in their hydrophilic size. The former has one glycerol molecule as a hydrophilic moiety and the latter has two molecules of glycerol. One can expect bigger particles with the $C_{14}G_2$ surfactant system. This is because the bigger area of the surfactant's headgroup results in a smaller critical packing parameter. Figure 2.27 shows the normalized $I(q)$ and the corresponding $p(r)$ for the 5 wt% $C_{14}G_1$ and $C_{14}G_2$ surfactants in tetradecane.

The forward scattering intensity $I(q = 0)$ of the $C_{14}G_2$ system is much higher than that of the $C_{14}G_1$ system and decays following a $I(q) \sim q^{-1}$ behavior, indicating that the aggregates have cylindrical structure in the $C_{14}G_2$ system. Minute observation of the scattering function reveals that the $I(q)$ in the cross-section region ($q \sim 2\,\mathrm{nm}^{-1}$) shifts towards the forward direction in the $C_{14}G_2$ system indicating changes in the internal structure of micelles. Namely, the cross-sectional diameter of the aggregates in the former system is higher than in the latter system. The effect of hydrophilic size of the surfactant on the reverse micellar structure is well reflected in the PDDF curves Figure 2.27b. A pronounced peak in the low-q side and an extended tail in the high-r side of the $p(r)$ curves indicate the presence of long

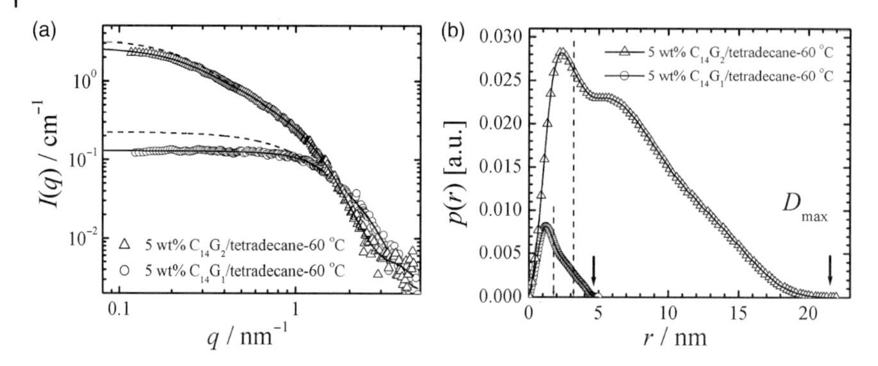

Figure 2.27 (a) The X-ray scattering intensities $I(q)$ of 5 wt% $C_{14}G_1$ and $C_{14}G_2$ in tetradecane in absolute unit at 60 °C, and (b) the corresponding $p(r)$-functions.

cylindrical micelles in the $C_{14}G_2$ system. On the other hand, a slightly elongated ellipsoidal prolate-type structure is observed in the $C_{14}G_1$ system. The position of the inflection point as indicated by the broken line (see Figure 2.27b after the maximum in the $p(r)$-curves increases in the $C_{14}G_2$ system. Minute observation of the $p(r)$ curves revealed that the cross-sectional diameter of the $C_{14}G_2$ system is nearly twice that of the $C_{14}G_1$ system. A small bump seen at $r \sim 6$ nm in the $p(r)$-function of the 5 wt% $C_{14}G_2$/tetradecane system (Figure 2.27b) most probably came from critical fluctuations, as the systems are close to the phase-separation point [45, 65]. Thus, the present SAXS data have highlighted the significance of the surfactant's hydrophilic size to the structural control of reverse micelles.

In order to complement the GIFT analysis of the SAXS data, model calculations for selected systems are carried out, testing different plausible structure models and the results are shown in Figure 2.28. The calculations are carried out based on the method reported in detail elsewhere [96]. For the model calculations, only two systems, 5 wt% $C_{14}G_1$/tetradecane and 5 wt% $C_{14}G_2$/tetradecane, were considered. The GIFT analysis of the SAXS data has confirmed slightly elongated ellipsoid prolate-type micelles with maximum size ~5 nm in the former system and long cylindrical micelles with maximum length ~22 nm in the latter system. The core radius R_c and the total axial length L for a cylinder or the short and long axes of an ellipsoidal prolate, a and b, obtained from the theoretical model calculations are consistent with the results obtained from the GIFT method despite a visible difference of the depth of the minimum in the experimental and theoretical $I(q)$ functions, see Figure 2.28a. Besides, the $p(r)$-functions derived from the model calculations are very similar to the $p(r)$-functions deduced from the GIFT method, except that the small bump at $r \sim 6$ nm is absent in the $p(r)$-function of the model for the 5 wt% $C_{14}G_2$/tetradecane system because no onset of critical fluctuations is considered in the model calculation. Thus, these model calculations confirm the general results obtained directly from the model-free data-analysis procedure.

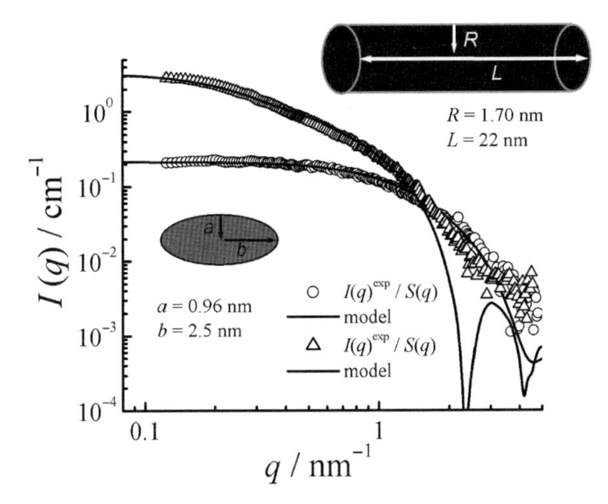

Figure 2.28 The model calculation (full lines) and experimental X-ray scattered intensities of 5 wt% $C_{14}G_1$/tetradecane (circles) and 5 wt% $C_{14}G_2$/tetradecane (square) as typical examples. The data were fitted using a model of a homogeneous ellipsoidal prolate and a homogeneous cylinder.

2.4 Conclusion

The wide range of applications and increasing interests on the studies of nonionic surfactant reverse micelles or W/O microemulsions has shown their significance in colloid and polymer sciences. Due to biocompatibility and biodegradability of the glycerol-based nonionic surfactants, studies on the self-assemblies of these surfactants in polar and nonpolar solvents offer various practical applications in the food, cosmetics, and pharmaceutical industries.

This chapter dealt with the studies of phase behavior and self-assembled structures of glycerol-based nonionic surfactants in a variety of nonpolar organic solvents, mainly focusing on the structure (shape and size) control of the reverse micelles in surfactant/oil binary systems. Contrary to the conventional poly(oxyethylene) type nonionic surfactants, glycerol-based nonionic surfactants depending on their head-group size have the potential to form various self-assembled structures in nonpolar solvents without external water addition. The monoglycerol-based surfactants could not form any liquid-crystalline structure at all temperatures and compositions in different oils such as liquid paraffin, squalane, squalene, and *n*-alkanes; however, as described in Section 2.2.2, the diglycerol-based surfactants spontaneously form various types of self-assembled structures in oils. The formation of lamellar liquid crystal in concentrated regions and the dispersion of reverse vesicles in the dilute regions advance the practical application of these surfactants.

Reverse micelles are expected to form in the ternary mixtures of surfactant/oil/water and are often in spherical shape. Despite this anticipation, this chapter showed a clear picture on the formation of a variety of reverse micellar structures (spherical, prolate, rod-like, cylindrical and planar-like) in surfactant/oil systems without external water addition. Of course, a small amount of water is always present in the glycerol-based surfactants as an impurity as they are hydroscopic. The structure parameters of the micelles are determined by *inverse Fourier transformation* of the SAXS data utilizing *indirect Fourier transformation* (IFT) and its updated version, the *generalized indirect Fourier transformation* (GIFT) methods, depending on the volume fraction of the surfactants. The results obtained from the IFT and GIFT are supported by theoretical model calculations. It has been found that irrespective of the nature of surfactant, composition always favors the 1D micellar growth and temperature always favors rod-to-sphere transition in the reverse micellar structure. The reverse micelles observed in the glycerol-based nonionic surfactants solubilize some amount of polar solvents such as water or glycerol. The reverse micelles swell with water and the maximum micellar diameter and cross section increase with water. In other words, water favors 2D micellar growth in reverse systems. In aromatic oils and in cyclohexane, small globular or ellipsoidal prolate types of reverse micelles are observed. As described in Section 2.3.2.2, the maximum size of the diglycerol surfactant-based reverse micelles increases in parallel to the chain length of *n*-alkanes. Formation of rod-like micelles in the short-chain alkane octane eventually transformed into planar-like micelles in the long-chain oil hexadecane. Similar oil effects are observed in monoglycerol-based surfactant, with a difference that in monoglycerol systems, the extent of micellar growth with chain length of alkanes is relatively smaller. A dramatic micellar growth both in terms of micellar size and cross section is observed on increasing the headgroup size of the surfactant.

This chapter has shown a possible route to the free structural control of glycerol-based nonionic surfactants reverse micelles, and it is found that the solvent properties, temperature, composition, chain length of surfactant and headgroup size, and water can be the tunable parameters for the structural variation of reverse micelles.

Dedication

With all due respect, I (LKS) would like to dedicate this chapter to my first PhD advisor, late Prof. Hironobu Kunieda. I happened to be the final student to join his group, and although, I could not complete my PhD under his supervision, he introduced me to several interesting features of surfactants in a short period of time. He also highlighted the importance of phase behavior study to understand fundamental science and also its significance in the practical applications. I still remember his common advice to his students "*You have to improve yourself, and be independent*". I hope many of his students recall these words and I am quite sure he will be remembered for a long time due to his important contribution to colloid and interface science society. I pray to the God to keep his soul in rest and peace.

Acknowledgment

LKS thanks the Japan Society for the Promotion of Science (JSPS) for a postdoctoral fellowship for foreign researchers. Many thanks to Dr. Takaaki Sato Shinshu University, Japan and Prof. Dr. Otto Glatter of the University of Graz, Austria for fruitful discussion. Thanks also go to Dr. Conxita Solans, CSIS Barcelona for encouragement to write this chapter. Without her support, the chapter would never have come to this shape. We also thank Dr. Tharwat Tadros for his precious time to edit this chapter.

References

1 Uddin, H.Md., Rodriguez, C., Watanabe, K., López-Quintela, M.A., Kato, T., Furukawa, H., Harashima, A., and Kunieda, H. (2001) *Langmuir*, **17**, 5169.
2 Kaneko, M., Matsuzawa, K., Uddin, H.Md., López-Quintela, M.A., and Kunieda, H. (2004) *J. Phys. Chem. B*, **108**, 12736.
3 Kunieda, H., Tanimoto, M., Shigeta, K., and Rodriguez, C. (2001) *J. Oleo Sci.*, **50**, 633.
4 Kunieda, H., Uddin, H.Md., Horii, M., Furukawa, H., and Harashima, A. (2001) *J. Phys. Chem. B*, **105**, 5419.
5 Kunieda, H., Shigeta, K., Ozawa, K., and Suzuki, M. (1997) *J. Phys. Chem. B*, **101**, 7952.
6 Forster, S., Zisenis, M., Wenz, E., and Antonietti, M. (1996) *J. Chem. Phys.*, **104**, 9956.
7 Zhong, X.F., Varsheny, S.K., and Eisenberg, A. (1992) *Macromolecules*, **25**, 7160.
8 Desjardins, A., van de Ven, T.G.M., and Eisenberg, A. (1992) *Macromolecules*, **25**, 2412.
9 Shrestha, L.K., Masaya, K., Sato, T., Acharya, D.P., Iwanaga, T., and Kunieda, H. (2006) *Langmuir*, **22**, 1449.
10 Rodriguez, C., Uddin, Md.H., Watanabe, K., Furukawa, H., Harashima, A., and Kunieda, H. (2002) *J. Phys. Chem. B*, **106**, 22.
11 Luisi, P.L., and Strab, B.E. (eds) (1987) *Reverse Micelles: Biological and Technological Relevance of Amphiphilic Structures in Apolar Media*, Plenum Press, New York.
12 Pileni, M.P. (1989) *Structure and Reactivity in Reverse Micelles*, vol. 65 (ed. M.P. Pileni), Elsevier, Amsterdam.
13 Barzykin, A.V., and Tachiya, M. (1996) *Heterog. Chem. Rev.*, **3**, 105.
14 Jones, M.N., and Chapman, D. (1995) *Micelles, Monolayers, and Biomembranes*, John Wiley & Sons, Inc., New York.
15 De, T., and Maitra, A. (1995) *Adv. Colloid Interface Sci.*, **59**, 95.
16 Boutonnet, M., Kizling, J., and Stenius, P. (1982) *Colloids Surf.*, **5**, 209.
17 López-Quintela, M.A., Tojo, C., Blanco, M.C., García Rio, L., and Leis, J.R. (2004) *Curr. Opin. Colloid Interface Sci.*, **9**, 264.
18 López-Quintela, M.A. (2003) *Curr. Opin. Colloid Interface Sci.*, **8**, 137.
19 Cushing, B.L., Kolesnichenko, V.L., and O'Connor, C.J. (2004) *Chem. Rev.*, **104**, 3893.
20 Lisiecki, I., and Pileni, M.P. (1993) *J. Am. Chem. Soc.*, **115**, 3887.
21 Pileni, M.P. (1997) *Langmuir*, **13**, 3266.
22 Sharma, S.C., Kunieda, H., Esquena, J., and Rodriguez-Aberu, C. (2006) *J. Colloid Interface Sci.*, **299**, 297.
23 Pileni, M.P. (1993) *Adv. Colloid Interface Sci.*, **46**, 139.
24 Angelico, R., Palazzo, G., Colafemmina, G., Cirke, P.A., Giustini, M., and Ceglie, A. (1998) *J. Phys. Chem. B*, **102**, 2883.
25 Riter, R.E., Kimmel, J.R., Undiks, E.P., and Levinger, N.E. (1997) *J. Phys. Chem. B*, **101**, 8292.
26 Cason, J.P., and Roberts, C.B. (2000) *J. Phys. Chem. B*, **104**, 1217.
27 Li, Q., Li, T., and Wu, J. (2000) *J. Phys. Chem. B*, **104**, 9011.

28 Kanamaru, M., and Einaga, Y. (2002) *Polymer*, **43**, 3925.

29 Tung, S.-H., Huang, Y.-E., and Raghavan, S.R. (2007) *Langmuir*, **23**, 372.

30 Tung, S.-H., Huang, Y.-E., and Raghavan, S.R. (2006) *J. Am. Chem. Soc.*, **128**, 5751.

31 Penfold, P., Staples, E., Tuckeer, I., and Cummins, P. (1997) *J. Colloid Interface Sci.*, **185**, 424.

32 Warnheim, T., Jonsson, A., and Sjoberg, M. (1990) *Prog. Colloid Polym. Sci.*, **82**, 271.

33 Jones, M.-C., Gao, H., and Leroux, J.-C. (2008) *J. Controll. Release*, **132**, 208.

34 Herrington, T.M., and Shali, S.S. (1988) *J. Am. Oil. Chem. Soc.*, **65**, 1677.

35 Rodriguez-Aberu, C., Acharya, D.P., Hinata, S., Ishitobi, M., and Kunieda, H. (2003) *J. Colloid Interface Sci.*, **262**, 500.

36 Krog, N. (2001) *Crystallization Processes in Fats and Lipid System* (eds N. Garti and K. Sato), Marcel Dekker, New York, pp. 505–526.

37 Richardson, G., Bergenstahl, B., Langton, M., Stading, M., and Hermansson, A.M. (2004) *Food Hydrocolloid*, **18**, 655.

38 Kunieda, H., Shrestha, L.K., Acharya, D.P., Kato, H., Takase, Y., and Gutiérrez, J.M. (2007) *J. Dispers. Sci. Technol.*, **28**, 133.

39 Shrestha, L.K., Saito, E., Shrestha, R.G., Kato, H., Takase, Y., and Aramaki, K. (2007) *Colloid Surf. A*, **293**, 262.

40 Shrestha, L.K., Shrestha, R.G., Solans, C., and Aramaki, K. (2007) *Langmuir*, **23**, 489.

41 Shrestha, L.K., Shrestha, R.G., Sharma, S.C., and Aramaki, K. (2008) *J. Colloid Interface Sci.*, **328**, 172.

42 Glatter, O. (1981) *J. Appl. Crystallogr.*, **14**, 101.

43 Glatter, O. (1979) *J. Appl. Crystallogr.*, **12**, 166.

44 Brunner, P.J., and Glatter, O. (1997) *J. Appl. Crystallogr.*, **30**, 431.

45 Glatter, O., Fritz, G., Lindner, H., Brunner, P.J., Mittelbach, R., Strey, R., and Egelhaaf, S.U. (2000) *Langmuir*, **16**, 8692.

46 Bergmann, A., Fritz, G., and Glatter, O. (2000) *J. Appl. Crystallogr.*, **33**, 1212.

47 Pusey, P.N., Fijnaut, H.M., and Vrijm, A. (1982) *J. Chem. Phys.*, **77**, 4270.

48 Salgi, P., and Rajagopolan, R. (1993) *Adv. Colloid Interface Sci.*, **43**, 169.

49 Fritz, G., and Bergmann, A. (2004) *J. Appl. Crystallogr.*, **37**, 815.

50 Glatter, O. (1991) *Progr. Colloid Polym. Sci.*, **84**, 46.

51 Glatter, O. (1980) *J. Appl. Crystallogr.*, **13**, 577.

52 Glatter, O. (1980) *J. Appl. Crystallogr.*, **13**, 7.

53 Glatter, O., and Hainisch, B. (1984) *J. Appl. Crystallogr.*, **17**, 435.

54 Ekwall, P., and Brown, G.H. (eds) (1975) *Advances in Liquid Crystals*, vol. 1, Academic Press. Exerowa D. R. and Krugliakov, P. M. (1998) *Foam and Foam Films: Theory, Experiment, Applications*, vol. 5, Elsevier, New York

55 Aramaki, K., Olsson, U., Yamaguchi, Y., and Kunieda, H. (1999) *Langmuir*, **15**, 6226.

56 Kunieda, H., Shigeta, K., and Suzuki, M. (1999) *Langmuir*, **15**, 3118.

57 Kunieda, H., Kabir, H., Aramaki, K., and Shigeta, K. (2001) *J. Mol. Liq.*, **90**, 157.

58 Shigeta, K., Olsson, U., and Kunieda, H. (2001) *Langmuir*, **17**, 4717.

59 Kunieda, H., Uddin, Md.H., Furukawa, H., and Harashima, A. (2001) *Macromolecules*, **34**, 9093.

60 Aramaki, K., Hossain, Md.K., Rodriguez, C., Uddin, Md.H., and Kunieda, H. (2004) *Macromolecules*, **36**, 9443.

61 Kunieda, H., Kaneko, M., Lopez-Quintela, M.A., and Tsukahara, M. (2004) *Langmuir*, **20**, 2164.

62 Rodríguez-Abreu, C., Acharya, D.P., Aramaki, K., and Kunieda, H. (2005) *Colloid Surf. A*, **269**, 59.

63 Sato, T., Hossain, Md.K., Acharya, D.P., Glatter, O., Chiba, A., and Kunieda, H.J. (2004) *Phys. Chem. B*, **108**, 12927.

64 Shrestha, L.K., Sato, T., Acharya, D.P., Iwanaga, T., Aramaki, K., and Kunieda, H. (2006) *J. Phys. Chem. B*, **110**, 12266.

65 Shrestha, L.K., and Aramaki, K. (2007) *J. Dispers. Sci. Technol.*, **28**, 1236.

66 Shrestha, L.K., Sato, T., and Aramaki, K. (2007) *Langmuir*, **23**, 6606.

67 Shrestha, L.K., Sato, T., and Aramaki, K. (2007) *J. Phys. Chem. B*, **111**, 1664.

68 Shrestha, L.K., Shrestha, R.G., Varade, D., and Aramaki, K. (2009) *Langmuir*, **25**, 4435.

69 Shrestha, L.K., Glatter, O., and Aramaki, K. (2009) *J. Phys. Chem. B*, **113**, 6290.

70 Schick, M.J. (1962) *J. Colloid Sci.*, **17**, 801.

71 Hey, M.J., Ilett, S.M., and Davidson, G. (1995) *J. Chem. Soc. Faraday Trans.*, **91**, 3897.

72 Tasaki, K. (1996) *J. Am. Chem. Soc.*, **118**, 8459.

73 Karlström, G. (1985) *J. Phys. Chem.*, **89**, 4962.

74 Kjellander, R., and Florin, E. (1981) *J. Chem. Soc., Faraday Trans. 1*, **77**, 2053.

75 Shrestha, L.K., Sato, T., and Aramaki, K. (2009) *Phys. Chem. Chem. Phys.*, **11**, 4251.

76 Krog, N.J. (1990) *Food Emulsion*, 2nd edn (eds K. Larsson and S.E. Friberg), Marcel Dekker, New York, p. 127.

77 Larsson, K., and Dejmek, P. (1990) *Food Emulsion*, 2nd edn (eds K. Larsson and S.E. Friberg), Marcel Dekker, New York, p. 97.

78 Hernqvist, L. (1988) *Crystallization and Polymorphism of Fats and Fatty Acids* (eds N. Garti and K. Sato), Marcel Dekker, New York, p. 97.

79 Larsson, K. (1967) *Nature*, **213**, 383.

80 Israelachvilli, J.N., Mitchell, D.J., and Ninham, B.W. (1976) *J. Chem. Soc. Faraday Trans. II*, **72**, 1525.

81 Mitchell, D.J., and Ninham, B.W. (1981) *J. Chem. Soc. Faraday Trans 2*, **77**, 601.

82 Ruckenstein, E., and Nagarajan, R. (1980) *J. Phys. Chem.*, **84**, 1349.

83 Kitahara, A. (1980) *Adv. Colloid Interface Sci.*, **12**, 109.

84 Mathews, M.B., and Hirschhorn, E. (1953) *J. Colloid Sci.*, **8**, 86.

85 Luisi, P.L., Scartazzini, R., Haering, G., and Schurtenberger, P. (1990) *Colloid Polym. Sci.*, **268**, 356.

86 Schurtenberger, P., Magid, L.J., King, S.M., and Lindner, P. (1991) *J. Phys. Chem.*, **95**, 4173.

87 Wu, G., Zhou, Z., and Chu, B. (1993) *Macromolecules*, **26**, 2117.

88 Alexandridis, P., Olsson, U., and Lindman, B. (1995) *Macromolecules*, **28**, 7700.

89 Alexandridis, P., and Andersson, K. (1997) *J. Colloid Interface Sci.*, **194**, 166.

90 Li, J., Zhang, J., Han, B., Gao, Y., Shen, D., and Wu, Z. (2006) *Colloid Surf. A*, **279**, 208.

91 Nagarajan, R., and Wang, C.C. (1995) *Langmuir*, **11**, 4673.

92 Serafin, J.M. (2003) *J. Chem. Edu.*, **80**, 1194.

93 Soda, K., and Hirashima, H. (1990) *J. Phys. Soc. Jpn.*, **59**, 4177.

94 Moitzi, C., Freiberger, N., and Glatter, O. (2005) *J. Phys. Chem. B*, **109**, 16161.

95 Strey, R., Glatter, O., Schubert, K.-V., and Kaler, E.W. (1996) *J. Chem. Phys.*, **105**, 1175.

96 Glatter, O. (1980) *Acta Phys. Austriaca*, **52**, 243.

3
Nonionic Microemulsions: Dependence of Oil Chain Length and Active Component (Lidocaine)

Joakim Balogh, Ulf Olsson, Skov Pedersen Jan, Helena Kaper, Håkan Wennerström, Karin Schillén, and Maria Miguel

3.1
Introduction

This chapter reviews and presents the latest developments on the work on the nonionic microemulsions stabilized by surfactants of ethylene oxide alkyl ether type, C_mE_n, where m is the number of carbons in the alkyl chain and n is the number of ethylene oxide groups.

The history of what is now called microemulsions started with the experimental studies of Schulman and coworkers [1–3], Windsor [4, 5] and, to some degree, with Ekwall [6, 7].

This was followed by theoretical work by Prince [8, 9], and by Friberg [10], and Shinoda [11–13], and later Shinoda together with Kunieda [14–20]. It soon became clear that droplets of either oil-in-water or water-in-oil were formed in the water-/oil-rich part of the phase diagram, respectively. A problem arose when Shinoda and Friberg showed that a continuous phase existed when going from the oil-rich to the water-rich area without any macroscopic phase separation [21, 22]. Various solutions to this problem, for example, based on the idea of simultaneously having two different types of droplets with "positive" and "negative" interface, were quickly discarded. Friberg suggested a structure containing different curvatures [21] and Shinoda proposed a structure with zero curvature [22]. The bicontinuous structure suggested by Scriven [23], having the properties suggested by Friberg and Shinoda, was later supported by experimental data. One of the techniques to show the presence of a bicontinuous structure was NMR self-diffusion [24]. This technique was later improved by Stilbs [25] and used to study microemulsions [26, 27]. It was further improved [28, 29] and became an important technique and widely used in Lund together with other NMR techniques [30–41].

The work on nonionic surfactants and microemulsions of the ethylene oxide type had initially two strong legs to stand on, Shinoda and Kunieda with coworkers [42–46] in Yokohama, and Kahlweit and Strey with coworkers [47–56] in Göttingen (later Strey and Sottmann with coworkers [57–62] in Cologne). These were followed by among others the group in Lund [30, 36, 39, 63–68]. Later, others active

Self-Organized Surfactant Structures. Edited by Tharwat F. Tadros

ISBN: 978-3-527-31990-9

in this field include Langevin, Hellveg and Gradzielski with coworkers [69–78] just to mention a few. The contribution by Lindman [79] gives more information about the early days of microemulsions. What is well established about microemulsions and nonionic surfactants can be summarized by the following few sentences with selected references. Microemulsions are thermodynamically stable one-phase systems of oil and water of a colloidal length-scale [80–84] (i.e. 10^{-9}–10^{-6} m) where the water and oil domains are separated by a monolayer of surfactant [85–87]. Nonionic microemulsions are stabilized by nonionic surfactants [88–91].

Microemulsions have been studied by several techniques, for example, small-angle neutron and X-ray scattering (SANS and SAXS, respectively), dynamic and static light scattering (DLS and SLS, respectively) and cryogenic transmission electron microscopy (cryo-TEM). The latest development of various techniques are described in the recent review by Gradzielski [92] and the book chapter by Hellweg [93] is recommended for a description of the developments in scattering techniques for studying microemulsions.

There is a lot of interest in microemulsions as is shown by this book [94] and by recent publications on microemulsion models [95, 96], as well as oil chain length dependence [97] and the use of poly(ethylene glycol) (PEG) and microemulsions for pharmaceutical purposes [98, 99]. Microemulsions are used in many industrial applications [100], especially within the area of cosmetics and detergents [101] as well as in pharmaceutical applications [102]; one of the more important ones being as drug-delivery systems [103–112].

In this chapter the recent developments in interpreting experimental data from SAXS [113, 114], SANS [115, 116], NMR [113, 117] DLS and SLS [113, 118] and in interfacial tension [116, 119] measurements on nonionic microemulsions are presented. Some of the work on microemulsion models and the scaling behavior of these systems are also presented [116, 119]. With this knowledge of microemulsions and their scaling properties, there was an interest in using these type of C_mE_n nonionic microemulsions as models for a drug-delivery system and to investigate the effect of the drug added to the system [114, 120–122].

3.2 Microemulsion Model

The microemulsion can be described by modeling the surfactant film. The properties of the surfactant film are governed by the curvature elasticity and the curvature free energy, G_c given by the surface integral $G_c = \int dA g_c$, of the curvature energy g_c and the area A. The commonly used model, which is referred to as the Helfrich model, has the curvature energy

$$g_c = 2\kappa(H - H_0)^2 + \bar{\kappa}K \tag{3.1}$$

where H_0 is the spontaneous curvature κ and $\bar{\kappa}$ is the bending rigidity, and saddle splay modulus, respectively. Using the principal curvatures, c_1 and c_2, H, the mean curvature, can be expressed as $(c_1 + c_2)/2$, and K, the Gaussian curvature, as $c_1 \times c_2$.

For $\bar{\kappa} < 0$ discrete aggregates are preferred and for $\bar{\kappa} > 0$ saddle-like structures are preferred. It has been found that κ and $\bar{\kappa}$ are on the order of a few $k_B T$ [123] and that $\bar{\kappa}$ is negative. Others have obtained κ and $\bar{\kappa}$ on the order of unity and –0.5 in units of $k_B T$ [60, 70, 71], respectively. There are also models where $\bar{\kappa}$ is positive [124] due to how the enthalpy has been taken into account. Small differences in droplet size result in big variations of the derived values for κ and $\bar{\kappa}$. Since they are often obtained from experimental results as the combination $(2\kappa + \bar{\kappa})$ [75, 76] there are several possibilities to interpret the results. It has been shown that the assumptions made in the model play an important role in determining the actual values of κ and $\bar{\kappa}$ [72].

When the spontaneous curvature H_0 is zero, either a lamellar flat structure, L_α, with both principal curvatures equal to zero, or a bilayer bicontinuous structure, L_3, saddle-shaped with both principal curvatures equal but of opposite sign, can be formed. Numerous investigations (both experimental and theoretical) that concern the bilayer bicontinuous phase have previously been carried out [123, 125–129].

For the C_mE_n microemulsions, it has been found that H_0 depends linearly on the temperature, T:

$$H_0 = \alpha(T_0 - T) \tag{3.2}$$

where α is a constant and T_0, the temperature where the spontaneous curvature is equal to zero, is also known in literature as the phase-inversion temperature, PIT. This approximation holds for at least $T = T_0 \pm 20\,\text{K}$ [62, 123, 125].

3.3 Phase Studies

It is always essential to perform phase studies in order to obtain a basic understanding of the general phase behavior of a system, to determine the phase boundaries, and to follow the kinetics of the structural changes in the system. Since there are at least three components, the phase behavior has to be determined as a function of these parameters. With these systems being temperature sensitive, a fourth parameter, the temperature, also has to be considered. The phase diagram can thus be illustrated as a phase prism, as displayed in Figure 3.1. To be able to study the effect of the different parameters different cuts are made in the phase prism. Among these, there are three classical cuts, which will be presented here. The first cut is known as the Shinoda cut [22], which, with a fixed amount of surfactant and by varying the ratio between oil and water, gives information about the system's effect on the surfactant. The second cut is the Kahlweit fish cut [50] (the name originates from the fact that the three-phase region can be symbolized by the body of a fish and the one-phase region by its tail), which corresponds to a fixed ratio between water and oil (often 1 : 1) with varying surfactant concentration. This cut gives information about the effectiveness of the surfactant. The third cut is the Lund cut [130] with a fixed ratio between surfactant and oil and varying the

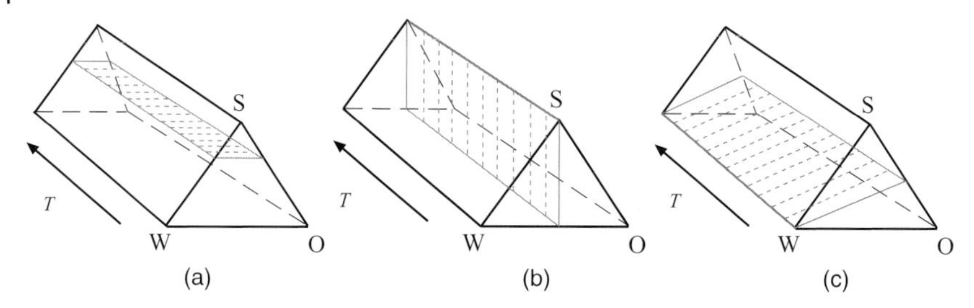

Figure 3.1 Phase prism with different cuts. (a) Shinoda cut, (b) Kahlweit cut, and (c) Lund cut. The phase triangle, consisting of surfactant (S), oil (O), and water (W), becomes a phase prism with the dependence of temperature (T) that these systems have.

water content. In the water-rich part this cut leads to an increase in the concentration of the aggregates without changing the ratio between the dispersed phase and the disperser (thus keeping the aggregates of fairly constant size for small or moderate changes in concentration).

The $C_{12}E_5$-decane-water system with the Lund cut with surfactant to oil ratio of 0.815 : 1 has been thoroughly investigated [34, 35, 130, 131]. There were also some indications that hexadecane did not act as expected [132] and therefore a systematic investigation was initiated. Comprehensive phase studies [113] of the $C_{12}E_5$-hexadecane-water system, where the ratio between the oil and the surfactant was 1 : 0.815, Lund cut, as well as the Kahlweit fish cut with equal parts of oil and water, were performed. The temperature boundaries of the microemulsion phase were established and the lower emulsification boundary temperature T_{EFB}, the minimum temperature of a one-phase microemulsion, as well as the upper emulsification boundary, T_{UEFB}, the maximum temperature for the microemulsion, were established. The phase diagrams of hexadecane and decane for the Lund cut are shown in Figures 3.2a and b and the phase diagrams from the Kahlweit fish cut are shown in Figure 3.2c. The structures have been determined by SAXS or NMR, as well as visual inspection and the use of polarized light to find heterogeneous liquid-crystalline phases like lamellar and hexagonal phases.

The more important boundary is the lower one since the microemulsions generally contain spherical droplets in the vicinity of this phase boundary [84]. The failure temperature T_{EFB} can be compared to T_0. The temperature differences from T_0 are fairly constant between the decane and hexadecane systems. It is around 13 K for the system containing water (H_2O), and it changed slightly for the system containing D_2O. By using Equations 3.1 and 3.2, it is shown that the systems are compared at equal curvature when studied at T_{EFB}. Differences between the systems are not due to trivial temperature effects and it is therefore possible to draw more detailed conclusions about the surfactant film properties (in terms of κ and $\bar{\kappa}$). The main differences (apart from the expected increases in temperature [84]) between the phase diagram of the decane system and the hexadecane system are that the hexadecane system displays a lamellar phase only at high concentrations and that no cubic phase exists. The T_{UEFB} increases with concentration for

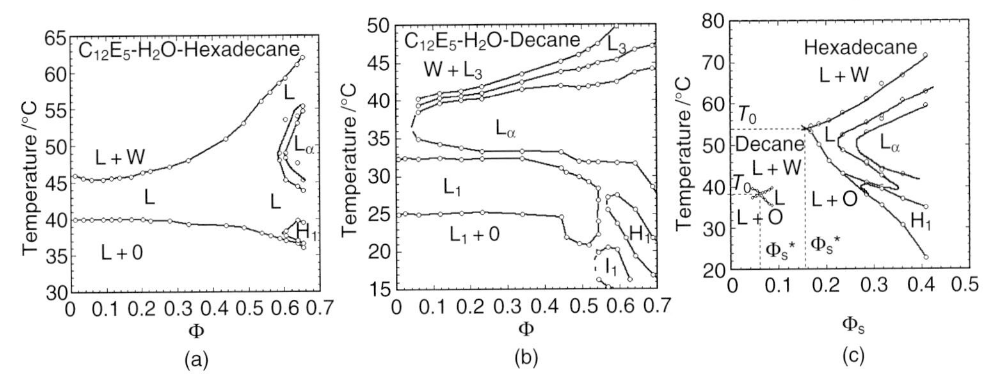

Figure 3.2 (a) The phase diagram of the system $C_{12}E_5$-water-hexadecane, (b) the phase diagram of the system $C_{12}E_5$-water-decane using the Lund cut with 0.815 : 1 surfactant to oil ratio on volume base, (c) the phase diagrams from the Kahlweit fish cut with oil-to-water ratio 1 : 1 on volume base of the system $C_{12}E_5$-water-hexadecane and a small part of the phase diagram of the system $C_{12}E_5$-water-decane. **L + O** is microemulsion with excess oil and **L + W** is the microemulsion and excess water and **L** is the microemulsion $\mathbf{L}_\alpha$ is the lamellar phase and $\mathbf{L}_3$ is the bicontinuous bilayer phase. Figures are adapted from ref. [113] and the data taken for **2a** from ref. [113], **2b** from ref. [131] and **2c** from ref. [113] with the decane data in **2c** coming from ref. [123].

the hexadecane system resulting in a microemulsion phase having a narrower temperature range for lower concentrations and wider range at higher concentration than that of the decane system.

3.4 Microemulsions at Emulsification Boundary

Initially, the decane and hexadecane microemulsion systems were studied at the emulsification boundary (T_{EFB}) to make an accurate comparison between the two systems. Deuterium-NMR relaxation using selectively deuterium-labeled surfactant is a useful method to detect variation of micelle size when changing composition and/or temperature. In a relaxation NMR experiment the combined motion of longitudinal ($1/T_1$) relaxation of the micelles and the transverse relaxation ($1/T_2$) of the surfactants inside the film (internal diffusion) are measured. The measured parameters are therefore sensitive to the micellar size but essentially insensitive to intermicellar interaction. More information can be found in papers by Wennerström [133] and Halle [33]. At T_{EFB} the microemulsion droplets are generally spherical at infinite dilution [84]. The results from the relaxation measurements (expressed as $1/T_2$–$1/T_1$) on the two systems are presented in Figure 3.3.

When examining these results, it is obvious that the droplets in the hexadecane system grow with increasing volume fraction of the droplets. The relaxation changes by about 50% and the droplets are therefore assumed to change size with

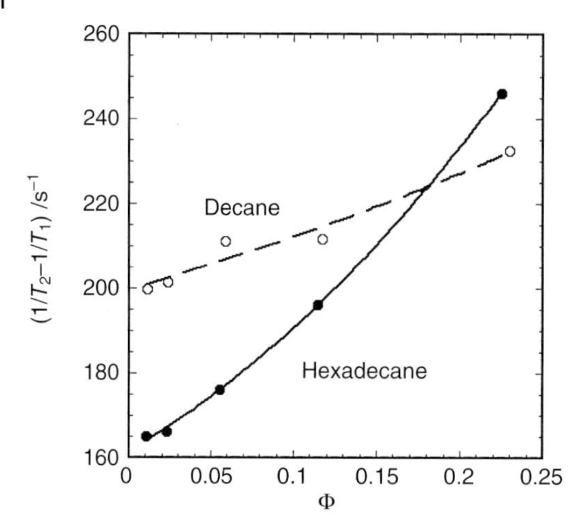

Figure 3.3 NMR relaxation data ($1/T_2 - 1/T_1$) for the systems with decane or hexadecane with $C_{12}E_5$ and water as a function of droplet volume fraction. The figure is adapted from ref. [113] and data taken for hexane from ref. [113] and for decane from ref. [34].

concentration. The droplets in the hexadecane system can be seen as growing with a factor of about $(1 + 4\Phi)$. The relaxation of droplets in the decane system changes only by 10%, in the same concentration range. This was initially interpreted as there was no significant growth with concentration [34]. In view of the result for the hexadecane system, the NMR relaxation data of the decane system could also be explained by a small size increase of the droplets in the decane system [113]. For both systems, the droplet growth is interpreted as a one-dimensional growth where the aggregates become ellipsoidal in shape. The model of Halle [134] for determining the axial ratios, ρ (ρ = long axis/short axis), has been used in this study.

The DLS results (apparent collective diffusion coefficients, D) of the decane system measured earlier [130] at T_{EFB} for this system was compared with the results of the hexadecane system [113] also measured at its T_{EFB} but using a different light-scattering setup [135, 136]. Whereas the concentration behavior of the collective diffusion coefficient of the droplets in the decane system follows that of hard spheres, it deviated from the hard-sphere behavior in the hexadecane system. However, without the previous NMR relaxation results (Figure 3.3), it would not have been possible to distinguish whether this deviation was due to the interactions between the aggregates or growth of the aggregates or a combination of the two. But in the light of the NMR relaxation results, the deviation was interpreted as a growth of the aggregates. Also, the SLS data of the decane system pointed towards interacting hard spheres [130]. There was only a small difference in the SLS results from the hard-sphere model indicating that the small growth observed in the NMR relaxation data (Figure 3.3) is slightly overestimated. For the hexadecane system, the prediction from the NMR relaxation data of a concentration

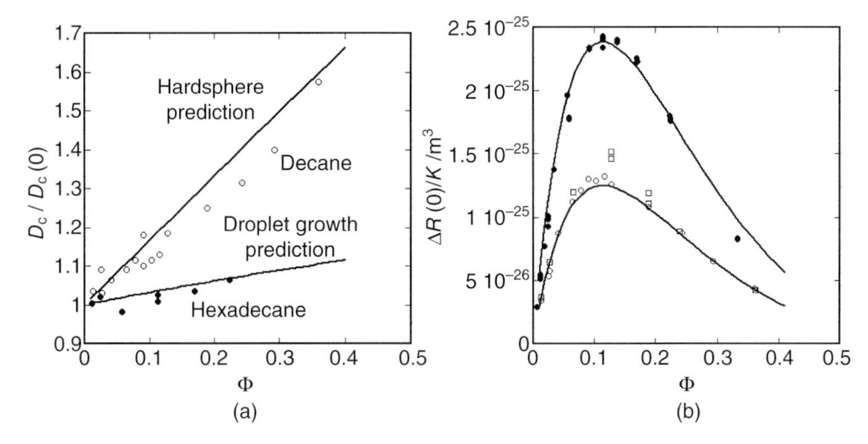

Figure 3.4 (a) Results from DLS and (b) SLS measurements on the hexadecane (filled circles) or decane (open circles) -$C_{12}E_5$-water systems at the respective emulsification boundary. The hard-sphere model gives a good description of the concentration dependence of the normalized apparent collective diffusion coefficient (D_c/D_0) of the decane system in (a) and the prediction from NMR relaxation data (with the growth following $(1 + 4\Phi)$) gives a good description for the concentration dependence of D_c/D_0 of the hexadecane system. The SLS data are presented in (b) as the excess Rayleigh ratio normalized with the constant K ($K = 4\pi^2 n^2 (dn/d\Phi)^2/\lambda^4$, where n is refractive index of water, $dn/d\Phi$ is the refractive index increment and λ is the wavelength of the laser light) to be able to compare the systems measured at different wavelengths. The SLS measurement performed in Lund for the decane system is shown as open squares in (b). Figures are adapted from and data taken from ref. [113]. The decane data originate from ref. [130].

growth with the factor of about $(1 + 4\Phi)$, led to a good description of both the DLS and the SLS data. The results from the DLS and SLS measurements on the decane and hexadecane systems are shown in Figure 3.4.

The SLS data in Figure 3.4b clearly show that the droplets of the hexadecane system are larger than those in the decane system. The hard-sphere radius of droplets in the hexadecane system is 91 Å compared to 76 Å for droplets in the decane system and the apparent hydrodynamic radii obtained from the data in Figure 3.4a are 105 Å and 86 Å, respectively.

The decane and the hexadecane systems were also studied using SAXS at laboratory instrument [137] in Aarhus, Denmark. Both were studied [138] at their respective T_{EFB} [113]. In this study both volume polydispersity and axis ratio were derives. These two parameters influence the scattering data in very similar ways, so it should in principle not be possible to determine both at the same time. However, the particle anisotropy has very little influence on the structure factor effects originating from interparticle interference and therefore it was possible to determine both. It was confirmed that both kinds of droplets grow with droplet concentration. The growth was higher for the droplets in the hexadecane system compared to droplets in the decane system, as shown in Figure 3.5a. At the highest droplet

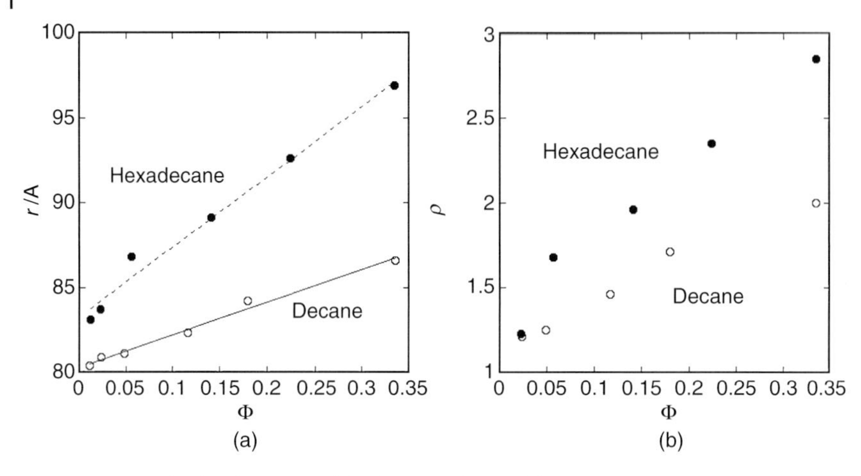

Figure 3.5 (a) The droplet size and (b) the axial ratio (b) versus droplet volume fraction for the -$C_{12}E_5$-water system with hexadecane (filled circles) or with decane (open circles) obtained from SAXS experiments. The lines are guides for the eye. The figures are adapted from and data taken from ref. [113].

volume fraction of 0.35, axis ratios of 2.5 and 2, respectively, were determined as can be seen in Figure 3.5b. At the lowest droplet volume fraction, Φ, =0.011, the volume-equivalent size difference was 5% between the droplets in the decane system (80 Å) and the droplets in the hexadecane system (83 Å).

From the scattering results above, it can be concluded that the differences between the two systems could not only be detected in the phase diagram but also in the behavior at the emulsification boundary. The droplets are essentially behaving as hard spheres but display a one-dimensional growth with concentration, especially in the hexadecane case. A size difference between the droplets in the decane and hexadecane systems was also observed, even at the same surfactant-to-oil volume ratio. It is worth noticing that using different techniques gives different radius partly due to the different definitions, so an "absolute" size to be used in calculations of κ and $\bar{\kappa}$ is hard to give.

3.5 Influence of Oil Chain Length

With the observed differences both in phase diagrams and microstructures between the decane and hexadecane microemulsion systems, it became interesting to study whether this was a sudden or a gradual change with oil chain length. Therefore, an investigation of the oil chain length (carbon length) dependence on the microemulsion systems was performed by varying the oil from octane to hexadecane. Both the Lund cut phase diagram, where only the microemulsion phase is shown, and the fish cut phase diagram, where also the two-phase regions were

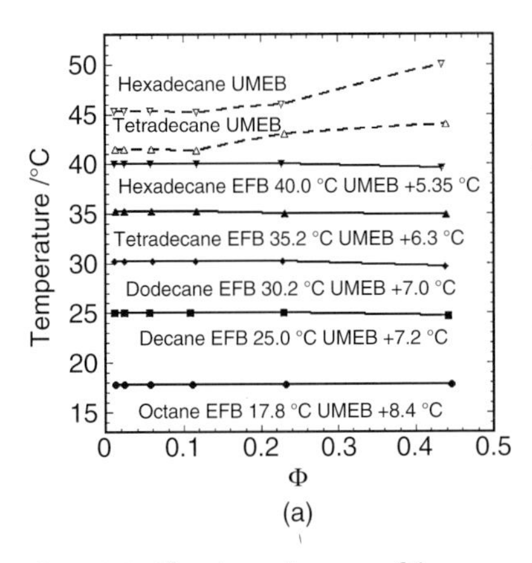

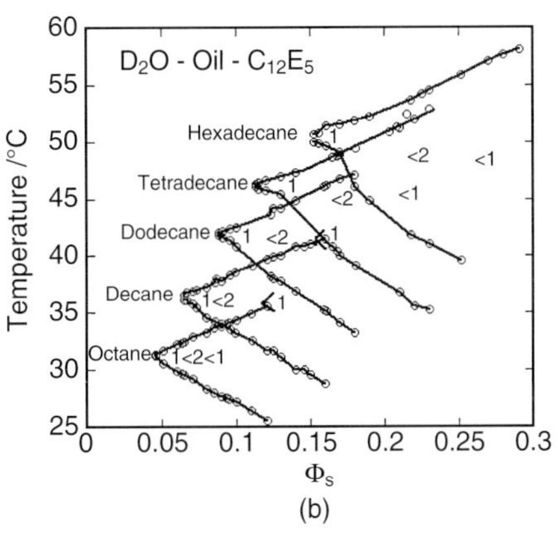

Figure 3.6 The phase diagrams of the systems of $C_{12}E_5$-alkane-water with the alkane varying from octane to hexadecane. In (a) the Lund cut with 0.815 : 1 surfactant-to-oil volume ratio using H_2O showing T_{EFB} for all the systems and only T_{UEFB} for tetradecane and hexadecane, since for the other alkanes the line describing T_{UEFB} is parallel to the one describing T_{EFB}. In (b) the Kahlweit fish cut with oil-to-water volume ratio (=1 : 1) using D_2O. "1" and "2" indicate one- and two-phase regions, respectively. The figures are adapted from ref. [118]. The data in (a) are taken from ref. [118] and in (b) from ref. [116].

determined, were determined by visual inspection and are presented for the two systems in Figure 3.6.

Note that the fish cut diagram in Figure 3.6b presents the different systems containing D_2O and not H_2O. This gives a difference of 2–3 K in temperatures for the fish cuts [113, 116] while the Lund cut is lowered with only 1–2 K, resulting in a different temperature difference between T_0 and T_{EFB} depending on whether H_2O or D_2O is used. From these results it can be concluded that the difference between T_0 and T_{EFB} is constant for the different oil chain lengths so the systems are compared at the same spontaneous curvature, H_0, according to Equation 3.2. The systematic behavior is that the microemulsion phase in the Lund cut has a narrower temperature range, the longer the oil, and that the "disappearance" of the lamellar phase is gradual.

The variation with oil chain length is well studied for the fish cut by Kahlweit, Strey and Sottmann with coworkers [55, 58, 139], but apart from what was already known, the phase boundaries for the two-phase region (i.e. the microemulsion lamellar phase) were also determined in [116]. In Figure 3.7a the minimum volume fraction of surfactant, Φ_{s*}, and the maximum volume fraction of surfactant, Φ_{su}, are shown as a function of the carbon length of the oil. The relative width of the microemulsion $((\Phi_{su} - \Phi_{s*})/2)$ is shown as a function of oil chain length in Figure 3.7b.

It is observed in Figure 3.7b that the relative width increases with increasing oil length. The decrease in relative width for the hexadecane system is only due to that the lowest concentration was used (independent of temperature) and not the one at T_0. This gives a bigger difference for longer oils since the phase diagrams

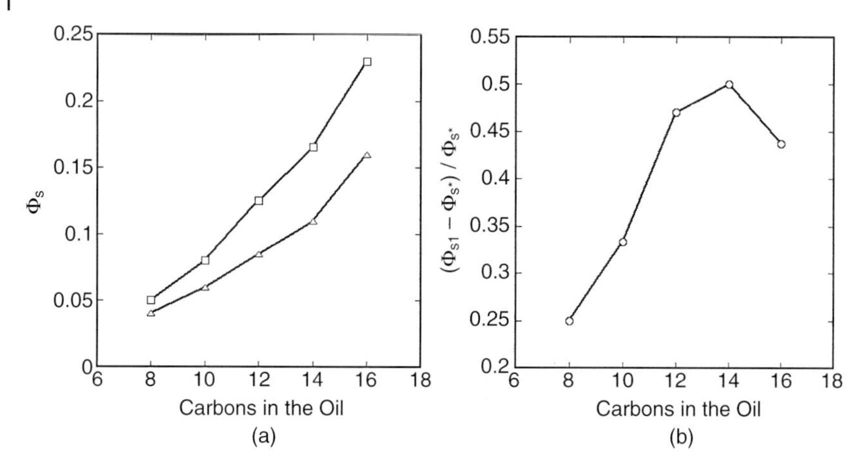

Figure 3.7 (a) Microemulsion boundaries with the surfactants, Φ_{su} (squares) and Φ_{s*} (triangles) plotted versus oil length. (b) The relative width $((\Phi_{su} - \Phi_{s*})/2)$ of the microemulsion phase for the systems of $C_{12}E_5$-alkane-D_2O (circles) plotted versus oil length. The figures are adapted from ref. [140] and data taken from ref. [113]. Note that the concentrations are taken at the minimum concentrations regardless of temperature and not at T_0.

became more twisted (less symmetrical) the longer the oils as seen in Figure 3.6b. Had the values at T_0 been used, there would have been an increase in relative width also for hexadecane.

To further investigate whether the change in the microstructure (from a spherical to an ellipsoidal shape) with volume fraction of droplets is gradual or sudden, both dynamic and static light-scattering experiments were performed on the different alkane systems. The results are presented in Figure 3.8. As observed from both the DLS (Figure 3.8a) and the SLS (Figure 3.8b) results, it is evident that the droplets grow gradually with increasing oil chain length.

To obtain further insight into these systems SANS measurements [84, 115] were performed at the Paul Scherrer Institut in Villigen, Switzerland, at the SANS1 beamline [141]. The contrast-matched balanced system had been studied by Strey *et al.* previously [57] using a Kahlweit cut (oil: water 1 : 1) for some of the systems, but here measurements on the different systems were performed under the same experimental conditions. A contrast-matched system can be considered as an effectively binary system assuming that the microemulsion system is symmetric with exchange of water and oil domains. In such a scattering experiment only the scattering from the surfactant film is detected. The sample concentrations were just above the minimum surfactant concentration for the different systems (see Figure 3.9a). It can be observed that the difference between scattering curves for the systems is gradual.

The axial ratios of the aggregates at fixed concentration resulting from the different techniques. NMR relaxation and self-diffusion, DLS and SANS (from the contrast variation, see later Figure 3.13) are presented in Figure 3.10 as a function

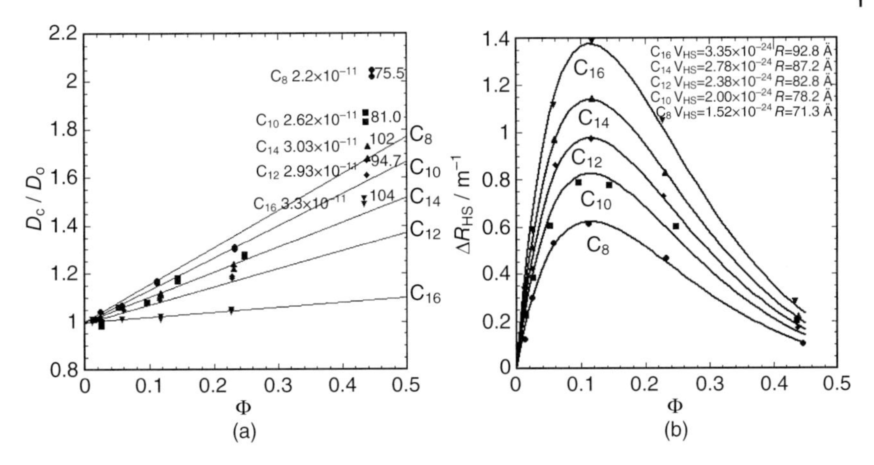

Figure 3.8 Results from (a) dynamic light-scattering and (b) static light-scattering measurements performed at the emulsification boundary for different $C_{12}E_5$-alkane-H_2O systems. The apparent hydrodynamic radii are shown in 8(a) and the hard sphere radii in 8(b). The figures are adapted and data taken from ref. [118].

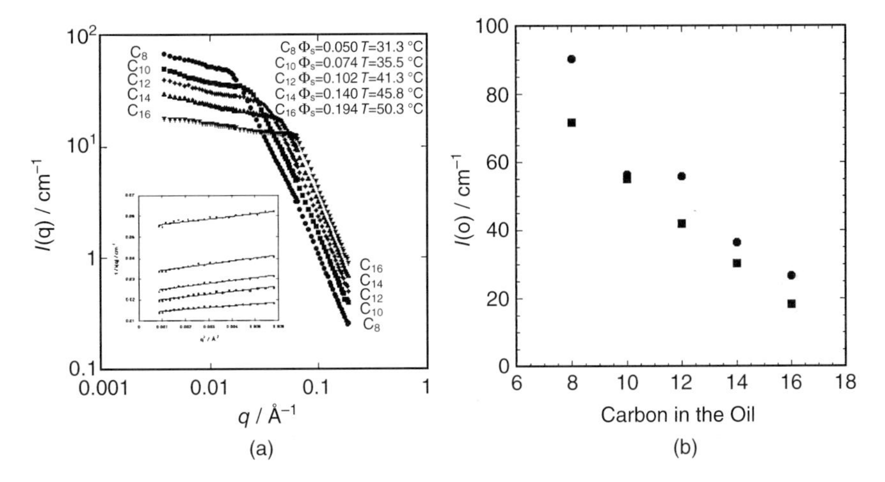

Figure 3.9 (a) The scattering curves from SANS measurements on the systems of $C_{12}E_5$-alkane-D_2O (the experimental conditions are shown in the figure) using a Kahlweit cut (oil : water 1 : 1). The inset shows the inverse intensity versus the squared scattering length, (b) the forward scattering (at zero angle) as a function of carbon length for two samples for each carbon length. The figures are adapted and data taken from ref. [116].

of oil length at corresponding T_{EFB}. It is observed that the growth of the droplets with concentration increases gradually with the oil carbon length.

The main conclusion is that the shape change observed when increasing the oil length is gradual. Additional conclusions are that the longer the oil the larger the droplets and the more pronounced is the growth with concentration.

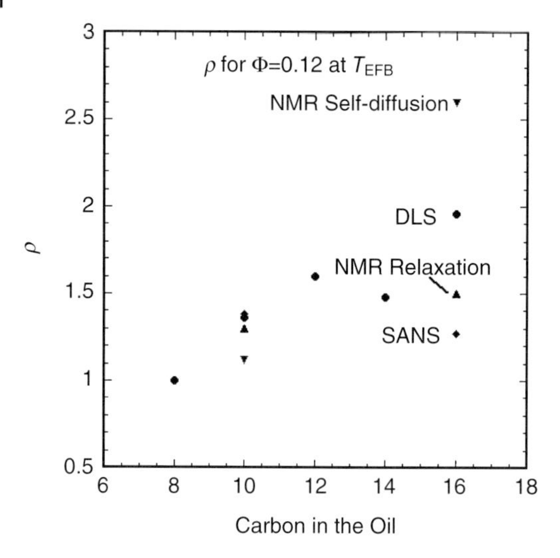

Figure 3.10 Axial ratios obtained for the different $C_{12}E_5$-alkane-H_2O systems at the emulsification boundary for samples with $\Phi = 0.12$ using NMR self-diffusion (downward phasing triangles), NMR relaxation (upward phasing triangles), DLS (circles), and SANS (diamonds). The figure is adapted from ref. [118] and data taken from refs [113, 115, 118].

3.6 The Effect of Temperature

With increasing temperature a more dramatic growth of the droplets takes place. This growth has been shown to be one-dimensional [35] as predicted by Safran and coworkers [142]. The results from the NMR relaxation measurements at elevated temperatures are shown in Figure 3.11; the decane system is presented in Figure 3.11a, and the hexadecane system in Figure 3.11b.

In Figure 3.11, it can be noticed that for the highest droplet concentration in both systems, the relaxation, and there by the aggregate size, passes through a maximum at elevated temperatures after which the size decreases at even higher temperatures or the upper phase boundary is reached. This is contrary to what is expected. A decrease in size with increasing temperatures should not be expected since the surfactant continues to change with temperature [84, 143]. Instead, this behavior can be interpreted as that the systems change to a bicontinuous phase [34]. For these temperatures the axial ratio has no meaning so they are not used in the further investigation. Crude values of transition temperatures into a bicontinuous structure can thus be determined from this maximum in Figure 3.11.

To confirm that the systems change from discrete aggregates to becoming bicontinuous, NMR self-diffusion experiments were performed. The results are shown in Figure 3.12. This technique [28, 29] measures the diffusion of the surfactants and the oils, and can thus be used to determine whether the studied structures consist of discrete aggregates or are bicontinuous.

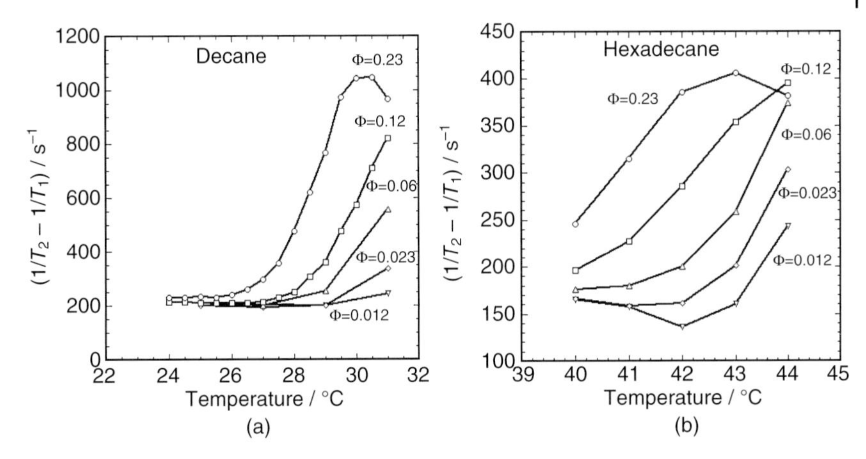

Figure 3.11 NMR relaxation as function of temperature for the (a) $C_{12}E_5$-water-decane system and for (b) the $C_{12}E_5$-water-hexadecane system at five droplet concentrations. The figures are adapted from ref. [117] and data for the decane system are taken from ref. [34] and the hexadecane system from ref. [117].

The aggregates in the hexadecane system start to grow at lower concentrations and at lower temperature than the aggregates in the decane system. Furthermore, the aggregates in the hexadecane system stop to grow at a lower temperature. This results in that the aggregates in the decane system grow more (i.e. to larger axial ratios). However, the growth starts both at higher concentration and at higher temperature increase.

To be able to study the structure of the system in more detail it would be interesting to investigate by scattering the microemulsion droplets with different contrast, that is, with a difference between the scattering length density of the scattered object and the solvent/background. SANS is very suitable for this [144] since deuterated compounds have scattering length densities that are very different from those of the corresponding ordinary compounds. There are also possibilities for contrast variation for SAXS by using for example, chlorinated oils for changing the contrast of the oil phase [145], however, the physical properties of chlorinated oils are very different from those of ordinary alkanes and it is in practice not possible to use this option.

In the SANS technique, the fact that different isotopes of the same element can have very different scattering length density is exploited. In the case of hydrogen, the value is negative, while deuterium has a positive value. This make it possible to vary the contrast so that the oil at the inside of the droplets and the water at the outside have the same contrast by using combination of deuterated oils and mixtures of H_2O and D_2O. What is really detected in the experiment is then the scattering of the surfactant film. Both a contrast where the oil and the water was matched [116] as well as a contrast-variation SANS study with ten different H_2O/D_2O ratios have been performed [115]. There has been an earlier study [146] on contrast variations that concerned reverse water-in-oil system with AOT

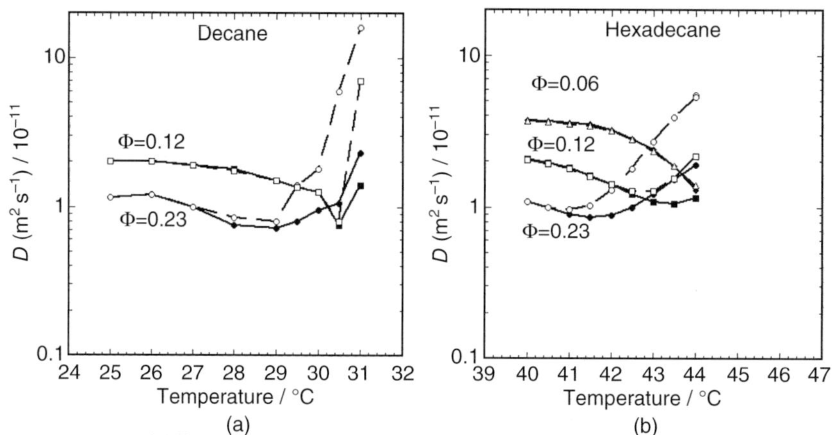

Figure 3.12 Self-diffusion coefficient plotted versus temperature for two (in a) respective three (in b) concentrations, (a) the $C_{12}E_5$-water-decane, system and (b) the $C_{12}E_5$-water-hexadecane system. The dotted lines are the oil diffusion and the permanent lines surfactant diffusion. When the two are different the system is bicontinuous. The figures are adapted from ref. [117] and data for the decane system are taken from ref. [35] and for the hexadecane system from ref. [117].

[bis(2-ethylhexyl)sulfosuccinate sodium salt] as surfactant and without varying temperature and concentration. This study showed that it is possible from an extensive contrast-variation series to derive both the volume/composition polydispersity and the axis ratio of the droplets.

In the extensive contrast-variation study [115] about 720 separate measurements in total were performed for the decane and hexadecane systems. The ten contrasts that these correspond to for each system were then fitted simultaneously to obtain the size distribution and the average shape of the microemulsion droplets. Some of the SANS scattering curves and their fits are presented in Figure 3.13.

The forward scattering minimum (match point) as shown in Figure 3.14 occurred at the same scattering length density for all temperatures and did not broaden with temperature or concentration, indicating that only one type of aggregate was present. If there were two types of aggregates present they would have different compositions and thus different scattering length and hence broaden the minimum.

The decane and the hexadecane systems were also studied using dynamic and static light scattering at temperatures above T_{EFB} [118] and the results are displayed in Figure 3.15. In line with the previous studies using other techniques, the DLS and SLS data also show that the droplets in the hexadecane system start to grow at a lower concentration and at a lower temperature increase than the droplets in the decane system. This conclusion is drawn from the observed deviation from the linear fit of D_c/D_0 *vs.* droplet volume fraction (Figures 3.15a and b) and from the deviation from the hard-sphere behavior of the excess Rayleigh ratio (Figures 3.15c and d). The earlier growth of droplets in the hexadecane system still leads to change into bicontinuous structure at a lower axial ratio, $p = 4$ rather than 6, as for

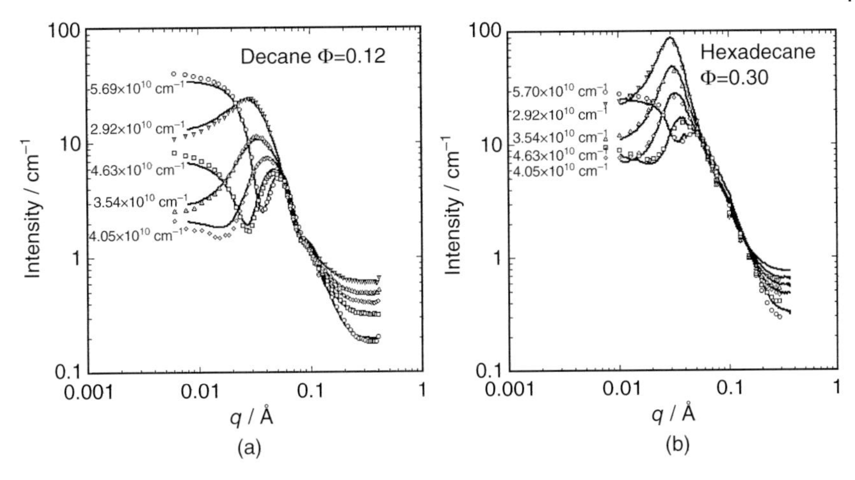

Figure 3.13 SANS data and their fits versus scattering vector q, at chosen scattering length densities for (a) the $C_{12}E_5$-water-decane system, and for (b) the $C_{12}E_5$-water-hexadecane system, at the respective emulsification boundary. The figures are adapted from and data taken from ref. [115].

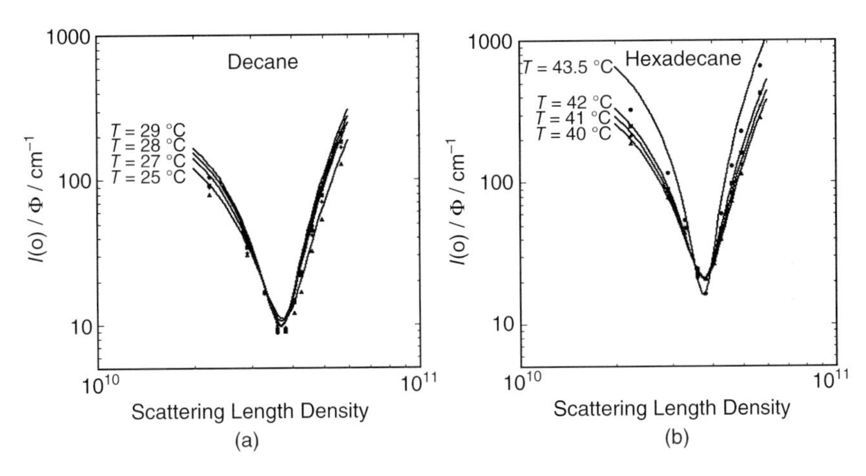

Figure 3.14 Intensity in the forward direction versus scattering length density of the water phase for (a) the $C_{12}E_5$-water-decane system, and for (b) the $C_{12}E_5$-water-hexadecane system, for four temperatures. The figures are adapted from and data taken from ref. [115].

droplets in the decane system, which can be seen in Figure 3.16, where the axial ratios of the aggregates obtained from the different techniques are presented.

From all the results presented above, it can be concluded that the droplets grow with temperature and that they do not grow to axial ratios >4–6 before they either reach the phase boundary or the system become bicontinuous. The growth increases with the oil length, but the axial ratio when system of discrete droplets

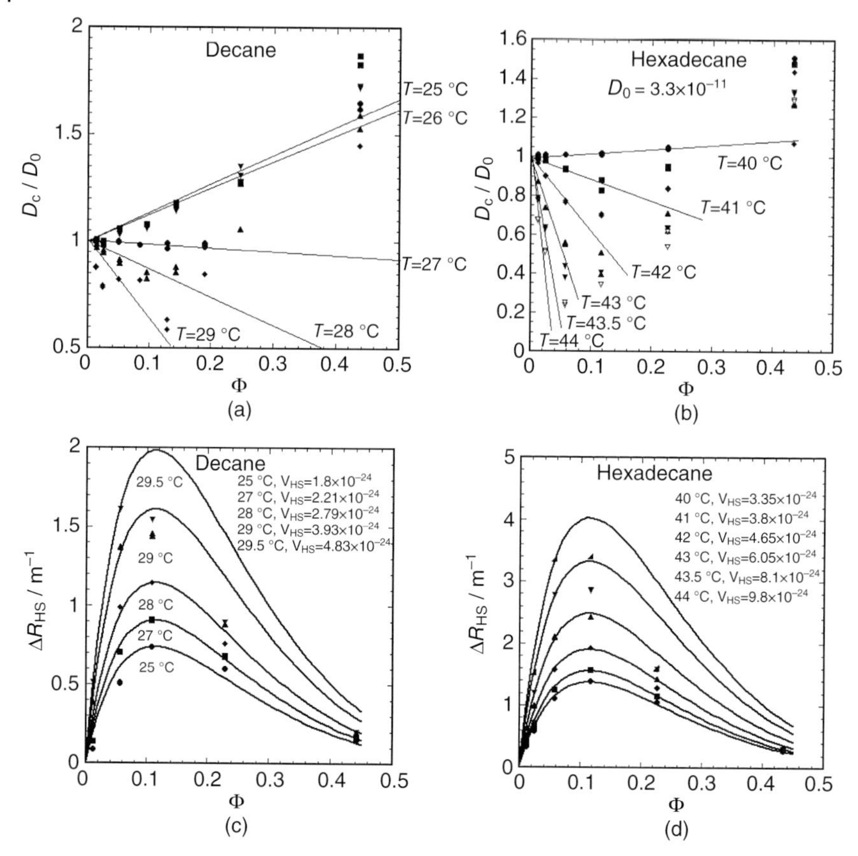

Figure 3.15 Results from dynamic light-scattering measurements on (a) the $C_{12}E_5$-water-decane system, and (b) the $C_{12}E_5$-water-hexadecane system. Results from the static light-scattering measurements on (c) the decane system and (d) the hexadecane system for different temperatures. The obtained fitting parameters of the SLS data are also shown in (c) and (d). The figures are adapted from and data taken from ref. [118].

changes into a bicontinuous system is lower the longer the oil length. It is also evident that there is only one population of droplets and not a mixture of spheres and rods giving a similar behavior as ellipsoidal droplets.

3.7 The Temperature at Which the Microemulsion Becomes Bicontinuous

Using the deviations of specific parameters measured with different techniques from hard-sphere trends, it is possible to make preliminary predictions on when the discrete aggregates change into a bicontinuous system. Since there is no macroscopic phase separation it is not possible to acquire this information from the

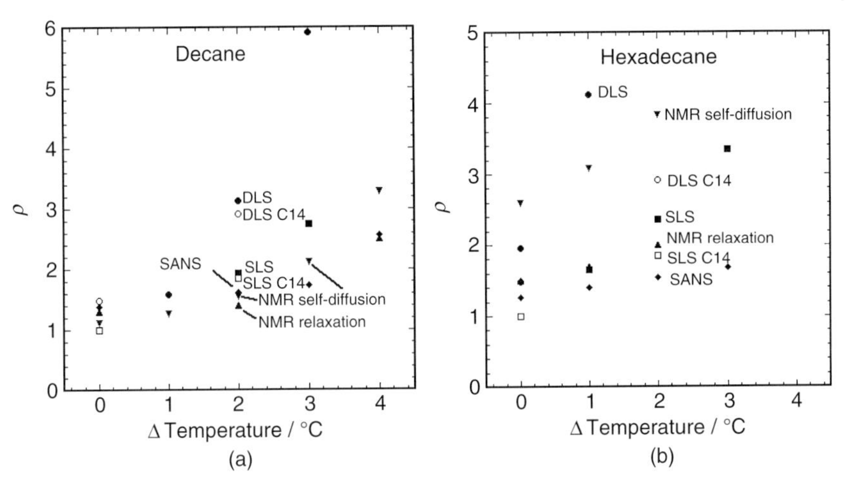

Figure 3.16 Axial ratios determined by DLS (circles), SLS (squares) NMR relaxation (upward phasing triangles), NMR diffusion (downward phasing triangles), and SANS (diamonds) for (a) the $C_{12}E_5$-water-decane system, and for (b) the $C_{12}E_5$-water-hexadecane system, as a function of temperature increase ΔT from T_{EFB}. The results for the tetradecane system are indicated as hollow symbols for DLS and SLS. Figures adapted from ref. [118] and data taken from refs [113, 115, 117, 118].

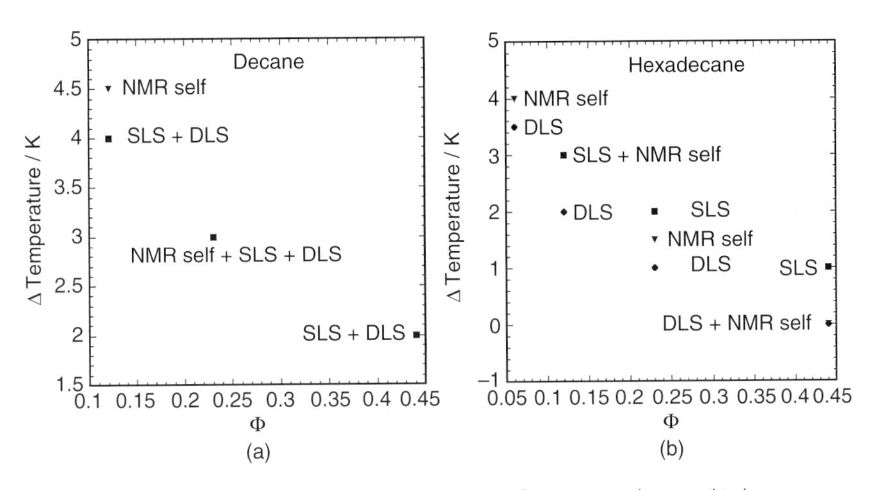

Figure 3.17 Estimates of the temperatures at which the systems become bicontinuous based on results from light-scattering and NMR self-diffusion techniques: (a) the system of $C_{12}E_5$-water-decane, (b) the system of $C_{12}E_5$-water-hexadecane. The figures are adapted from ref. [118] and data taken from refs [35, 117, 118].

phase diagrams. Good predictions of when the systems becomes bicontinuous can be made using DLS, SLS and NMR relaxation as well as SANS. In Figure 3.17, DLS, SLS and NMR results are compared in order to estimate the temperature of this transition.

It is shown that the decane system needs a higher temperature increase from T_{EFB} to change into bicontinuous structure compared to the hexadecane system. The temperature increase is inversely proportional to the concentration. It can be concluded that light scattering can be used to get good estimates of temperatures of the transition from discrete droplets to a bicontinuous structure.

3.8 Interfacial Tension: Investigating the Microemulsion Model and Scaling

Investigations of the microemulsion model can be done by combining the dataset from phase diagrams, SANS and interfacial tension data to retrieve values of κ and $\bar{\kappa}$. However, the datasets can also be used individually to determine which of the two models for the Gibbs energy, G, works the best. The two models are based on, respectively, the polynomial expansion [126] of the curvature: $G/V = a_3\Phi_S^3 + a_5\Phi_S^5$, and the renormalization [147–150] of the bending rigidity: $G/V = a\Phi_S^3 + (1 + b\ln\Phi_S)$, where V is the volume of the system. The information needed from phase diagrams is the minimum surfactant concentration, Φ_{s^*}, the maximum concentration of microemulsion, Φ_{su}, and the minimum concentration to have a lamellar phase, $\Phi_{sL\alpha}$, all at T_0. From the SANS measurements, the forward scattering data of the contrast-matched samples is needed and from surface tension measurement the minimum surface tension is needed. It was concluded that both the polynomial expansion and the renormalization models gave a good description of the data, and it was not possible to determine which model is the best one.

It has been shown by Kunieda and Shinoda that there is a correlation between ultralow surface tension and critical solution phenomena [151]. The interfacial tension of the balanced system has been studied by Sottmann and Strey [60] using spinning-drop measurements. The authors explored the full temperature range and varied both the oil chain length and the surfactant. For the $C_{12}E_5$ system the oil was varied from octane to tetradecane. From the data it could be shown that the interface tension of a given system obeys the parabolic relation $\gamma = \varepsilon\,(T_0 - T)^2 + \gamma_0$ over a large temperature range. Here, ε is a constant and the value of the interfacial tension at $T = T_0$, denoted as γ_0, is numerically small compared to values obtained a few degrees away from γ_0. The surface tension versus the square of the temperature difference from T_0 for the different oils is shown in Figure 3.18.

Figure 3.18 shows that all the data follow one universal straight line for this surfactant system regardless of the oil component. The different alkane systems also have the same value of ε within experimental accuracy. The interfacial value at $T = T_0$, γ_0, is on the order of $1\,\mu N/m$ and this value changes by an order of magnitude between octane and tetradecane. For general information about surface tensions of colloidal systems see the book edited by Hartland [152]. The interpretation of the interfacial data was extended using more surfactants [119]. Figure 3.19 shows that the minimum surface tension versus Φ_{s^*} nearly follows the same dependence for different surfactants and oils.

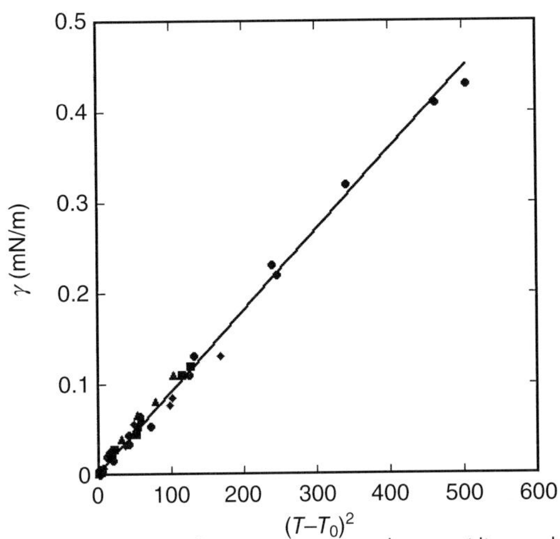

Figure 3.18 The surface tension versus the temperature difference squared from T_0 for the system of $C_{12}E_5$-water-alkane. The symbols are for the different oils, octane (circles), decane (squares), dodecane (diamonds) and tetradecane (triangles). All of the data follow one universal straight line. The figure is adopted from ref. [116] and data taken from ref. [116] originating from ref. [60].

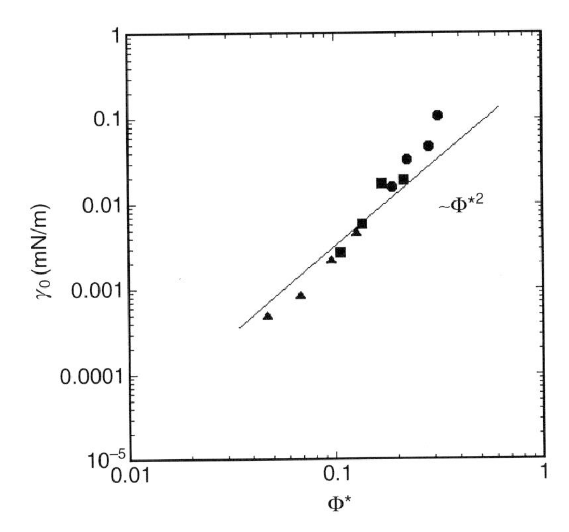

Figure 3.19 The surface tension at the minimum surfactant concentration for the different systems of C_mE_n-water-alkane with several different oils and three surfactants, C_8E_3 (circles), $C_{10}E_4$ (squares), and $C_{12}E_5$ (triangles). All data show a similar behavior. The line scales with minimum surfactant squared as a guide for the eye. The figure is adapted from and data taken from ref. [119].

From the results of the interpretation of the surface tension we may conclude that the scaling (the fact that these systems show similar behavior) of the systems is indeed valid, as has also been shown by Strey and Sottmann [58]. It could also be observed that all the SANS curves of the matched system can fit on top of each other if the concentration differences is taken into consideration when the intensity and the scattering vector is in log scale [116].

3.9
Microemulsions as Models for Drug-Delivery Systems

Having all this information about these microemulsions, we have proposed to use them as possible models for drug-delivery systems. Lidocaine (2-diethylamino-N-(2,6-dimethylphenyl)acetamide) was here chosen as the drug since there are applications where lidocaine is delivered through application directly on the skin (topical drug-delivery systems). The present system could also be a model of a system delivering through the skin (dermal drug-delivery system). Note that ethylene oxide has also been reported to promote drug penetration through the skin [153–155].

First, the solubility of lidocaine in oil (the alkanes mentioned above, octane to hexadecane) and water was studied. At 298 K it was possible to make a 10% lidocaine mixture in the oils but for the longer oils it was crucial that the temperature was 298 K rather than 293 K. Not much of the 1% lidocaine added to water was dissolved, not even at high temperatures. Lidocaine was thus assumed to have low water solubility and that it would be more hydrophobic than hydrophilic even though it is slightly amphiphilic. Assuming that the lidocaine would be in the oil phase, close to the surfactant, made it necessary to lower the amount of oil equally, in order to have the same amount of hydrophobic compound to the surfactant. The lidocaine concentration chosen was 0%, 1% and 10% based on oil volume, where oil volume is defined as the volume of lidocaine and oil. The droplets were expected to be of equal size at T_{EFB} since the surfactant-to-oil ratio would be the same. In the initial work [122] it was shown that the phase boundaries change slightly with 1% substitution but for 10% substitution the T_{EFB} is lowered substantially. However, T_{UEFB} does not decrease as much, resulting in a wider microemulsion phase, as seen in Figure 3.20. From these results it is not possible to conclude about the size of the microemulsion droplets. It is thus not possible to explain the difference just as resulting from a change in the surfactant film curvature. For determining sizes, dynamic and static light scattering was employed to study [120] the systems. Figure 3.21 presents the results from these measurements on the system with various amount of lidocaine. It is observed that the droplets become smaller for the system with 10% oil substituted by lidocaine. This is more evident when examining the SLS results (see Figure 3.21b). The difference between 0 and 1% substitution with lidocaine is small but quite significant for 10%. The size of droplets (i.e. the apparent hydrodynamic radius) decreases from 80 Å to 70 Å when 10% of the oil is substituted with lidocaine. The change in shape was confirmed

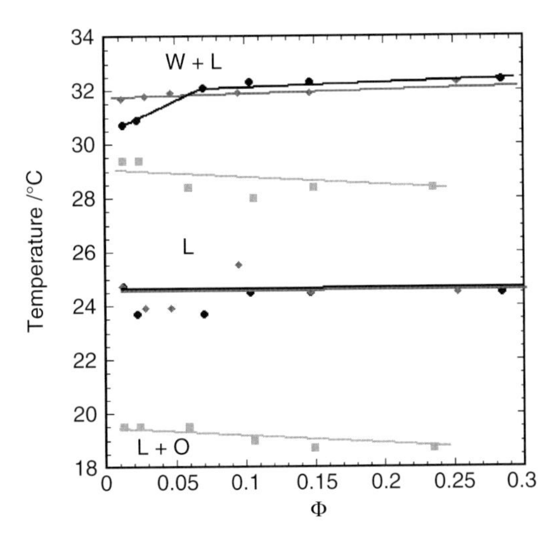

Figure 3.20 The phase diagrams for the Lund cut with 0.85 : 1 surfactant-to-oil volume ratio for the $C_{12}E_5$-water-decane system, where part of the oil is substituted with lidocaine. Without lidocaine (circles), with 1% (diamonds), and with 10% (squares). **W + L** indicates the two-phase with excess water, **L** the microemulsion, and **L + O** the two-phase with excess oil. The figure is adapted from and data taken from ref. [121].

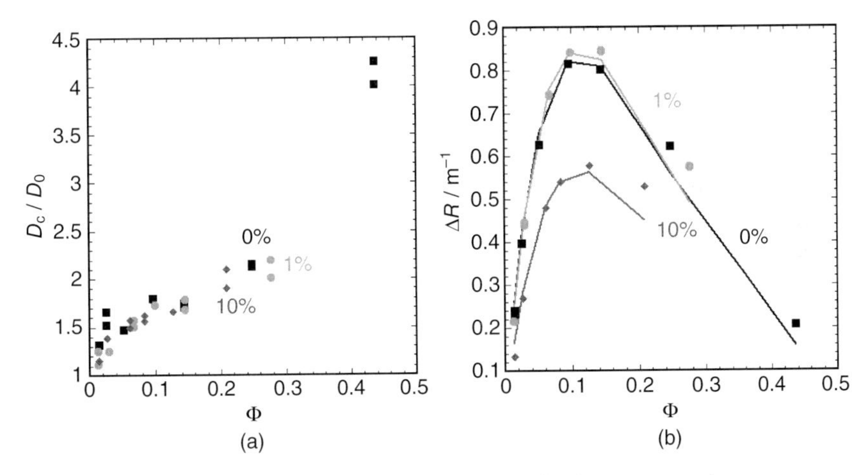

Figure 3.21 Results from (a) dynamic light scattering (D_c/D_0 *vs.* volume fraction of droplets), and (b) static light scattering (excess Rayleigh ratio *vs.* volume fraction of droplets) measurements at the emulsification boundary for the $C_{12}E_5$-water-decane system, where part of the oil is substituted with lidocaine. The figures are adapted from and data taken from ref. [120].

by SAXS measurements at MAX-Lab in Lund, Sweden using the I711 beam line [156] at the MAX II ring. The aggregates became more elongated (ellipsoidal) with droplet concentration and with the concentration of lidocaine [114], as shown in Figure 3.22.

By increasing the temperature significant differences were observed using light-scattering techniques. The droplet growth can be seen as an increased deviation from the hard-sphere model when fitting the DLS and the SLS data. In Figure 3.23, the concentration dependence of the normalized collective diffusion coefficient (D_c/D_0) obtained from DLS measurements on the three systems with lidocaine at selected temperatures is displayed.

The change in the size and shape of the droplets with temperature was confirmed by using SAXS. The SAXS data obtained for the system with 1% of the oil substituted with lidocaine at 3 temperatures is presented in Figure 3.24. The

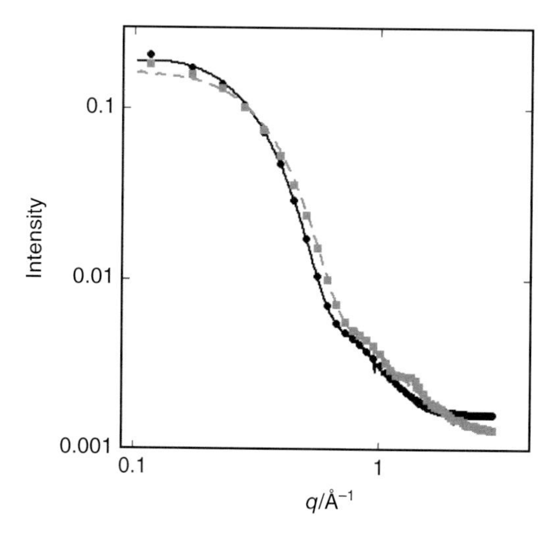

Figure 3.22 The SAXS data at the emulsification boundary for the $C_{12}E_5$-water-decane system, where part of the oil is substituted with lidocaine of 1% lidocaine (circles) and 10% lidocaine (squares). The figure is adapted from and data taken from ref. [114].

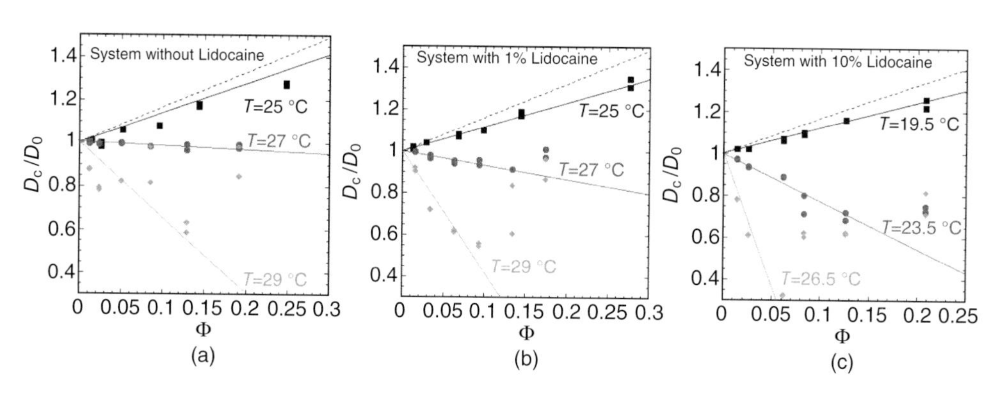

Figure 3.23 Normalized collective diffusion coefficient at three temperatures for the $C_{12}E_5$-water-decane system, where part of the oil is substituted with lidocaine. (a) Without lidocaine, (b) with 1% of the oil substituted with lidocaine, and (c) 10% substituted with lidocaine. The figures adapted from and data taken from ref. [121].

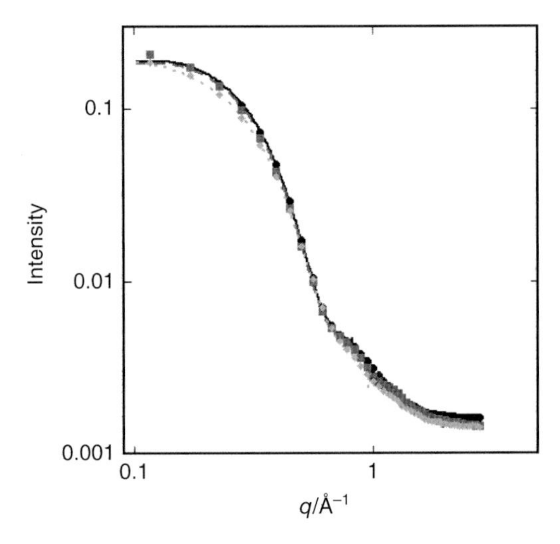

Figure 3.24 The difference with temperature for the system $C_{12}E_5$-water-decane, where 1% of the oil is substituted with lidocaine at 298 K (circles), 300 K (squares) and 302 K (diamonds). The figure adapted from and data taken from ref. [114].

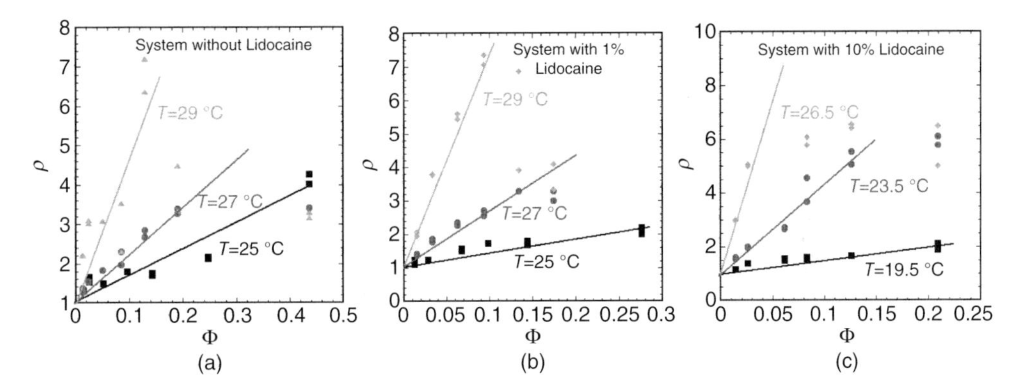

Figure 3.25 The axial ratio determined from DLS as a function of droplet volume fraction for three temperatures for the $C_{12}E_5$-water-decane system, where part of the oil is substituted with lidocaine. (a) Without lidocaine, (b) with 1% of the oil substituted with lidocaine, and (c) 10% substituted with lidocaine. The lines are guides for the eye. The figures are adapted from and data taken from ref. [121].

derived droplet growth with temperature displayed in Figure 3.25 can again be interpreted as an axial-ratio growth with temperature.

Self-diffusion NMR studies [121] confirms that the microemulsion systems initially consist of discrete aggregates that at higher temperatures change into bicontinuous microemulsions. The temperature increase needed to change the system into a bicontinuous microemulsion also changes with the amount of added

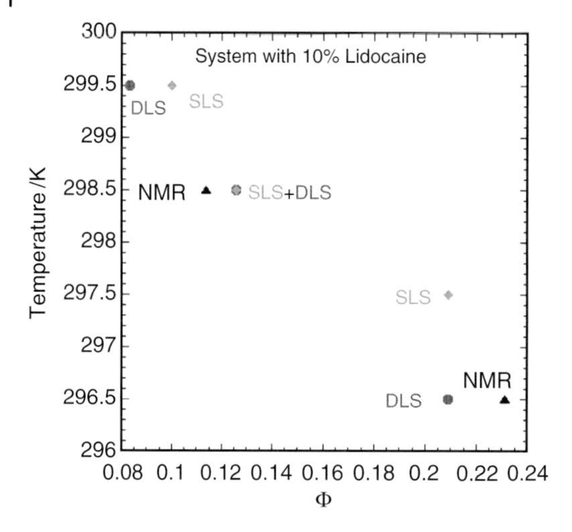

Figure 3.26 Temperature at which the system becomes bicontinuous versus droplet volume fraction for the $C_{12}E_5$-water-decane system, where 10% of the oil is substituted with lidocaine. The figure is adapted from and data taken from ref. [121].

lidocaine. The temperature for the system with 10% of the oil substituted with lidocaine where it changes from discrete droplets to bicontinuous as derived from DLS/SLS and from self-diffusion NMR is shown in Figure 3.26. The values from DLS and SLS lay close to those from NMR, which confirm that the above-described procedure, where the deviation from the hard-sphere model is evaluated, can be used to study this structural transition. From the experimental results, it can be concluded that the model drug-delivery system changes both microscopically and macroscopically when a drug is added to the microemulsion at higher doses.

3.10 Conclusion

Throughout this work the growth of the microemulsion droplets has been interpreted as the droplets becoming ellipsoidal in shape with a defined ratio between the two axis, p. The SAXS and the NMR data could be well fitted by the ellipsoidal model. This does not rule out that the aggregates may be peanut shaped, that might be more energetically favored. The microemulsion systems of $C_{12}E_5$ with a ratio of surfactant to oil in the range of 0.815–0.85 show some consistent features. The droplets grow to one-dimensionally elongated objects, and if one assumes that they are ellipsoidal, one gets an axial ratio of 4–6 before the system become bicontinuous or reaches the upper phase boundary. Results show that there is no formation of long worm-like micelles, only a formation of a bicontinuous microemulsion structure. There is a systematic gradual change in size when changing the oil. The droplets grow at both lower temperature increase from T_{EFB} and at lower droplet concentration, but also change into a bicontinuous microemulsion at lower axial ratios, the longer the oils. In addition, a droplet growth at T_{EFB} can also be observed for the longer oils.

Some of the difference between the systems can be explained by a numerically smaller (negative) value of $\bar{\kappa}$ the longer the oil.

With the microemulsions it was possible to study the effects of the addition of the drug lidocaine on a model drug-delivery system, the $C_{12}E_5$-decane-water system. Substituting oil with lidocaine results in changes of the phase boundaries, as well as the droplet size and growth with temperature and concentration. With changes of both the macroscopic and microscopic properties of the system when adding a drug it is obviously necessary to take the effect of the drug into account when formulating a drug-delivery system.

Another conclusion is that the nonionic microemulsion systems with C_mE_n are useful since they are so well characterized. The scaling properties of their behavior make them extra useful since the results can be generalized. Even if a lot is known about these nonionic microemulsions far from all the aspects are fully understood.

References

1 Hoar, T.P., and Schulman, J.H. (1943) *Nature*, **152**, 102.

2 Schulman, J.H., and Riley, D.P. (1948) *J. Colloid Interface Sci.*, **3**, 383.

3 Schulman, J.H., and Friend, J.A. (1949) *J. Colloid Interface Sci.*, **4**, 497.

4 Windsor, P.A. (1954) *Solvent Properties of Amphiphilic Compounds*, Butterworths, London.

5 Windsor, P.A. (1968) *Chem. Rev.*, **68**, 1.

6 Ekwall, P., Danielsson, I., and Mandell, L. (1960) *Angew. Chem. Int. Ed. Engl.*, **79**, 119.

7 Ekwall, P. (1967) *Svensk Kemisk Tidskrift*, **79**, 605.

8 Prince, L.M. (1967) *J. Colloid Interface Sci.*, **23**, 165.

9 Prince, L.M. (1969) *J. Colloid Interface Sci.*, **29**, 216.

10 Gillberg, G., Lehtinen, H., and Friberg, S.E. (1970) *J. Colloid Interface Sci.*, **33**, 40.

11 Shinoda, K., Nakagawa, T., Tamamushi, B., and Isemura, T. (1963) *Colloidal Surfactants*, Academic Press, New York.

12 Shinoda, K. (1968) Principles of solution and solubility, in *Solvent Properties of Surfactant Solutions*, vol. 2 (ed. K. Shinoda), Marcel Dekker, New York.

13 Shinoda, K. (ed.) (1968) *Solvent Properties of Surfactant Solutions*, vol. 2, Marcel Dekker, New York.

14 Shinoda, K., and Kunieda, H. (1971) *Abstr. Pap. Am. Chem. S.*, 51.

15 Shinoda, K., and Kunieda, H. (1973) *J. Colloid Interface Sci.*, **42**, 381.

16 Kunieda, H., and Shinoda, K. (1972) *Nippon Kagaku Kaishi*, 2001.

17 Kunieda, H., and Shinoda, K. (1976) *J. Phys. Chem.*, **80**, 2468.

18 Kunieda, H., and Shinoda, K. (1978) *J. Phys. Chem.*, **82**, 1710.

19 Kunieda, H., and Shinoda, K. (1979) *J. Colloid Interface Sci.*, **70**, 577.

20 Kunieda, H., and Shinoda, K. (1980) *J. Colloid Interface Sci.*, **75**, 601.

21 Friberg, S.E., Lapczynska, I., and Gillberg, G. (1976) *J. Colloid Interface Sci.*, **56**, 19.

22 Shinoda, K. (1983) *Prog. Colloid Polym. Sci.*, **68**, 3.

23 Scriven, L.E. (1976) *Nature*, **263**, 123.

24 Bull, T., and Lindman, B. (1975) *Mol. Cryst. Liq. Cryst.*, **28**, 155.

25 Stilbs, P., and Moseley, M.E. (1979) *Chem. Scr.*, **13**, 26.

26 Stilbs, P., Moseley, M.E., and Lindman, B. (1980) *J. Magn. Reson.*, **40**, 401.

27 Lindman, B., Stilbs, P., and Moseley, M.E. (1981) *J. Colloid Interface Sci.*, **83**, 569.

28 Stilbs, P. (1987) *Prog. Nucl. Magn. Reson. Spectrosc.*, **19**, 1.

29 Söderman, O., and Stilbs, P. (1994) *Prog. Nucl. Magn. Reson. Spectrosc.*, **26**, 445.

30 Lindman, B., Söderman, O., and Wennerström, H. (1987) *Ann. Chim.-Rome.*, **77**, 1.

31 Kahlweit, M., Strey, R., Haase, D., Kunieda, H., Schmeling, T., Faulhaber, B., Borkovec, M., Eicke, H.F., Busse, G., Eggers, F., Funck, T., Richmann, H., Magid, L., Söderman, O., Stilbs, P., Winkler, J., Dittrich, A., and Jahn, W. (1987) *J. Colloid Interface Sci.*, **118**, 436.

32 Halle, B., and Wennerström, H. (1981) *J. Magn. Reson.*, **44**, 89.

33 Halle, B., and Wennerström, H. (1981) *J. Chem. Phys.*, **75**, 1928.

34 Leaver, M.S., Olsson, U., Wennerström, H., and Strey, R. (1994) *J. Phys. II*, **4**, 515.

35 Leaver, M., Furo, I., and Olsson, U. (1995) *Langmuir*, **11**, 1524.

36 Lindman, B., Söderman, O., and Wennerstöm, H. (1987) NMR of surfactant system, in *Surfactant Solution: New Methods of Investigation*, vol. 22 (ed. R. Zana), Marcel Dekker, New York.

37 Lindman, B., Olsson, U., and Söderman, O. (1995) Surfactant solutions: aggregation phenomenon and microheterogeneity, in *Dynamics of Solutions and Fluid Mixes by NMR* (ed. J.-J. Delpuech), John Wiley & Sons, Inc., New York.

38 Söderman, O., Henriksson, U., and Olsson, U. (1987) *J. Phys. Chem.*, **91**, 116.

39 Söderman, O., and Olsson, U. (1995) Micellar solutions & microemulsions, in *Encyclopedia of Nuclear Magnetic Resonance* (eds D.M. Grant and R.K. Harris), John Wiley & Sons, Ltd, Chichester, p. 3046.

40 Jonströmer, M., Jönsson, B., and Lindman, B. (1991) *J. Phys. Chem.*, **95**, 3293.

41 Jonströmer, M., Nagai, K., Olsson, U., and Söderman, O. (1999) *J. Dispersion Sci. Technol.*, **20**, 375.

42 Kunieda, H., and Shinoda, K. (1982) *J. Dispersion Sci. Technol.*, **3**, 233.

43 Shinoda, K., and Lindman, B. (1987) *Langmuir*, **3**, 135.

44 Shinoda, K., Kunieda, H., Obi, N., and Friberg, S.E. (1981) *J. Colloid Interface Sci.*, **80**, 304.

45 Kunieda, H., Hanno, K., Yamaguchi, S., and Shinoda, K. (1985) *J. Colloid Interface Sci.*, **107**, 129.

46 Kunieda, H., Asaoka, H., and Shinoda, K. (1988) *J. Phys. Chem.*, **92**, 185.

47 Kahlweit, M., Lessner, E., and Strey, R. (1983) *Abstr. Pap. Am. Chem. S.*, **186**, 62.

48 Kahlweit, M., Lessner, E., and Strey, R. (1983) *Colloid Polym. Sci.*, **261**, 954.

49 Kahlweit, M., Lessner, E., and Strey, R. (1984) *J. Phys. Chem.*, **88**, 1937.

50 Kahlweit, M., and Strey, R. (1985) *Angew. Chem. Int. Ed. Engl.*, **24**, 654.

51 Kahlweit, M., and Strey, R. (1986) *J. Phys. Chem.*, **90**, 5239.

52 Kahlweit, M., and Strey, R. (1987) *J. Phys. Chem.*, **91**, 1553.

53 Kahlweit, M., Strey, R., and Firman, P. (1986) *J. Phys. Chem.*, **90**, 671.

54 Kahlweit, M., Strey, R., Firman, P., and Haase, D. (1985) *Langmuir*, **1**, 281.

55 Kahlweit, M., Lessner, E., and Strey, R. (1983) *J. Phys. Chem.*, **87**, 5032.

56 Strey, R., Winkler, J., and Magid, L. (1991) *J. Phys. Chem.*, **95**, 7502.

57 Sottmann, T., Strey, R., and Chen, S.H. (1997) *J. Chem. Phys.*, **106**, 6483.

58 Sottmann, T., and Strey, R. (1996) *J. Phys. Condens Matter*, **8**, A39.

59 Sottmann, T., and Strey, R. (1996) *Ber. Bunsenges. Phys. Chem.*, **100**, 237.

60 Sottmann, T., and Strey, R. (1997) *J. Chem. Phys.*, **106**, 8606.

61 Sottmann, T., and Strey, R. (1998) *Tenside Surfact. Det.*, **35**, 34.

62 Strey, R. (1994) *Colloid Polym. Sci.*, **272**, 1005.

63 Olsson, U., Shinoda, K., and Lindman, B. (1986) *J. Phys. Chem.*, **90**, 4083.

64 Olsson, U., Nagai, K., and Wennerström, H. (1988) *J. Phys. Chem.*, **92**, 6675.

65 Lindman, B., and Wennerstöm, H. (1991) *J. Phys. Chem.*, **95**, 6053.

66 Lindman, B., Shinoda, K., Jonströmer, M., and Shinohara, A. (1988) *J. Phys. Chem.*, **92**, 4702.

67 Lindman, B., Shinoda, K., Olsson, U., Anderson, D., Karlström, G., and Wennerström, H. (1989) *Colloid Surf.*, **38**, 205.
68 Strey, R., Schomäcker, R., Roux, D., Nallet, F., and Olsson, U. (1990) *J. Chem. Soc. Faraday Trans.*, **86**, 2253.
69 Hellweg, T., and Langevin, D. (1997) *Opt. Methods Phys. Colloidal Dispersions*, **104**, 155–156.
70 Hellweg, T., and Langevin, D. (1998) *Phys. Rev. E*, **57**, 6825.
71 Hellweg, T., and Langevin, D. (1999) *Phys. A*, **264**, 370.
72 Gradzielski, M. (1998) *Curr. Opin. Colloid In.*, **3**, 478.
73 Gradzielski, M. (1998) *Langmuir*, **14**, 6037.
74 Gradzielski, M., Langevin, D., Sottmann, T., and Strey, R. (1997) *J. Chem. Phys.*, **106**, 8232.
75 Gradzielski, M., and Langevin, D. (1996) *J. Mol. Struct.*, **383**, 145.
76 Gradzielski, M., Langevin, D., and Farago, B. (1996) *Phys. Rev. E*, **53**, 3900.
77 Gradzielski, M., Langevin, D., Magid, L., and Strey, R. (1995) *J. Phys. Chem.*, **99**, 13232.
78 Gradzielski, M., Langevin, D., Sottmann, T., and Strey, R. (1996) *J. Chem. Phys.*, **104**, 3782.
79 Lindman, B. (2009) Some thoughts about microemulsions, in *Microemulsion: Background, New Concepts, Applications, Perspective* (ed. C. Stubenrauch), John Wiley & Sons, Ltd, Oxford, p. XV.
80 Shah, D.O. (1980) *Introduction to Colloids and Surface Chemistry*, 3rd edn, Butterworths, London.
81 Hiemenz, P.C. (1986) *Principles of Colloid and Surface Chemistry*, 2nd edn, Marcel Dekker Inc, New York.
82 Bourrel, M., and Schechter, R.S. (eds) (1988) *Microemulsion and Related Systems: Formulation, Solvency and Physical Properties*, vol. 30, Marcel Dekker, New York.
83 Kumar, P., and Mittal, K.L. (eds) (1999) *Handbook of Microemulsion Science and Technology*, Marcel Dekker, New York, p. 849.
84 Evans, D.F., and Wennerström, H. (1999) *The Colloidal Domain: Where Physics, Chemistry, Biology and Technology Meet*, 2nd edn, John Wiley & Sons, Inc., New York, NY.
85 Rosen, M.J. (1989) *Surfactants and Interfacial Phenomena*, 2nd edn, John Wiley & Sons, Inc., New York.
86 Holmberg, K. (2002) *Handbook of Applied Surface and Colloid Chemistry*, vol. 1, John Wiley & Sons, Ltd, Chichester.
87 Holmberg, K., Jönsson, B., Kronberg, B., and Lindman, B. (2003) *Surfactants and Polymers in Aqueous Solution*, 2nd edn, John Wiley & Sons, Ltd, Chichester.
88 Schick, M.J. (ed.) (1967) *Nonionic Surfactants*, vol. 1, Marcel Dekker, New York.
89 Mitchell, D.J., Tiddy, G.J.T., Waring, L., Bostock, T., and McDonald, M.P. (1983) *J. Chem. Soc. Faraday Trans.*, **79**, 975.
90 Schick, M.J. (ed.) (1987) *Nonionic Surfactants : Physical Chemistry*, vol. 23, Marcel Dekker, New York.
91 Cross, J. (ed.) (1986) *Nonionic Surfactants: Chemical Analysis*, vol. 19, Marcel Dekker, New York.
92 Gradzielski, M. (2008) *Curr. Opin. Colloid Interface Sci.*, **13**, 263.
93 Hellweg, T. (2009) Scattering techniques to study the microstructure of microemulsions, in *Microemulsion: Background, New Concepts, Applications, Perspective* (ed. C. Stubenrauch), John Wiley & Sons, Ltd, Oxford, p. 48.
94 Stubenrauch, C. (ed.) (2009) *Microemulsion: Background, New Concepts, Applications, Perspective*, John Wiley & Sons, Ltd, Oxford.
95 Komura, S. (2007) *J. Phys. Condens Matter*, **19**, 463101.
96 Wennerstöm, H., and Olsson, U. (2009) *Comptes Rendus Chimie*, **12**, 4.
97 Fanun, M. (2008) *J. Dispersion Sci. Technol.*, **29**, 289.
98 Djekic, L., Ibric, S., and Primorac, M. (2008) *Int. J. Pharm.*, **361**, 41.
99 Djekic, L., and Primorac, M. (2008) *Int. J. Pharm.*, **352**, 231.
100 Solans, C., and Kunieda, H. (eds) (1997) *Industrial Application of Microemulsions*, vol. 66, Marcel Dekker, New York.
101 Rybinski, W.v., Hloucha, M., and Johansson, I. (2009) Microemulsions in

cosmetics and detergents, in *Microemulsion: Background, New Concepts, Applications, Perspective* (ed. C. Stubenrauch), John Wiley & Sons, Ltd, Oxford, p. 230.

102 Patravale, V.B., and Date, A.A. (2009) Microemulsions: pharmaceutical applications, in *Microemulsion: Background, New Concepts, Applications, Perspective* (ed. C. Stubenrauch), John Wiley & Sons, Ltd, Oxford, p. 259.

103 Bagwe, R.P., Kanicky, J.R., Palla, B.J., Patanjali, P.K., and Shah, D.O. (2001) *Crit. Rev. Ther. Drug.*, **18**, 77.

104 Trotta, M., Gasco, M.R., and Morel, S. (1989) *J. Control Release*, **10**, 237.

105 Kreilgaard, M. (2002) *Adv. Drug Deliv. Rev.*, **54**, S77.

106 Lawrence, M.J., and Rees, G.D. (2000) *Adv. Drug Deliv. Rev.*, **45**, 89.

107 Vandamme, T.F. (2002) *Prog. Retin. Eye Res.*, **21**, 15.

108 Spernath, A., and Aserin, A. (2006) *Adv. Colloid Interface*, **128**, 47.

109 Krafft, M.P., Chittofrati, A., and Riess, J.G. (2003) *Curr. Opin. Colloid Interface Sci.*, **8**, 251.

110 Malmsten, M. (2006) *Soft Matter*, **2**, 760.

111 Malmsten, M. (2007) *J. Dispersion Sci. Technol.*, **28**, 63.

112 Trotta, M., Morel, S., and Gasco, M.R. (1997) *Pharmazie*, **52**, 50.

113 Balogh, J., Olsson, U., and Pedersen, J.S. (2006) *J. Dispersion Sci. Technol.*, **27**, 497.

114 Balogh, J., and Pedersen, J.S. (2008) Investigating the effect of adding drug (lidocaine) to a drug-delivery system using small-angle X-ray scattering, in *Colloids for Nano- and Biotechnology* (ed. Z.D. Horvolgyi, and E. Kiss), Siofok, Hungary, pp. 101–106.

115 Balogh, J., Olsson, U., and Pedersen, J.S. (2007) *J. Phys. Chem. B*, **111**, 682.

116 Balogh, J., Kaper, H., Olsson, U., and Wennerström, H. (2006) *Phys. Rev. E*, **73**, 041506.

117 Balogh, J., and Olsson, U. (2007) *J. Dispersion Sci. Technol.*, **28**, 223.

118 Balogh, J. (2010) *Adv. Colloid Interface Sci.*, **159**(1), 22–31.

119 Wennerström, H., Balogh, J., and Olsson, U. (2006) *Colloid Surf. A*, **291**, 69.

120 Balogh, J., Schillén, K., and Miguel, M. (2007) Investigating a nonionic oil-in-water microemulsion containing a hydrophobic drug (Lidocaine), in *II Iberic Meeting of Colloids and Interfaces (RICI2)* (ed. A. Valente, and J. Sexas de Melo), Coimbra, Portugal, pp. 311–319.

121 Balogh, J., Schillén, K., Miguel, M., and Pedersen, J.S. (2009) To be submitted 2010.

122 Balogh, J., Schillén, K., and Miguel, M.D. (2007) *AAPS J.*, **9**, W4179.

123 Le, T.D., Olsson, U., Wennerström, H., and Schurtenberger, P. (1999) *Phys. Rev. E*, **60**, 4300.

124 Bergström, L.M. (2008) *Colloid Surf. A*, **316**, 15.

125 Olsson, U., and Wennerström, H. (1994) *Adv. Colloid Interfac*, **49**, 113.

126 Wennerström, H., and Olsson, U. (1993) *Langmuir*, **9**, 365.

127 Wennerström, H., Daicic, J., Olsson, U., Jerke, G., and Schurtenberger, P. (1997) *J. Mol. Liq.*, **72**, 15.

128 Le, T.D., Olsson, U., Wennerström, H., Uhrmeister, P., Rathke, B., and Strey, R. (2001) *Phys. Chem. Chem. Phys.*, **3**, 4346.

129 Le, T.D., Olsson, U., Wennerström, H., Uhrmeister, P., Rathke, B., and Strey, R. (2002) *J. Phys. Chem. B*, **106**, 9410.

130 Olsson, U., and Schurtenberger, P. (1993) *Langmuir*, **9**, 3389.

131 Leaver, M.S., Olsson, U., Wennerström, H., Strey, R., and Wurz, U. (1995) *J. Chem. Soc. Faraday Trans.*, **91**, 4269.

132 Evilevitch, A. (2001) Molecular exchange in colloidal dispersions. Doctoral thesis. Lund University.

133 Wennerström, H., Lindman, B., Söderman, O., Drakenberg, T., and Rosenholm, J.B. (1979) *J. Am. Chem. Soc.*, **101**, 6860.

134 Halle, B. (1991) *J. Chem. Phys.*, **94**, 3150.

135 Jansson, J., Schillen, K., Olofsson, G., da Silva, R.C., and Loh, W. (2004) *J. Phys. Chem. B*, **108**, 82.

136 Schillén, K., Jansson, J., Löf, D., and Costa, T. (2008) *J. Phys. Chem. B*, **112**, 5551.

137 Pedersen, J.S. (2004) *J. Appl. Crystallogr.*, **37**, 369.

138 Orthaber, D., Bergmann, A., and Glatter, O. (2000) *J. Appl. Crystallogr.*, **33**, 218.

139 Kahlweit, M., Strey, R., Firman, P., Haase, D., Jen, J., and Schomäcker, R. (1988) *Langmuir.*, **4**, 499.
140 Kaper, H. (2004) Thermodynamics of balanced microemulsions. Master thesis. Lund University, CAU Kiel.
141 Kohlbrecher, J., and Wagner, W. (2000) *J. Appl. Crystallogr.*, **33**, 804.
142 Safran, S.A., Turkevich, L.A., and Pincus, P. (1984) *J. Phys. Lett.-Paris*, **45**, L69.
143 Karlström, G. (1985) *J. Phys. Chem.*, **89**, 4962.
144 Stuhrmann, H.B. (1982) Contrast variation, in *Small Angle X-Ray Scattering* (eds O. Glatter and O. Kratky), Academic Press, London, p. 197.
145 Deen, G.R., and Pedersen, J.S. (2008) *Langmuir*, **24**, 3111.
146 Arleth, L., and Pedersen, J.S. (2001) *Phys. Rev. E*, **63**, 061406.
147 Safran, S.A., Roux, D., Cates, M.E., and Andelman, D. (1986) *Phys. Rev. Lett.*, **57**, 491.
148 Andelman, D., Cates, M.E., Roux, D., and Safran, S.A. (1987) *J. Chem. Phys.*, **87**, 7229.
149 Cates, M.E., Andelman, D., Safran, S.A., and Roux, D. (1988) *Langmuir*, **4**, 802.
150 Gompper, G., Endo, H., Mihailescu, M., Allgaier, J., Monkenbusch, M., Richter, D., Jakobs, B., Sottmann, T., and Strey, R. (2001) *Europhys. Lett.*, **56**, 683.
151 Kunieda, H., and Shinoda, K. (1982) *Bull. Chem. Soc. Jpn.*, **55**, 1777.
152 Fowler, C.E., Li, M., Mann, S., and Margolis, H.C. (2002) *J. Dent. Res.*, **81**, A490.
153 Sarpotdar, P.P., and Zatz, J.L. (1986) *J. Pharm. Sci.*, **75**, 176.
154 Sarpotdar, P.P., and Zatz, J.L. (1986) *Drug Dev. Ind. Pharm.*, **12**, 1625.
155 Sarpotdar, P.P., and Zatz, J.L. (1987) *Drug Dev. Ind. Pharm.*, **13**, 15.
156 Cerenius, Y., Stahl, K., Svensson, L.A., Ursby, T., Oskarsson, A., Albertsson, J., and Liljas, A. (2000) *J. Synchrotron. Radiat.*, **7**, 203.

4
Some Characteristics of Lyotropic Liquid-Crystalline Mesophases

Idit Amar-Yuli, Abraham Aserin, and Nissim Garti

4.1
Introduction

The present review is dedicated to the memory of Professor Kunieda Hironobu and his fundamental scientific contribution in the study of lyotropic liquid crystals.

Various liquid-crystal structures have been observed in many amphiphilic systems, such as block copolymers in selective solvents [1, 2], membrane lipids [3], and surfactants in aqueous solution [4]. These structures have found widespread application in the preparation of mesoporous templates [5], the colloidal structure design of nanomaterials, including semiconducting assemblies [6], and the formation of microstructured polymeric gels [7] and vesicles for drug delivery [8].

The lyotropic liquid crystals have been studied as a separate category of liquid crystals since they are mostly composed of amphiphilic molecules and water. The lyotropic liquid-crystal structures exhibit the characteristic phase sequence from normal micellar cubic (I_1) to normal hexagonal (H_1), normal bicontinuous cubic (V_1), lamellar (L_α), reverse bicontinuous cubic (V_2), reverse hexagonal (H_2), and reverse micellar cubic (I_2). These phase transitions can occur, for instance, when increasing the apolar volume fraction [9], or decreasing the polar volume fraction of the amphiphilic molecule, for example, poly(oxyethylene) chain length in nonionic poly(oxyethylene) alkyl (oleyl) or cholesteryl ether-based systems [10, 11].

Israelachvili *et al.* developed a simple theory that led to useful insights and understanding regarding the generic sequence of mesophases expected on water dilution or temperature changes [12]. The effective critical packing parameter (CPP) $\mathrm{CPP} = \frac{V_s}{a_0 \cdot l}$ describes, in general, the geometric features of the lipid. V_s is the hydrophobic chain volume, a_0 is the cross polar headgroup area, and l is the chain length of the molecule in its molten state [12, 13]. The packing parameter is useful for predicting which phases can be preferentially formed by a given lipid, since it links the molecular shape and properties to the favored curvature of the polar/apolar interface. Therefore, it raises perceptions concerning the topology based on the shape of the aggregate. The magnitude of the headgroup area depends

Self-Organized Surfactant Structures. Edited by Tharwat F. Tadros

ISBN: 978-3-527-31990-9

on the pure amphiphilic molecule as well as the degree of hydration, which is sensitive to water concentration and temperature. As a general rule, headgroup area increases with hydration and decreases with an increase in temperature. An increase in temperature affects the hydrophobic surfactant tail as well, by increasing its thermal motions. As a result, its length will decrease while its volume is expected to increase.

A hypothetical lyotropic phase diagram, which exhibits the phase transitions that can be induced by varying the composition (water content) or the temperature, is displayed in Figure 4.1 [12, 14–17]. This phase diagram qualitatively condenses the major factors affecting the mesophase behavior and transitions, for example, the molecular shape of the surfactant, the composition (water content), and temperature [12, 15–17].

It can be seen that when CPP equals unity, the preferred mesophase is lamellar. Since the cross sections of the polar head and the hydrophobic tail are similar, the curvature is zero, flat. When the surfactant head cross section is larger than that

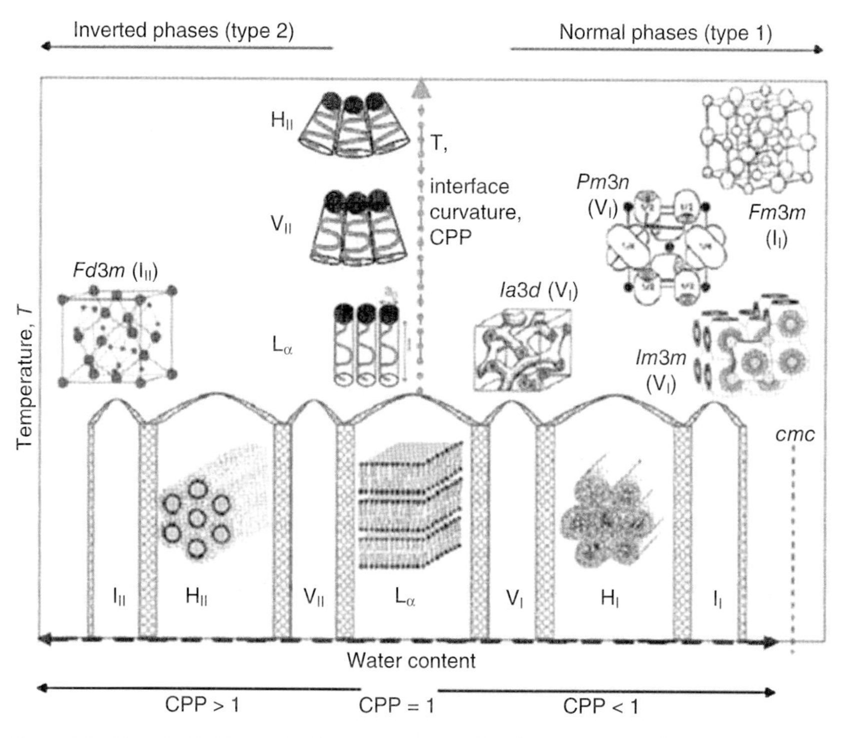

Figure 4.1 Hypothetical lyotropic binary phase diagram where phase transitions can be induced by varying water content or temperature. The indicated mesophases are: L_α–lamellar, Ia3d, Pn3m or Im3m–direct and inverted bicontinuous cubic (V_I and V_{II}, respectively), Fd3m–direct and inverted micellar/discontinuous cubic (I_I and I_{II}, respectively), H_I and H_{II}–direct and inverted hexagonal phase. The cmc (far right in the figure) represents the critical micellar concentration of surfactant from which it creates micelles [14].

of the tail, the preferred self-assembly will be of direct type. From this point, increasing the water content (*X*-axis) will decrease the CPP values and consequently mesophases more highly curved (toward the hydrophobic region, type 1) can be formed – direct bicontinuous cubic, hexagonal, micellar cubic, and eventually spheres (microemulsion). A similar mesophase transition sequence is theoretically expected if the surfactant has the opposite geometric properties, that is, larger tail than polar head cross section; however, with inverted/reversed type and hence, higher CPP values. Furthermore, tentatively, starting from the lamellar mesophase (CPP = 1) and increasing temperature (*Y*-red axis), the polar and apolar volume fractions are expected to decrease and increase, respectively, due to enhanced thermal motion of both water and surfactant tails. As a result, the CPP value (CPP > 1) will increase and type-2 mesophases will be formed with the same sequence.

Numerous attempts have been made to control these self-organized structures and to characterize their macro- as well as microscopic properties [18]. The effect of modifications in the surfactant structure (hydrophilic or hydrophobic group) and the impact of an additional component on mesophase behavior have been examined [19–22]. It has been suggested that the interfacial curvature that determines the phase behavior depends on the relationship between three elements: tail volume, tail length, and area per headgroup.

Kunieda and coworkers investigated diverse colloidal systems including spherical and elongated micelles, and liquid-crystalline systems on the basis of polymer or other amphiphilic molecules mixed with water [4, 10, 18, 23, 24].

They identified the influence of the addition of hydrophobic molecules on self-organizing structures as two-fold [23, 24]: (1) the penetration effect, in which molecules penetrate into the surfactant palisade layer and expand the effective cross-sectional area. As a result, the spontaneous curvature becomes more negative (or less positive); (2) the swelling effect, in which hydrophobic molecules are solubilized in the core of the aggregates and enlarge the volume of the aggregates. In this case, the cross-sectional area remains almost constant and the volume of the hydrophobic part is increased; thus the curvature tends to be more positive. The effect of the added molecules was assumed to be dependent on the hydrophilic–hydrophobic balance (HLB) between the added molecules and the surfactant that constructed the liquid crystal. For example, when decane was added to a relatively hydrophilic $C_{12}EO_7$-based system, the H_1–I_1 transition took place (curvature more positive), whereas the L_α–H_2 transition occurred (curvature more negative) in the $C_{12}EO_3$ system [23, 24].

Furthermore, Kunieda and coworkers were interested in replacing the traditional surfactants with environmentally friendly molecules to overcome biodegradation processes and aquatic toxicity [25]. The main environmentally friendly surfactants that were explored were poly(oxyethylene) cholesteryl ethers (ChEO*n*, where *n* is the number of oxyethylene, EO, units) with a bulky hydrophobic cholesteric group of natural origin [25–27]. Due to the distinct segregation tendency between the hydrophilic and hydrophobic groups, compared to the conventional alkyl ethoxylated surfactants, their phase behavior as a function of ethylene oxide chain length, water content, and temperature was found to be different [25–27]. The binary mixtures, and surely the ternary systems based on this type of surfactant, produced a variety of phases. Moreover, due to the effects of a bulky and nonflexible hydrophobic part of the surfactants and intricate hydrophobic–hydrophilic balance controlled by the different EO chain lengths, these systems exhibited novel intermediate phases [25–27]. Other nonionic surfactants that recently enabled the formation of unique structures, such as glycerol monooleate, required the incorporation of additional molecules [28, 29].

The current review displays an assortment of studies from Prof. Kunieda's research group describing unique liquid-crystalline systems and novel phases, which represent their contribution to this topic. Finally, modern studies focusing on the formation of novel and modified structures on the basis of nonionic surfactant, monoolein, will be discussed.

4.2 Phase Transitions Within Poly(oxyethylene) Cholesteryl Ethers-Based Systems

Rodŕiguez *et al.* [26] studied the phase behavior of short ethylene oxide chain length, trioxyethylene cholesterol ether, $ChEO_3$, mixed with water. The authors reported on a highly thermally stable lamellar liquid-crystalline phase above 75 wt% surfactant that consisted of rigid bilayers attributed to the rigid multiring skeleton cholesterol group (Figure 4.2). This phase exhibited conformational characteristics between the L_α phase and the L_β phase.

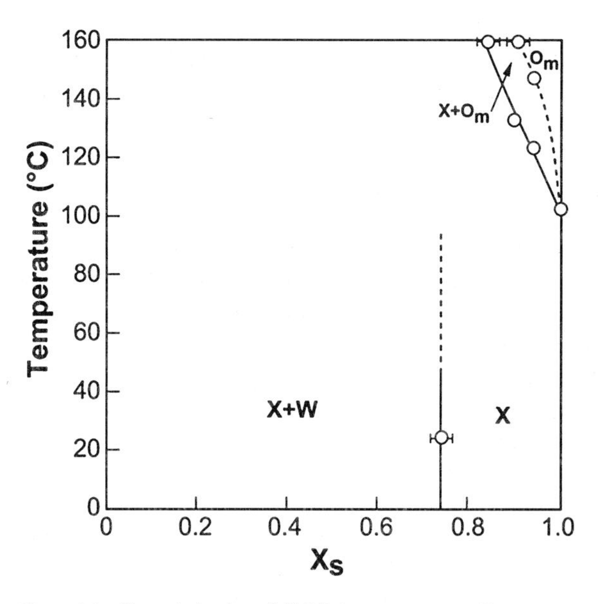

Figure 4.2 Phase behavior of $ChEO_3$/water systems. O_m, reverse micellar surfactant phase or surfactant liquid; W, excess aqueous phase; X, lamellar structure; X_S, the weight fraction of $ChEO_3$ in the system [26].

Analysis of SAXS measurements and calculations of the lipophilic chain length, d_L, revealed a partially extended conformation of the hydrophobic chain, that is, $d_L = 1.75$ nm (the extended conformation is $d_L max = 1.9$ nm) [26]. However, when the relative extension of the hydrophobic chain (d_L: $d_L$max) was compared to that of $C_{12}EO_3$ lamellar systems, it was found to be larger. The authors concluded that alkyl ethoxylated surfactants promote a more compact arrangement of the bilayer [26]. DSC measurements were also performed and exhibited only one peak, corresponding to the transition of lamellar liquid crystal to isotropic liquid. The entropy change during the transition from ordered liquid-crystalline structure to the disordered, fluid state was larger than that of typical L_α phases [30], but was smaller than that of previously reported L_β phases [31]. In agreement with the SAXS outcomes, it was indicated that the intercalation of cholesterol in the bilayers can decrease the enthalpy of the transition in comparison to transition from L_β phase to a liquid phase. Finally, the relatively large enthalpy (~10 kJ/mol) that was found for this transition compared with other nonionic surfactants, may explain the unusually high thermal stability of the L_α phase (Figure 4.2).

In order to improve the understanding of these systems, Kunieda and coworkers examined the thermotropic behavior of poly(oxyethylene) cholesteryl ethers with different chain lengths, ChEO*n*, mixed with water at a fixed concentration (~25 wt%) [32]. This study focused on the different fusion mechanisms that were involved in the solid–liquid phase transition. The solid–liquid transition temperature for ChEO*n* as a function of *n* is shown in Figure 4.3 (for comparison, the transition temperature for polyethylene glycol is also shown). In both cases, the transition temperature decreased when the chain length was diminished. However, for the cholesterol surfactant, when $n \leq 10$, a birefringent phase appeared between

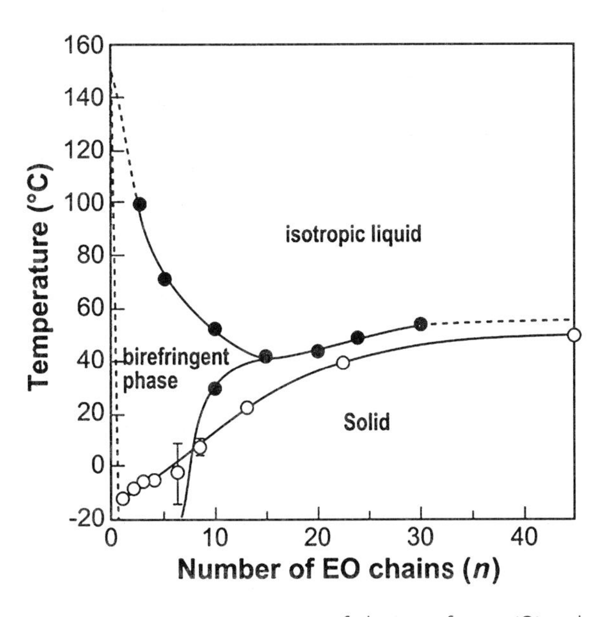

Figure 4.3 Transition temperature of $ChEO_X$ surfactant (●) and polyethylene glycol (○) [32].

the two phases, and the temperature of the transition to the isotropic liquid phase drastically increased. As the number of EO units increased, the flexibility of the surfactant increased and the melting temperature of the liquid-crystalline phase decreased.

López-Quintela *et al.* concluded that in case of large chain lengths the behavior was dominated by the EO chain, while for $n \leq 10$ the rigidity of the cholesterol group dominated, imposing its properties on the surfactants [32].

SAXS data analysis and calculation of the distance d as a function of the number of ethoxyethylene groups, n, revealed a linear increase of the double layer spacing (d) and suggested that the chains are perpendicular to the plane of the bilayer. The obtained fitting equation was $d = (36.9 \pm 0.5) + (3.18 \pm 0.04)n$ and the cholesterol moiety size was estimated as 18.5 ± 0.3 Å, corresponding to the length of the extended cholesterol molecule [32]. Hence, it was assumed that the double layer is an antiparallel structure with a complete overlapping of the oxyethylene chains, as depicted in Figure 4.4 [32]. The length of the oxyethylene unit was also

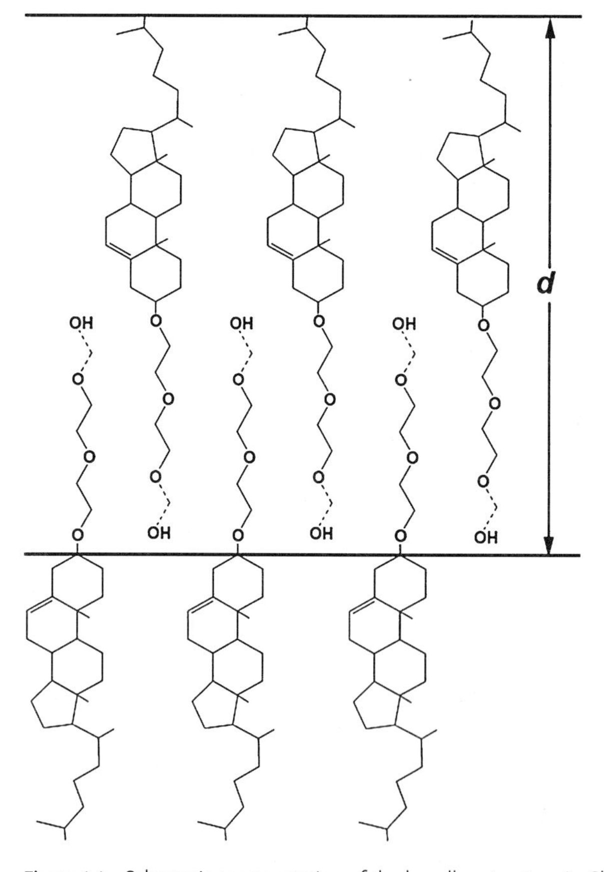

Figure 4.4 Schematic representation of the lamellar structure in $ChEO_n$ surfactant [32].

deduced (3.18 ± 0.04 Å) and implied a zigzag configuration for the oxyethylene unit (Figure 4.4).

WAXS measurements demonstrated that the two lateral distances for the long-chain surfactants were similar to those of other polyoxyethylene surfactants (3.87 and 4.72 Å). Thus, the authors concluded that the molecules were packed in the solid state similar to other polyoxyethylenes and that the thermodynamic behavior of the surfactants was mainly governed by the polyoxyethylene chain. Since the lateral distances among the polyoxyethylene chains are smaller than in the cholesterol, the polyoxyethylene chains could be more closely packed in the solid state only by an antiparallel arrangement, in order that the more voluminous cholesterol moieties do not interfere with each other. However, when the chain was reduced, below $n = 10$, the cholesterol group rigidity was pronounced, the lateral distances became similar to the lateral distances in the cholesterol, and transformation into a liquid-crystalline phase occurred [32].

In follow-up research, Sato *et al.* investigated the temperature- and composition-dependent phase behavior of water/$ChEO_{15}$ and water/$ChEO_{10}$ systems, representing relatively long and short EO chain length [33]. It was shown that nonionic surfactants, with both short and long EO chain length, when mixed with water were able to form a variety of self-organized structures as a function of composition and/or temperature (Figure 4.5). In general, the surfactant layer curvature gradually became more positive (convex surfactant layer toward water) with increasing water content. These systems exhibited a number of novel phase behaviors due to the effects of a bulky and nonflexible hydrophobic part of the surfactants and intricate hydrophobic–hydrophilic balance controlled by the different EO chain length. In the water/$ChEO_{15}$ system, aqueous micellar (W_m), discontinuous micellar cubic ($Fd3m$ space group, I_1), hexagonal (H_1), rectangular ribbon (R_1), and lamellar (L_α) phases were formed, whereas in the water/$ChEO_{10}$ mixture, W_m, R_1, defected lamellar (L_α^H), and L_α phases were produced [33].

In several surfactant-rich systems, especially when the hydrophobic part of the surfactant molecule was large or rigid, a number of intermediate phases, mainly between hexagonal (H_1) and lamellar (L_α) phases, have been identified. These intermediate phases were generally characterized by their inhomogeneous interfacial curvature and classified into the following types: rectangular ribbon (R_1), layered mesh (Mh_1), sponge phase (L_3), and possibly bicontinuous (V_1) structures [33–36]. Ionic systems more often showed rich behavior of intermediate phases and D_2O/nonionic surfactants such as $C_{22}EO_6$ and $C_{16}EO_6$ exhibited layered mesh structures [36, 37].

In this study, both water/$ChEO_{15}$ and water/$ChEO_{10}$ systems exhibited an extensive intermediate phase region, including R_1 and L_α^H (Figure 4.5). One should note that in the water/$ChEO_{10}$ system, although H_1 phase is missing and instead an unknown phase that could not be fitted to any known symmetry appeared, both R_1 and L_α^H were formed. In this system, in a poor-surfactant region, a W_m phase formed, in which short rod-like micelles were produced. With an increase in the weight fraction of the surfactant (W_S), the W_m phase transformed into a transparent, birefringent, and viscous liquid-crystalline phase at $W_S = 0.38$ via an unknown

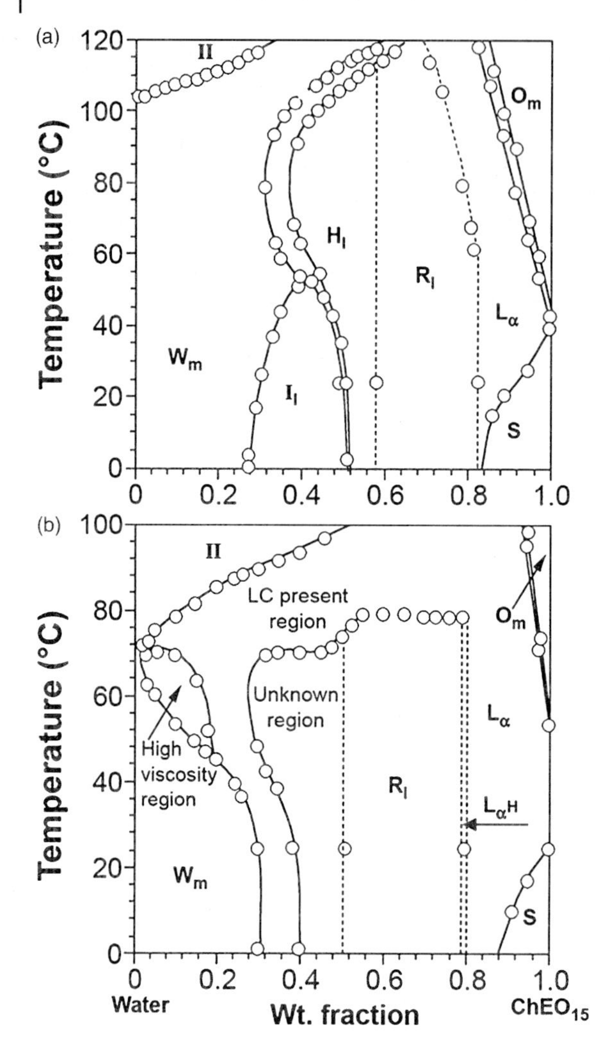

Figure 4.5 Phase diagram of binary water/$ChEO_n$ (n = 10 and 15) systems as a function of temperature and surfactant weight fraction W_S. W_m, I_1, H_1, R_1, L_α^H, L_α, and O_m denote aqueous micellar, micellar cubic, hexagonal, intermediate ribbon, distorted lamellar, lamellar, and reverse micellar phases, respectively. S is a solid-present region, and II indicates a two-phase region [33].

liquid-crystalline phase. Further increase in W_S led to the formation of R_1 phase at $W_S = 0.5$ (the detail of this phase will be discussed in the next section). It was surmised that this unknown liquid-crystalline phase, which is located between the W_m phase of rod-like micelles and the R_1 phase, is composed of long rod-like aggregates. Moreover, due to the bulky sterol group, the cross section of the rod-like aggregate in this phase was assumed to be distorted to an elliptical shape with an unknown spatial arrangement, instead of being circular as in the aggregates of the H_1 phase [33]. The R_1 phase changed into the L_α phase at higher surfactant weight fraction ($W_S = 0.8$) and in a narrow region close to the R_1–L_α boundary, L_α^H phase appeared (with a bump on the low-q side of the first-order reflection from the lamellar structure) [38].

To the contrary, in the water/$ChEO_{15}$ system, the I_1 phase was formed where the unknown phase was produced in the $ChEO_{10}$-based system. Upon further increase in surfactant concentration up to ~52 wt%, H_1 phase was formed and transformed into the intermediate R_1 phase at ~58 wt% of $ChEO_{15}$. The H_1–R_1 phase transformation was attributed to the increasing packing constrains caused by the bulky sterol moiety in the lipophilic core, due to which the circular cross section of the lipophilic core of the H_1 phase became elongated [33]. SAXS evolution in the H_1 to R_1 phase transformation suggested a gradual elongation of the cross section of the lipophilic core of the rod-like aggregate with increasing W_S. Finally the L_α phase was produced at ca. 83 wt%.

On the basis of the SAXS data, the interlayer spacing, d (Figure 4.6a), the effective cross-sectional area per surfactant molecule, a_S (Figure 4.6b), and the length of hydrophobic part, d_L (Figure 4.6c), were calculated and displayed as a function of the weight fraction of the surfactant, W_S (at 25 °C). Both systems showed similar d values despite the difference of the EO chain length, probably due to compression of the EO chain and packing modifications of surfactants. The water/$ChEO_{10}$ system exhibited smaller a_S due to its shorter EO chain and hence with more pronounced hydrophobicity, resulting in tighter packing than with the $ChEO_{15}$ molecule. Finally, d_L was found to be smaller in the $ChEO_{15}$-based system, only when L_α structure was formed, compared to the water/$ChEO_{10}$ system. The authors presumed that the hydrophobic part of $ChEO_{15}$ molecules was slightly tilted against the perpendicular axis to the lamellar layers [33].

For more profound understanding of the phase behavior in these systems, Sato *et al.* evaluated the hydration levels of the oxyethylene chain in each system [33]. The complex dielectric spectra of water/$ChEO_{10}$ and water/$ChEO_{15}$ binary systems (at 5, 10, and 15 wt% water) were determined at 25 °C by time-domain reflectometry (frequency range of 0.1–20 GHz, [39]). The low-frequency process was assigned to the kinetics of the hydrophilic layer of micelles, including the motion of hydrated oxyethylene chain and hydrated water. Additionally, the relaxation time of the high-frequency process was attributed to the cooperative rearrangement of the H-bond network of bulk water. Following various calculations, which are reported in the article, the effective hydration number of ethylene chain Z_{EO} was estimated. Z_{EO} corresponded to the number water molecules that could not contribute to the bulk water process due to hydration of the surfactant molecules. The Z_{EO} values for $ChEO_{10}$ and $ChEO_{15}$ were estimated as 40–45 and 55–65 molecules, respectively [33]. Correspondingly, ~4–4.5 water molecules were found to be hydrated per oxyethylene unit. Based on the analysis of Z_{EO}, fully hydrated micelles were present with nearly no bulk water when $W_S > {\sim}0.5$, which was consistent with the boundaries of the I_1 and H_1 phases. A further reduction of the water content resulted in reduction of hydrated water and incomplete hydration of the oxyethylene chains,

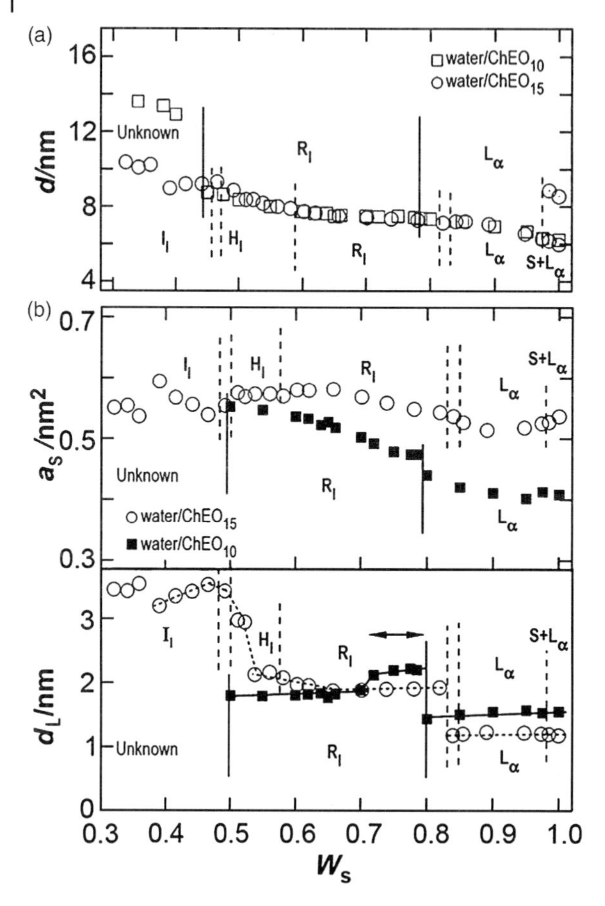

Figure 4.6 Concentration dependence of (a) the interlayer spacing, *d*, for liquid-crystalline phases; and (b) the effective cross-sectional area per surfactant molecule, a_s, and the length of the hydrophobic part of a surfactant molecule in $ChEO_n$/water ($n = 10$ and 15, respectively) systems as a function of the weight fraction of surfactant W_S at 25 °C [33].

leading to a reduction of the headgroup repulsion and hence to the formation of R_I and eventually L_α structures.

4.3 Nonconventional Liquid-Crystalline Structures

Some unique mesostructures, sometimes identified as ill-defined or defected lyotropic liquid crystals, such as ribbon (R_I), sponge (L_3), and low-viscosity, modified micellar cubic (Q_L), and hexagonal (H_2) phase, have been detected [29, 30, 33, 35, 36].

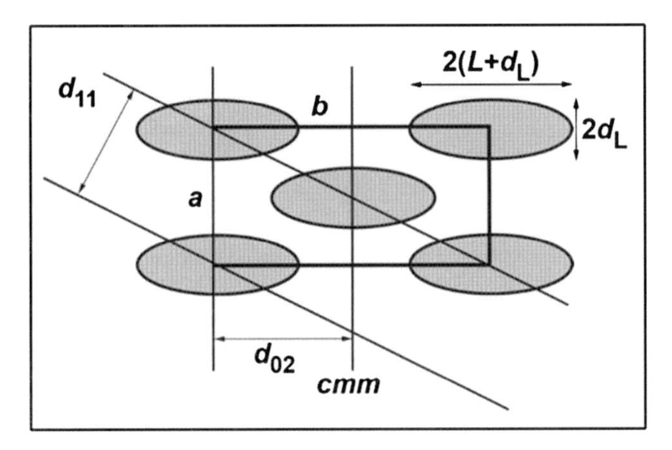

Figure 4.7 Schematic structure of the R_1 phase with a *cmm* symmetry with the unit cell parameters *a* and *b* [33].

4.3.1
Intermediate Ribbon (R_1) Phase

Liquid-crystalline mesophases with rhombohedral (ribbon, R_I), monoclinic, and tetragonal symmetry have been identified in some surfactant–water systems [15, 33]. In the binary mixtures of water/$ChEO_{10}$ and water/$ChEO_{15}$, in the poor-surfactant area and close to the L_α phase, an extremely wide intermediate phase of rectangular ribbon (R_1) structure with the centered rectangular symmetry, *cmm* or *cm*, has been identified (Figure 4.5; [33]). A schematic structure of the R_1 phase with a *cmm* symmetry suggested by the authors is depicted in Figure 4.7 (including the lattice parameters, *a* and *b*, for the ribbon-like aggregates that were evaluated from the *q*-vectors).

Sato *et al.* utilized the packing parameter (*P*, also mentioned above as CPP) approach for analyzing the structure in each system [12, 13]. It is known that for the L_α and H_1 phases, *P* values are 1 and 1/2, respectively. In the range of $1 > P > 2/3$, bicontinuous cubic (V_1) phase can be formed, and when $\frac{1}{2} > P > 2/3$ discontinuous intermediate phases, such as mesh (Mh_1) and ribbon (R_1) phases, are likely to be constructed. Using a hexagonal rod model, the axial ratio of the ribbon-like aggregate, ρ_A, and the packing parameter, *P*, were calculated. In Figure 4.8, ρ_A and *P* are plotted against W_S, together with the dimensions of the rectangular unit cell, *a* and *b*, and their ratio, *b*/*a*. In the R_1 region of the water/$ChEO_{15}$ system, with decreasing W_S, ρ_A decreased. This behavior corresponds to the gradual increase of the mean interfacial curvature, from $\rho_A = 1.7$ at $W_S = 0.82$ (close to the L_α–R_1 boundary) down to $\rho_A = 1$ at $W_S = 0.58$ (the R_1–H_1 boundary), indicative of an almost circular cross section of the cylindrical aggregates. The calculated packing parameters for the R_1 phase were in the range of $\frac{1}{2} < P < 2/3$, which was consistent with discontinuous intermediate structures. *P* monotonically decreased with decreasing W_S, and became very close to ~0.5 once more, exhibiting a signature of

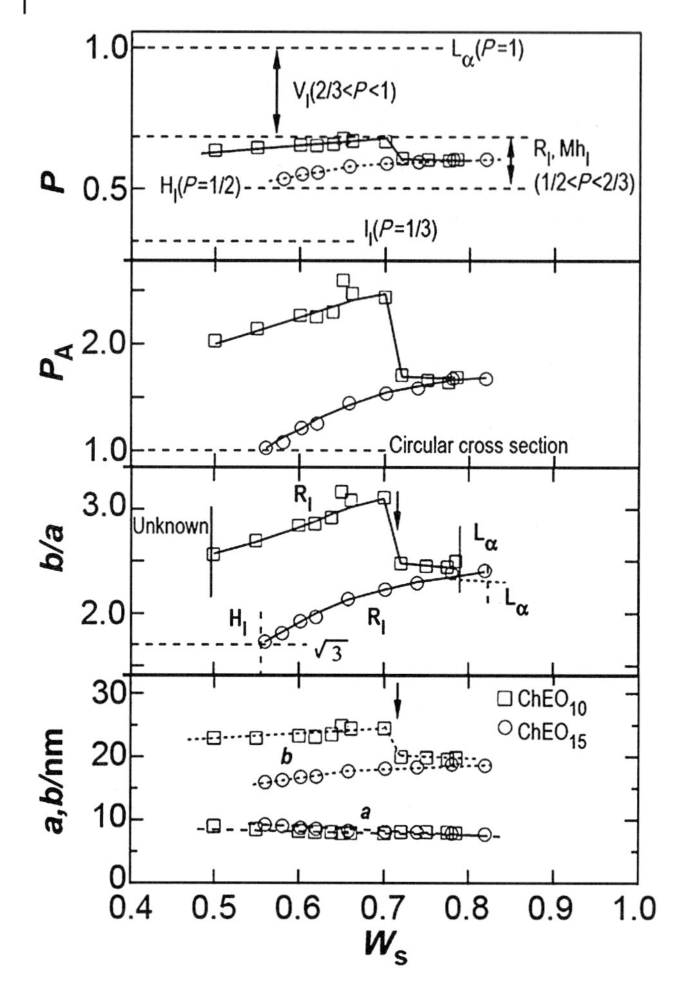

Figure 4.8 Concentration dependence of the structural parameters characterizing the ribbon (R_1) phase, the dimensions of the rectangular unit cell, *a* and *b*, the ratio *b*/*a*, the axial ratio of the ribbon-like aggregate, ρ_A, and the packing parameter, *P*, plotted against the surfactant weight fraction, W_S.

the H_1 phase, at $W_S = 0.58$. Thus, it was concluded that the R_1 phase in the water/$ChEO_{15}$ system, which covered a substantial region, from *ca.* 58 wt% to *ca.* 83 wt%, acted as a "distorted" hexagonal phase. With increasing water content, the interfacial curvature increased and became less inhomogeneous, and the "distortion" gradually disappeared up to the complete transformation to H_1 structure.

In the water/$ChEO_{10}$ system, the behavior of all the parameters characterizing the unit cell was not straightforward compared with that of the water/$ChEO_{15}$ system. The axial ratio of the aggregates, ρ_A, shown in Figure 4.8, was nearly constant despite the decrease of W_S, in $0.78 \geq W_S \geq 0.72$, and in parallel the lattice parameters *a* and *b* were almost constant. In this region, as shown in Figure 4.6c,

d_L exhibited larger values than that in the remaining R_1 region. Interestingly, at $W_S = 0.72$, ρ_A and P demonstrated an incremental increase, indicating a sudden growth of the aggregates in the direction of the long axis of the ribbon structure. This phenomenon was also confirmed by a sudden positional shift of the reflection from the (02) plane to the lower-q value (data not shown). At $W_S = 0.7$, P reached the maximum value of ~0.67, which was consistent with the threshold between bicontinuous and discontinuous intermediate regions as shown in Figure 4.8.

In contrast to the behavior of the water/$ChEO_{15}$ system, the peaks from the (11) and (02) planes did not merge into one peak but remained separate for the entire R_1 region. Furthermore, the reflection from the (02) plane gradually faded when W_S approached ~0.5, and a background of the SAXS spectrum became larger, indicative of the formation of highly distorted inhomogeneous structures. In the region of $0.72 \geq W_S \geq 0.5$, once more ρ_A decreased with decreasing W_S, and finally an unknown phase was formed at $W_S = 0.5$. In general, the surfactant layer curvature increased from negative to positive with decreasing surfactant content, so the increment of ρ_A and P, corresponding to a sudden decrease of the surfactant layer curvature, despite decreasing surfactant content, was a unique phenomenon in a water/surfactant binary system. In $0.78 \geq W_S \geq 0.72$, less elongated and "loosely" packed ribbon aggregates with a large d_L were formed, and at $W_S = 0.72$, a sudden compression in the direction of the short axis and elongation in that of the long axis of aggregates occurred. However, the packing parameter, P, could not exceed 2/3 and it decreased with decreasing W_S, as expected [33].

4.3.2 Novel Micellar Cubic Phase (Q_L)

The phase diagram of glycerol monooleate (GMO)/ethanol/water mixture studied by Efrat *et al.* revealed the existence of essentially four regions, lamellar, cubic bicontinuous liquid crystals, and micellar and sponge phases (Figure 4.9; [28]). An additional unique isotropic region that is located in the vicinity of the cubic liquid-crystal phase region was identified and termed Q_L. This new isotropic phase contained 49–54 wt% water, 41–33 wt% GMO, and 10–13 wt% alcohol, and was found to consist of discrete discontinuous micelles arranged in a cubic array.

The Q_L samples were transparent, nonbirefringent, low-viscosity fluid of high stability (minimum of 9 months) at room temperature. The rheological measurements showed that the Q_L phase was of low viscosity (36.6 Pa s in comparison to around 10^5 Pa s of the bicontinuous cubic phase [28]). Moreover, the flow curve had distinct nonhomogeneous flow regions and the frequency sweep diagram exhibited viscoelastic behavior that indicated the intermicellar network formation was held together by intermicellar structural forces.

SAXS measurements of the Q_L phase revealed five major peaks at 0.0583, 0.0923, 0.1099, 0.1299, and 0.1451 Å that have been translated into spacing ratios of $\sqrt{2}:\sqrt{5}:\sqrt{7}:\sqrt{10}:\sqrt{12}$ and were interpreted into a $Pm3n$, $P4_232$, or $P4_343$ space group of cubic symmetry. It was concluded that the ethanol mainly affected the polar region of the surfactant, since it was accommodated in the region of the headgroups.

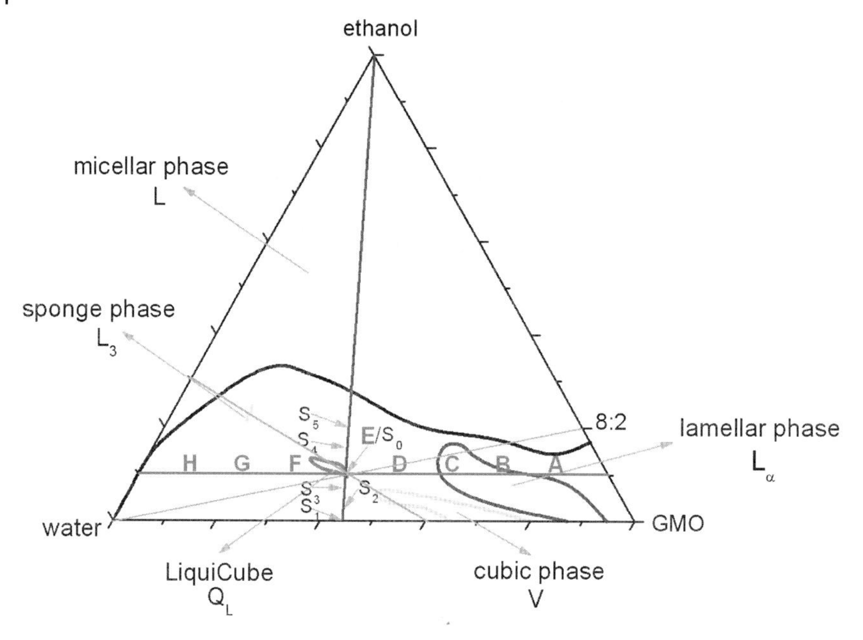

Figure 4.9 The phase diagram of the GMO/ethanol/water ternary system at 25 °C. The phase boundaries of the one-phase regions are drawn with solid lines. The phases indicated are lamellar phase (L_α), bicontinuous reverse cubic phase (V), and three isotropic phases: micellar isotropic phase (L), sponge phase (L_3), and the new Q_L phase. The marked samples A–H contain constant 10 wt% ethanol and they lie in a parallel line starting at 9 : 1 GMO/water and are diluted with water. Samples S_1 to S_6 have a constant GMO/water weight ratio (W_s/W_w) of 0.82, where W_s and W_w are the weight fractions of GMO and water, respectively. The water dilution line starting at 6 : 4 GMO/water axis and crossing the bicontinuous cubic phase runs along the Q_L phase and the sponge phase (L_3) marked by a dashed line [28].

This unique phase was detected also by cryo-TEM, which provides direct structural symmetry information. The cryo-TEM image of the Q_L sample exhibited highly ordered domains with cubic symmetry (arrows in the image) and was further confirmed by the fast Fourier transformations (FFT) (Figure 4.10; [28]). The FFT transformations were fitted to the bcc or fcc lattices, with points *a*, *b*, and *c* in the figure indexed as (2 0 0), (1 1 0), and (2 2 0) for the bcc structure, or (2 2 0), (0 2 0), and (0 4 0) for the fcc phase. The lattice parameter was found to be 103 ± 2 Å.

It was concluded that the incorporated ethanol, as a third component, strongly affected the lipid headgroups, causing a moderate effect on the lateral forces acting within the polar and apolar regions of the membrane, bending the layers and enabling the formation of cubic phase with very low viscosity.

In a further study, the authors demonstrated that the Q_L phase can solubilize high concentrations of the water-insoluble anti-inflammatory drug sodium diclofenac (Na-DFC) [40]. Up to 0.4 wt% Na-DFC, the Q_L structure remained intact

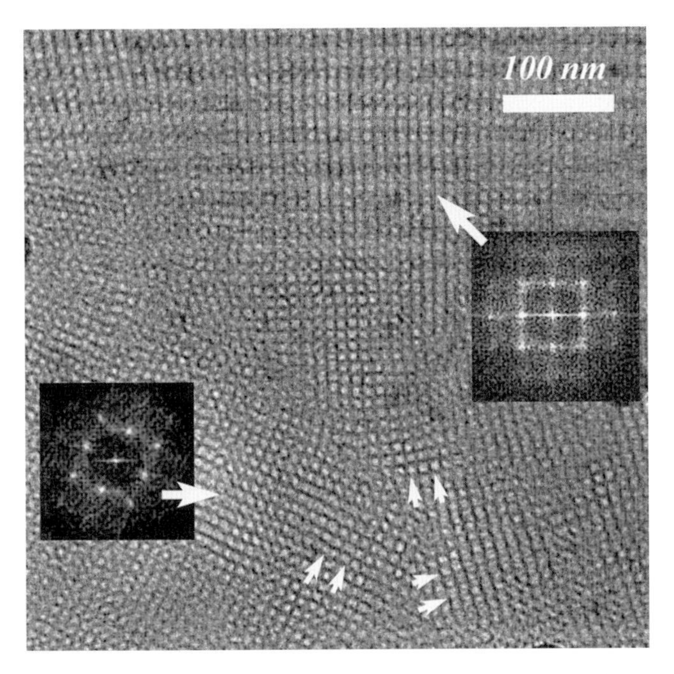

Figure 4.10 A typical cryo-TEM image of the Q_L sample. The two fast Fourier transformations clearly indicate cubic symmetry in the phase. Pairs of arrows in the image point to three liquid-crystalline regions that exist next to each other, all displaying a cubic phase [28].

with some influence on the hydration of the headgroups and on the intermicellar forces. However, at 0.8 to 1.2 wt% Na-DFC, the discontinuous micellar cubic phase was transformed into a more condensed mesophase of a bicontinuous cubic phase. At >1.2 wt% Na-DFC, the cubic phase was converted into a lamellar phase (L_α). Within 5.5 to 7.3 wt% Na-DFC, the mesophase was progressively transformed into a less ordered lamellar structure, and above 12 wt% Na-DFC crystals tended to precipitate out [40]. Here, we will concentrate solely on the Q_L structure, although an extensive and more expanded study is presented in reference [40].

It was concluded that at low Na-DFC concentrations the drug acted as a kosmotropic salt, it therefore salted-out the surfactant from its water layer; however, at higher levels it behaved like a hydrotropic, chaotropic salt and thus enabled the surfactant to salt-in.

SAXS measurements showed that once Na-DFC was solubilized (even in low levels) the diffraction-scattering reflection intensities were higher (and better resolved) than those of the empty samples. The first diffraction peak appeared at the higher q values (0.0583, 0.0622, 0.0629, and 0.0661 for 0, 0.1, 0.2, and 0.4 wt% Na-DFC, respectively). These results indicated a decrease of d value due to partial dehydration of the headgroups as the Na-DFC content increased without a significant change in the cubic space group.

Using the following equations (Equations 4.1 and 4.2 were established by Kunieda *et al.* [24, 33, 41–46], yet with some changes) the radius of the hydrophilic core of micelles, r_H, and the effective cross-sectional area per molecule, a_s, were calculated.

$$r_H = d\left\{\frac{3\varphi H}{4\pi n_m}\right\}^{1/3} C \tag{4.1}$$

$$a_S = \frac{3V_H}{r_H N_A} \tag{4.2a}$$

or

$$a_S = \frac{3v_H}{r_H} \tag{4.2b}$$

where ϕ is the volume fraction of the hydrophilic part of the surfactant, d is the measured interlayer spacing, C represents the Miller indices (hkl) assigned to the first peak of the X-ray spectra ((hkl) = (100) for primitive cubic, (hkl) = (110) for body-centered cubic, (hkl) = (110) for face-centered cubic, structures), and n_m is the number of micelles per unit cell (n_m = 1 for primitive cubic, n_m = 2 for body-centered cubic, and n_m = 4 for face-centered cubic structures). V_H (4.2a) is the molar volume of the hydrophilic moiety and v_H (4.2b) is the surfactant volume of the hydrophilic moiety.

It was found that in the Q_L structure two micelles comprise the unit cell, with an interlayer spacing of 107.8 Å. The radius of the hydrophilic core of micelles, r_H, was estimated as 3.48 nm and the effective cross-sectional area per molecule, a_s, was 0.13 nm^2 (see Figure 4.11; [40]).

Up to 0.4 wt% the added Na-DFC had a relatively small effect on the micelle structure parameter, for example, r_H (3.484, 3.264, 3.227, and 3.072 nm for 0, 0.1, 0.2, and 0.4 wt% Na-DFC, respectively), the radius of the lipophilic moiety, r_L (2.138, 2.009, 1.981, and 1.885 nm for 0, 0.1, 0.2, and 0.4 wt% Na-DFC, respectively), and a_s, (0.1319, 0.1397, 0.1413, and 0.1484 nm^2) (see Figure 4.12). The small quantities of solubilized Na-DFC influenced the headgroups by shrinking both the hydrophilic core and the effective hydrocarbon chain length. As a result, a_s increased, yet without a noticeable change in the CPP values. Furthermore, the Na-DFC affected the d values (107.8 to 101, 99.9, and 95.1 Å for 0.1, 0.2, and 0.4 wt% Na-DFC, respectively), hence, it was concluded that these low Na-DFC levels influenced mostly the *intermicellar force*.

These conclusions were further confirmed by the diffusion coefficients of the ingredients in the mixture. In general, upon addition of Na-DFC the diffusion coefficients of the GMO, ethanol, and Na-DFC were almost constant, yet the Na-DFC diffusion was higher than that of the GMO. These results indicated that the GMO did not increase its mobility at the interface and that the Na-DFC was located at the outer part of the interfacial layer within the lipophilic tails of the surfactant. Moreover, the water average diffusion coefficient values were high and decreased only slightly with increasing Na-DFC concentration. Hence, it was concluded that at lower solubilization levels the Na-DFC was a "water-structure

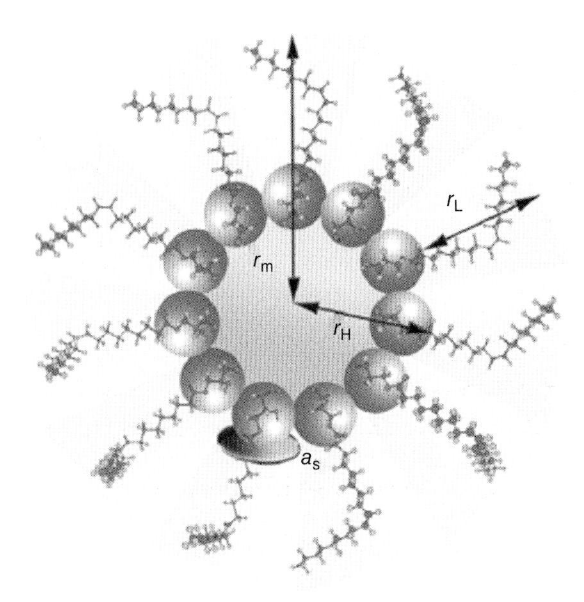

Figure 4.11 Schematic representation of spherical reverse micellar structure formed in discontinuous cubic (I_2) phase. The radius of the micelle is r_m, the radius of the hydrophilic moiety is r_H, the radius of the lipophilic moiety is r_L, and the effective cross-sectional area per surfactant molecule is a_s [40].

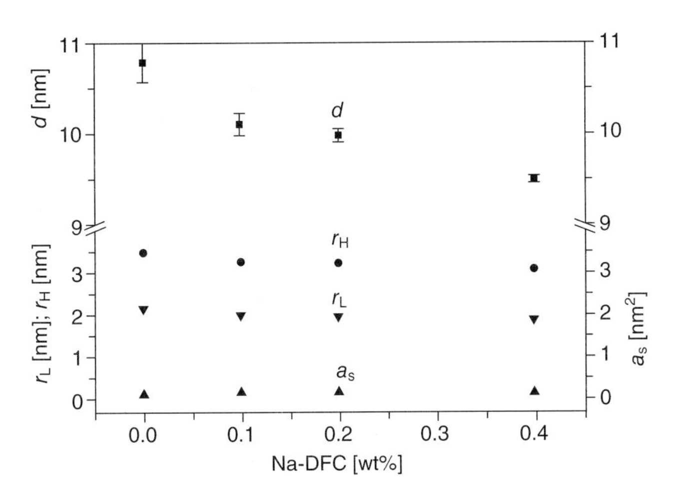

Figure 4.12 (●) Change in hydrophilic radius r_H, (▲) effective cross-sectional area a_s, (▼) lipophilic chain length r_L, and (■) interlayer spacing d, with Na-DFC added up to 0.4 wt% [40].

maker" (kosmotropic salt), that is, it stabilized the structure of the bulk water, remained nonpolarized, and nonionized and behaved like a lipophilic ion-pair [40].

The Na-DFC location and position within the interface as well as its polarization and partial ionization were strongly affected by its solubilization contents and the liquid-crystal structure. In the cubic phase the drug was located less close to the hydration layer; when the transition occurred it was exposed more to the water layer and the surfactant headgroups.

4.3.3
Low-Viscosity Reverse Hexagonal Phase (H_{II})

In the GMO/tricaprylin(TAG)/water mixtures a wide area of H_{II} liquid crystal was formed at room temperature and exhibited diverse structural properties when changing the water content. Below 20 wt% water the H_{II} were characterized by smaller lattice parameters (52.3 ± 0.5 Å) and effective crystallite sizes (772 ± 50 Å) relative to the more ordered, water-diluted (25 wt%) mixtures (55.4 and 1014, respectively) [47, 48]. Moreover, measurement of the room-temperature viscoelastic properties of these less-ordered H_{II} structures revealed that a decrease in water content was accompanied by a decrease in viscosity (zero shear rate viscosity of the less ordered was 4 times lower than the more ordered one). Hence, to effectively extend the range of the low-viscosity H_{II} phase, the temperature was varied (−10–40 °C) [48]. Yuli-Amar *et al.* found that upon heating the water-poor mixture (12.5 wt% water) above 35 °C it became fluid, yet the hexagonal symmetry was maintained [48]. The X-ray scattering patterns of the GMO/TAG/12.5 wt% water mixture at: 3 °C (2D WAXS), and at 7, 16, 25, 36, 40, and 50 °C (1-D SAXS) are shown in Figure 4.13a–g. At 3 °C (Figure 4.13a), a 2D WAXS image revealed four low-angle partially oriented reflections that were indexed as the first four orders of a lamellar phase with interlamellar spacing of 49 Å. In addition, there are a number of wide angle peaks at 4.59, 4.32, and 4.06 Å that are characteristic of a crystalline low-temperature phase, L_c. At 7 °C (Figure 4.13b), the SAXS profile shows a single Bragg Peak at 49 Å, which was associated with the lamellar phase observed in the WAXS image. At temperatures above 16 °C, three peaks were observed, and indexed as the (10), (11), and (20) reflections of a 2D hexagonal liquid-crystalline phase. Above 45 °C, the H_{II}–L_2 transformation took place (Figure 4.13g).

The mean lattice parameter (a) and the effective crystallite size (L_H) of the hexagonal structures were calculated, based on the three peak positions and the (10) peak line breadth (Table 4.1). In general, upon increasing the temperature the lattice parameter decreased. Between 20 and 28 °C the decrease was small, while at higher temperatures (36–40 °C) it was more pronounced (Table 4.1). The decrease in the lattice parameter with increasing temperature was mainly due to dehydration of the surfactant polar headgroups and by an increase in the hydrocarbon chain mobility. Additionally, the decrease in L_H upon increasing temperature (634–385 Å) was most pronounced between 26 and 40 °C. It was assumed that the elevated temperature that increased the mobility of both water and GMO chains resulted in increasing the CPP value and dehydrating part of the hydration water. Hence, the hexagonal phase, which is strongly dependent on hydrogen bonding, gradually shrank to complete disruption, while the curvature and CPP values increased. It was also assumed that the cylinder length of the H_{II} phase decreased as well, as a prologue to the H_{II}–L_2 transformation.

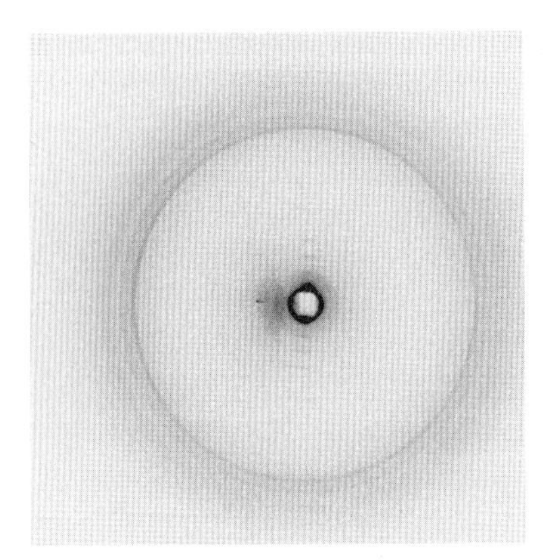

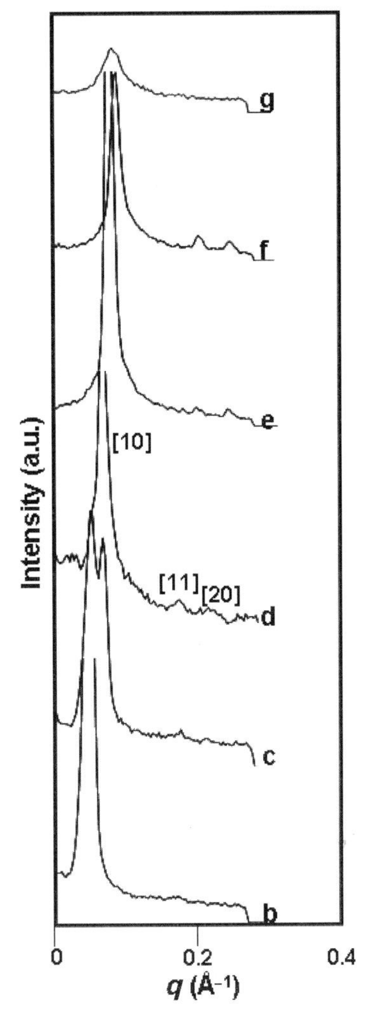

Figure 4.13 X-ray diffraction patterns of the GMO/tricaprylin 90/10 mixture with 12.5 wt% water after overnight incubation at −20 °C and measured at: (a) 3 °C (2D WAXS) and (b) 7, (c) 16, (d) 25, (e) 36, (f) 40, and (g) 50 °C (1-D SAXS) [48].

NMR (self-diffusion and ^{2}H-NMR) and ATR-FTIR measurements were used to evaluate conformational modifications at the molecular level and demonstrated that the GMO/TAG/12.5 wt% water mixture undergoes interrelated structural changes as a function of temperature that led to the formation of a fluid H_{II} phase.

On the basis of the results of the experimental techniques mentioned above, a picture of the different modifications as a function of temperature was drawn (Figure 4.14). Following low-temperature incubation, the ternary mixture self-associated into the L_c phase, which was stable between −10 and 5 °C (Figure 4.14

Table 4.1 Lattice parameters and the effective crystallite sizes (a and L_H, respectively) of the hexagonal structures in GMO/TAG/water mixture (12.5 wt% water concentration) as a function of temperature.

Temperature (°C) ± 0.5 °C	a (Å) ± 0.5 Å	L_H (Å) ± 50 Å
10.0	50.1	Not detectable
16.0	48.4	625
20.5	46.6	616
25.3	46.5	634
28.1	46.4	557
35.7	45.4	372
39.5	44.3	385
45.0	Not detectable	315
50.0	Not detectable	220

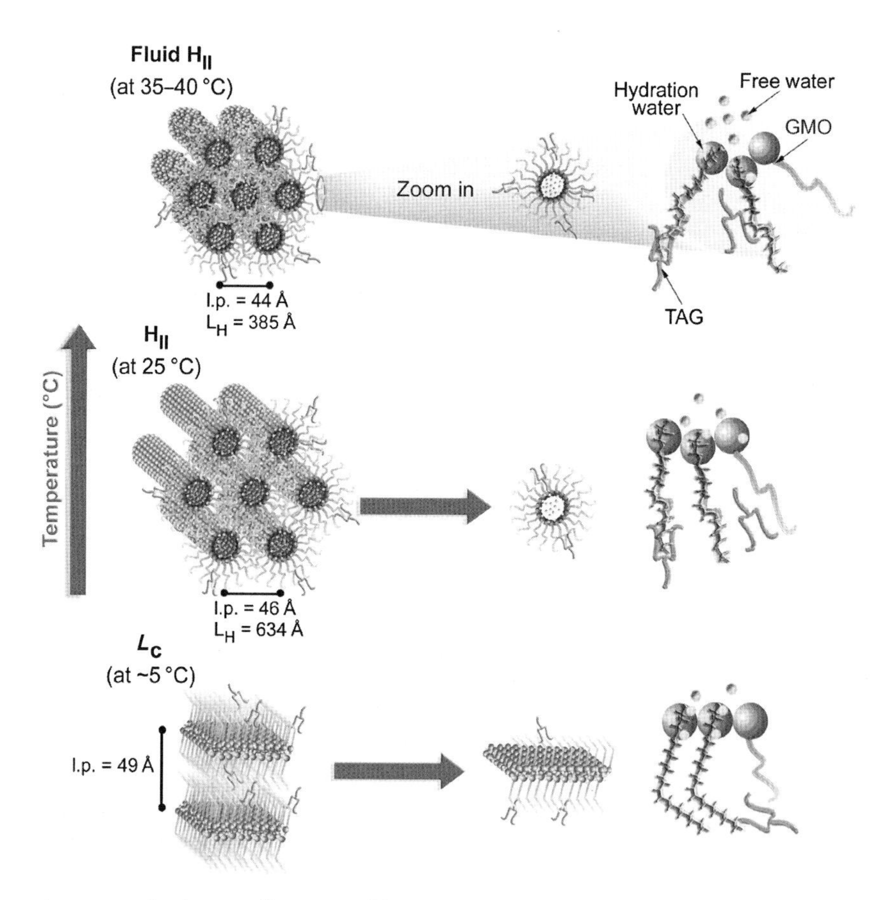

Figure 4.14 A schematic illustration of the structural and physicochemical properties of the mesophases as a function of temperature. l.p. and L_H correspond to the mean lattice parameter and effective crystalline size of the mesophases [48].

lower part). In the L_c structure the hydrocarbon chains were shown to be confined and tightly packed (WAXS), and hence their diffusion coefficients were the smallest (SD-NMR). The CH_2 asymmetric and symmetric stretching vibrations (FTIR) revealed an increase in the *gauche* conformer fraction at 5 and 15 °C, respectively, indicating the melting temperature of the GMO hydrocarbon chains (also detected by DSC).

With increasing temperature, the lamellar crystal structure was gradually transformed into a reverse hexagonal liquid-crystalline phase (>10 °C, Figure 4.14 center). The hexagonal phase was formed due to increasing curvature of the surfactant interfacial region (higher CPP values). The presence of TAG molecules assisted in the L_c–H_{II} transformation; hence TAG lowered the transformation temperatures (~20 and ~80 °C lower than for L_α and H_{II} phases in the GMO/water system, respectively). The hexagonal mesophase exhibited relatively rapid flow of water in the inner channels within the densely packed cylindrical-shaped aggregates of GMO. With increasing temperature, the surfactant headgroups were dehydrated and the hydrocarbon chain mobility was enhanced, both resulting in reduction of the lattice parameter. Above 35 °C additional structural changes occurred. The sample was liquefied yet the hexagonal symmetry was maintained (Figure 4.14, upper part). The fluidity of the hexagonal phase was explained by the significant reduction in domain size and cylinder length. This phenomenon was characterized by greater mobility of the GMO (as was detected by SD-NMR and FT-IR) and lesser mobility of the water (SD-NMR), and a significant dehydration process (FTIR).

In an additional study by Yuli-Amar *et al.*, in order to achieve low-viscosity reverse hexagonal phases at room temperature, ethanol and diethylene glycol monoethyl ether (Transcutol) were added to the ternary GMO/TAG/water mixture [29]. These studies were based on findings showing that alcohols can destroy liquid-crystal phases, and ethanol and PEG were shown to form discontinuous micellar cubic and sponge phases instead of bicontinuous phases [49–51]. It was shown that the addition of Transcutol or ethanol to the GMO/TAG/water mixture enabled the formation of a room temperature fluid H_{II} phase.

The authors followed the thermotropic behavior of the GMO/TAG/water mixture (Figure 4.15, thermogram I) during the heating scan from −20 to +40 °C and revealed the existence of two broad endothermic events with maxima at -1.0 ± 0.2 (peak A) and 6.1 ± 0.2 °C (peak B) [47, 48]. Peak A was found to be due to water fusion (with enthalpy of 24 J gr^{-1}) and peak B (with enthalpy of 36 J gr^{-1}) was related to the fusion of the hydrophobic moieties of the GMO solvated by the TAG [47, 48]. Upon addition of Transcutol or ethanol the fusion temperature and enthalpy of the water (ice) decreased significantly (Figure 4.15, thermograms II and III). The water fusion maxima decreased to -5.8 ± 0.2 and -8.8 ± 0.2 °C upon addition of Transcutol or ethanol, respectively. The subzero temperature of peak A (Figure 4.15, thermogram I) indicated the presence of weakly bound water. It was concluded that incorporation of the water-binding molecules such as Transcutol or ethanol led to stronger water binding with lower melting temperatures (Figure 4.15, thermograms II and III).

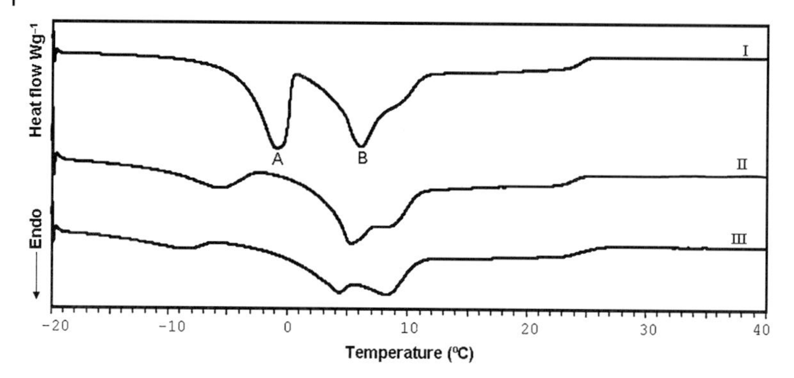

Figure 4.15 DSC thermograms of (I) GMO/tricaprylin/water mixture with two endothermic events at −1.0 ± 0.2 °C (peak A) and 6.1 ± 0.2 °C (peak B), (II) GMO/(tricaprylin + Transcutol)/water mixture, and (III) GMO/(tricaprylin + ethanol)/water mixture [29].

The incorporation of guest molecules affected the fusion of the hydrophobic moieties of the GMO molecules (peak B) as well. In the presence of Transcutol or ethanol the broad endothermic peak was split into two partially overlapping peaks with maxima at 5.2 ± 0.2 and 8.9 ± 0.2 °C, and 4.2 ± 0.2 and 8.1 ± 0.2 °C, respectively. Additionally, both molecules increased the fusion temperature of the hydrophobic GMO tails by 1–2 °C. These two guest molecules were found to compete for water with the GMO polar headgroups and led to dehydration of the surfactant headgroups. The ethanol was more efficient in binding water than Transcutol. Additionally, since the beginning of the fusion process of the hydrophobic moieties (peak onset) was slightly induced, less tightly packed hydrophobic chains were surmised in the low-viscosity H_{II} structure.

X-ray scattering was used to identify and confirm the structure of the different phases as well as their compositional boundaries as defined by the endothermic peaks of the DSC thermograms. For both samples (GMO/(TAG + Transcutol)/water or GMO/(TAG + ethanol)/water mixtures) measurements were made during heating from 3 to 40 °C, after overnight incubation at subzero temperature (Figures 4.16a–g). The X-ray scattering patterns (WAXS and SAXS) of GMO/(TAG + Transcutol)/water and GMO/(TAG + ethanol)/water mixtures as a function of temperature were similar.

At approx. 3 °C, that is, below the maximum temperature of peak B, the 2D WAXS images of both mixtures (with addition of Transcutol or ethanol) revealed four low-angle reflections that were indexed as the first four orders of a lamellar phase with interlamellar spacing of 52 Å. In addition, there were three wide angle peaks at 4.74, 4.49, and 4.14 Å of the crystalline low-temperature phase, L_c [51]. At 10 °C (Figure 4.16b), the SAXS profile of GMO/(TAG + Transcutol)/water exhibited two broad Bragg peaks indicative of the lamellar phase, with interlamellar spacing of 52 Å and domain size (based on the width of the first peak) of 300 ± 50 Å. At temperatures above 20 °C (25 °C in Figure 4.16e), three peaks were observed

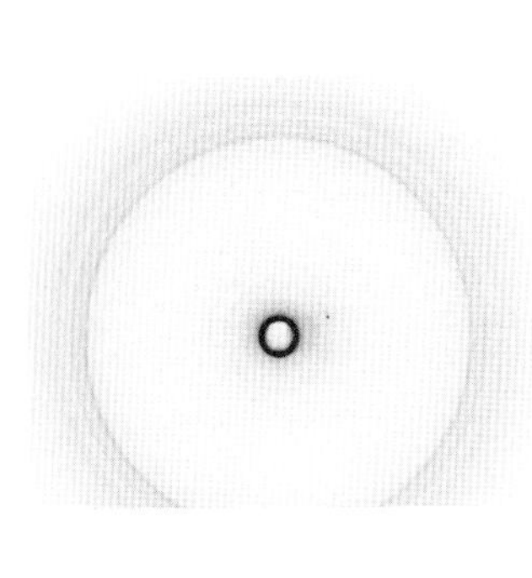

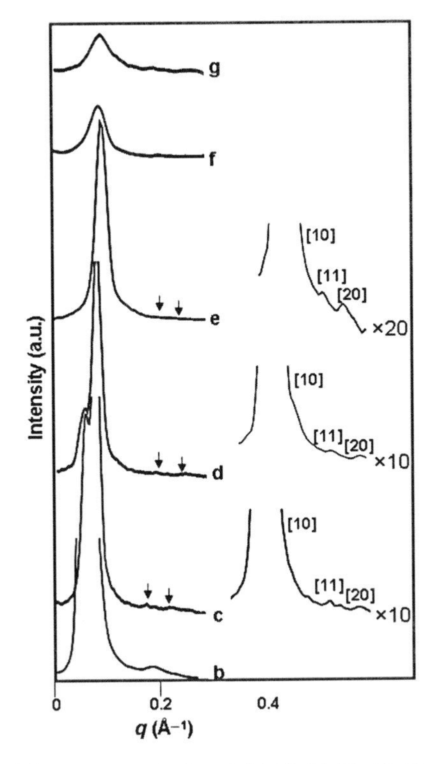

Figure 4.16 X-ray diffraction patterns of the GMO/(tricaprylin + Transcutol) 90/10 mixture with 12.5 wt% water after overnight incubation at −20 °C, measured at: (a) 3 °C (2D WAXS) and (b) 10, (c) 15, (d) 20, (e) 25, (f) 35, and (g) 40 °C (1D SAXS). The arrows on traces c, d, and e indicate the locations of the [11] and [20] diffraction peaks [29].

and indexed as the (10), (11), and (20) reflections of a 2D H_{II} phase. Upon increasing the temperature from 15 to 30 °C the lattice parameter of the H_{II} structure decreased (Table 4.2). In addition, there was a decrease in the effective crystallite size (430–180 Å), which was most pronounced between 25 and 40 °C. Between 35 and 40 °C (Figures 4.16f and g) only a single broad peak can be observed, implying random distribution of micelles.

It was concluded that the addition of either Transcutol or ethanol had structural effects similar to those of increasing temperature, in that there was an increase in the mobility of both the water and the GMO chains [29, 48]. These two small molecules, which compete for water with the GMO polar headgroups, increased the curvature and partially dehydrated the headgroups. As a result, less-ordered, fluid H_{II} structures with small domains were formed at room temperature (up to 30 °C).

The study was summarized in the following illustration that represents the modifications that took place in the H_{II} microstructure with each additive

Table 4.2 Lattice parameters and the effective crystallite sizes (a and L_H, respectively) of the hexagonal structures in GMO/TAG/Transcutol/water mixture (12.5 wt% water concentration) as a function of temperature.

Temperature (°C) ± 0.5 °C	a (Å) ± 0.5 Å	L_H (Å) ± 50 Å
15.0	49.8	446
20.1	49.9	411
25.0	45.8	430
30.2	44.8	286
35.4	N.A.	217
40.1	N.A.	180

(Figure 4.17). In the absence of additives, the typical H_{II} phase was formed and exhibited free flow of water in the core of the cylindrical GMO micelles (Figure 4.17, upper part). Upon addition of Transcutol or ethanol, the hexagonal phase was liquefied with reduced domain size, yet retained its symmetry. The presence of the strong hydrogen-bonding acceptors (hydrophilic guest molecules) in the water core of the cylinders and at the interface decreased the number of water molecules available to hydrate the GMO headgroups (Figure 4.17, lower part). This effect was more prominent in the presence of ethanol than of Transcutol. The dehydration process enhanced the mobility of the hydrocarbon chains, resulting in increased curvature and concomitant reduction of the lattice parameter.

The typical reverse hexagonal LLC composed of GMO/tricaprylin and water (Typ), and fluid hexagonal systems containing either 2.75 wt% Transcutol or ethanol (Tra and Eth, respectively) as a fourth component, were further explored as solubilization reservoirs for several bioactive molecules with different polarities [29, 52].

For example, the impact of ascorbic acid (AA) or ascorbyl palmitate (AP) guest molecules on the macrostructure of the three H_{II} phases was demonstrated. Up to 5 and 6 wt% AA and AP were solubilized in the well-ordered H_{II} (termed Typ + AA and Typ + AP, respectively). In the less ordered fluid systems containing 2.75 wt% Transcutol or ethanol, lower quantities of AA and AP guest molecules, 4 and 5 wt%, respectively, were solubilized (termed Tra + AA, Eth + AA and Tra + AP, Eth + AP, respectively).

Following the thermotropic behavior of the three systems as empty and loaded with the guest molecules revealed modifications in the melting points of the water and/or the hydrophobic peaks as discussed above. The melting temperatures of both water and hydrophobic moieties of the GMO in the three H_{II} phases (Typ, Tra, and Eth) empty and loaded with 3 wt% AA or AP guest molecules (Typ + AA, Typ + AP, Tra + AA, Tra + AP, and Eth + AA, Eth + AP), are summarized in Table 4.3 [52].

The incorporation of 3 wt% AA or AP molecules composed of five hydrogen-bond acceptor sites (four hydroxyl + one ester group and three hydroxyl + two ester

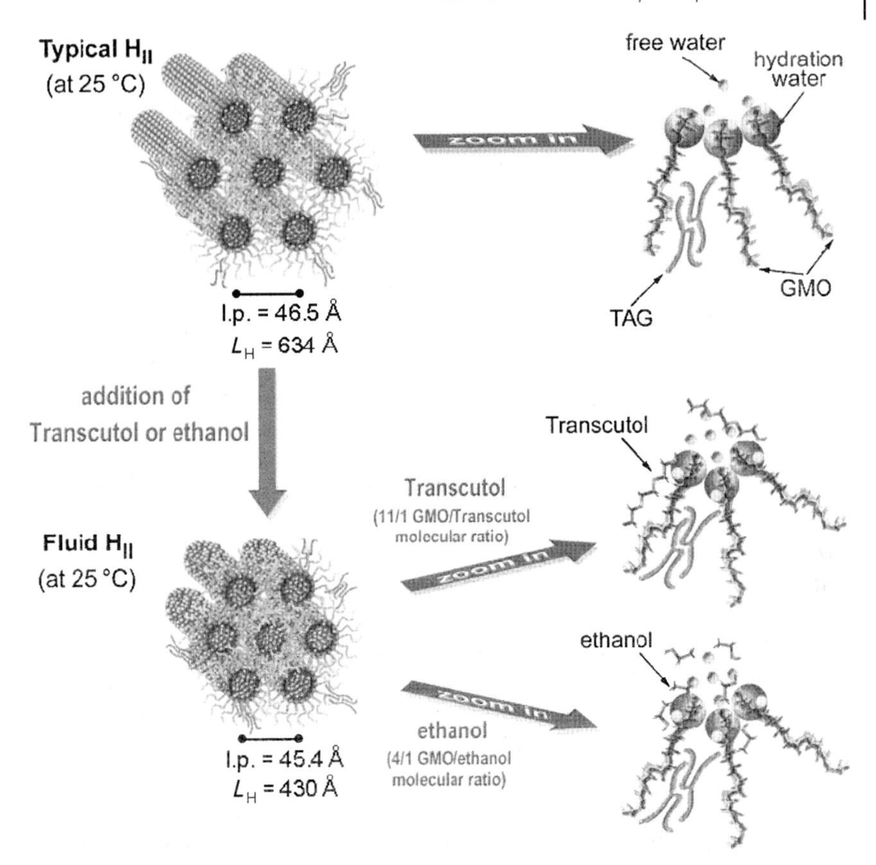

Figure 4.17 A cartoon illustrating the room-temperature structural and physicochemical properties of the fluid mesophases containing Transcutol and ethanol compared to those in their absence. l.p. and L_H correspond to the mean lattice parameter and the effective crystallite size in the hexagonal mesophase. The calculated ratios are: GMO/TAG 14/1 in the binary system and 12/1 in the ternary [29].

groups, respectively) had a significant effect on the thermal process of the mesophases (Table 4.3). An increase in content of the hydrophilic molecule, AA, decreased the water-thawing temperature in all three hexagonal systems. However, the water fusion decrease was the most pronounced (by 3 °C) in the Typ + AA H_{II} system (Table 4.3) and least pronounced (by 0.5 °C decrease) in the Tra + AA mixture. AP consists of water-binding sites similar to those of AA except for a single hydroxyl group that transformed into an ester group. Therefore, it had H-binding capability with water similar to that of the AA, and its accommodation in the three H_{II} phases decreased the water melting temperature, as seen in the presence of the AA molecule (Table 4.3). On the other hand, it was also detected that its long alkyl tail, which was solvated by the GMO hydrophobic tails, resulted in lower tail-melting temperatures in all three hexagonal systems (Table 4.3).

Table 4.3 Melting temperatures (°C) of three H_{II} systems (Typ, Tra, and Eth) empty and loaded with 3 wt% AA or AP. Typ represents the typical H_{II} system composed of GMO/TAG/water, Tra–GMO/(TAG + Transcutol)/water and Eth–GMO/(TAG + ethanol)/water.

Melting temperature (°C ± 0.2 °C)	Typ H_{II} system			Tra fluid H_{II} system			Eth fluid H_{II} system		
	empty	+AA	+AP	empty	+AA	+AP	empty	+AA	+AP
Water (peak A)	−2.1	−5.0	−5.4	−8.4	−8.9	−8.3	−9.9	−12.1	−12.3
Tails (peak B)	5.9	5.9	4.8	5.1	5.4	2.9	4.3	4.0	1.6
				6.5	6.8	5.1	5.5	5.1	5.0
				9.8			8.3	7.6	

It was concluded that although the mole quantity of the two guest molecules, AA and AP, compared to water was extremely small, they induced major competition for water binding with the GMO polar headgroups. The AA and AP molecules interfered with the water inter-hydrogen bonding, hence decreasing mainly the ice (water) melting temperature. However, in the presence of additional competitors for water binding, such as Transcutol or ethanol, the function of the AA and AP as water-binding agents was less distinct, particularly in the system containing Transcutol. Furthermore, once Transcutol and ethanol were present in the mixture, the contribution of the palmitic tail in solvating the GMO hydrophobic tails was marked, suggesting an additional effect that was later elucidated by the FTIR [52].

SAXS analysis in the presence of AA and AP molecules revealed an increase in the lattice parameter of the H_{II} phases of all three systems. In the typical H_{II} phase, upon addition of 3 wt% AA, a increased by 4.0 Å, and in the presence of Transcutol or ethanol a more moderate, yet noticeable, increase was observed (by 1.7–1.5 Å). Further increase in the solubilization loads of AA did not further alter the lattice parameter values. In the presence of the AP molecule only a slight increase in a (~1.5 Å) was detected in all three systems.

It was concluded that both AA and AP may increase the hydration level of the headgroups, thereby swelling the cylinders and significantly increasing the lattice parameter. However, the additional tail of the ascorbyl palmitate was effectively solvated by the GMO tail; hence there was a decrease in the extension of the acyl chains, particularly in the Typ + AP H_{II} system. Furthermore, in the fluid system, the Transcutol and ethanol moderated the swelling by causing dehydration of the GMO, thus decreasing parameter a. AA performed as a chaotropic agent effectively inducing cylinder swelling, mainly in the absence of the kosmotropic compounds (Transcutol or ethanol) while AP exhibited a more moderate chaotropic effect due to its opposite geometric contribution of additional tail volume [52].

FTIR spectroscopy was used to characterize the microstructure of the H_{II} phases and to spot the guest molecules' location. Each system was compared with and without AA or AP (3 wt%) while focusing on four major bands (Figure 4.18) to analyze the conformation of the surfactant molecule and its interaction with water and guest molecules in the different mesophases.

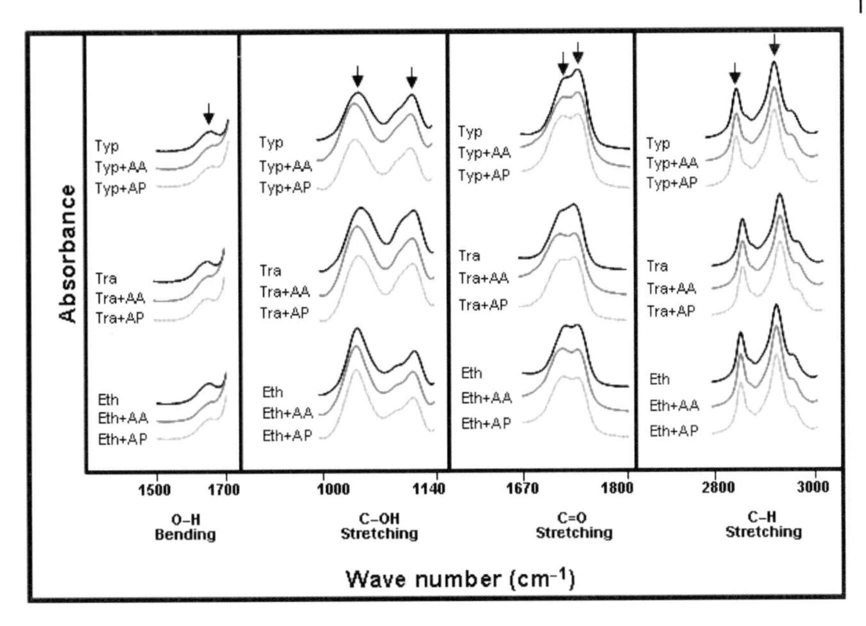

Figure 4.18 FTIR spectra, at 25 °C, of the three empty H_{II} systems (Typ, Tra, and Eth) and of 3 wt% AA or AP in each system (Typ + AA or AP, Tra + AA or AP, and Eth + AA or AP) in the frequency ranges: (from left to right) O–H bending 1500–1700, C–OH stretching 1000–1140, C=O stretching 1670–1800, and C–H stretching 2800–3000 cm^{-1}. Typ – GMO/TAG/water, Tra – GMO/(TAG + Transcutol)/water, and Eth – GMO/(TAG + ethanol)/water [52].

The bond vibrations in three regions of the mesophase were explored: the water-rich core (the water channels), the water/surfactant interface, and the lipophilic acyl chains.

The H–O–H bending band at ~1650 cm^{-1} was used to characterize the competitive interactions of the OH of the water with the GMO headgroups and the guest molecules (Figure 4.18). It is known that the OH bending vibration frequency increases when other moieties (e.g., hydroxyls) compete for the hydrogen bonding [31, 32]. Solubilizing AA or AP molecules augmented the frequency of the water bending vibrations (Figure 4.18, first segment). The frequency effect was more pronounced in the Typ and Eth systems (upward shifts of 4 and 2–3 cm^{-1}, respectively) and negligible (1 cm^{-1}) in the Tra mixture (Figure 4.18, first block). The hydrophilic guest molecules distorted the water structure and decreased the mean water–water H-bond angle, in the first hydration shell. The AA and AP guest molecules were recognized as solutes that destabilized the structure of bulk water (chaotropic effect) and decreased the root mean square of the H-bond angle (the smallest O· · · ·O–H angle formed by two neighboring water molecules) between water molecules [53].

At the interface, three vibrational modes of the GMO headgroups that reflect the interfacial arrangement of the lipid headgroups, were monitored – the stretching of the bonds C–OH (β, ~1121 cm^{-1}), C–OH (γ, ~1046 cm^{-1}), and C=O (carbonyl at the α position, 1720–1740 cm^{-1}) (Figure 4.18; [29, 48, 53]). The carbonyl band consists of two components, one originating from "free" (freely rotating) carbonyl (1740 cm^{-1}) and the second, from intramolecular hydrogen-bonded carbonyl groups (1730 cm^{-1}) [29, 48, 53].

The stretching frequencies of the C–OH, at β and γ positions, were affected by the guest molecules' solubilization in the H_{II} phases (Figure 4.18). The stronger the hydrogen-bonding between the surfactant and the water (or the guest molecules), the lower the stretching frequency of C–OH group (ν_{C-OH}). It should be noted that the low ν_{C-OH} vibration was attributed to the more hydrated or H-bonded OH group (γ position), hence more changes upon interactions with water or other hydrogen-bond acceptor groups are expected.

In the Typ H_{II} phase, upon addition of AA (Typ + AA) the C–OH stretching frequency shifted downward by 4.5 cm^{-1} only in the γ position. However, by solubilizing AP molecules the ν_{C-OH} shift was more moderate, 3 cm^{-1} in the γ position (Figure 4.18, second block). These results indicated stronger hydrogen-bonding at the hydroxyl groups, mainly in the γ position upon incorporation of AA compared to AP. It was surmised that the AA molecule is located closer to the hydroxyl group in the γ position while AP, due to its alkyl chain, was located deeper within the surfactant tails, closer to the hydrophobic region. In the Tra H_{II} phase, the peak position (γ) shifted toward a lower wave number (5 cm^{-1}, Figure 4.18, second block) although Transcutol, which is known to dehydrate the hydroxyl groups, was present. It was concluded that part of the Transcutol molecules that were H-bonded to the interface were exchanged by the AA or AP solubilizates (1/1 or 0.5 mole ratio of Transcutol/AA or AP). Consequently, more hydrogen bonds between the AA or AP molecules and the OH groups of the GMO were formed. Thus, the kosmotropic additive, Transcutol, was strongly attached to the water molecules and the added chaotropic guest molecules, AA and AP, efficiently interact with the interface hydroxylic groups. Once more, the effect was more pronounced in the γ position upon incorporation of AA compared to AP, showing that AA molecules were located closer to the hydroxyl group than AP.

On the contrary, in the system containing the hydrotrope, ethanol, the C–OH stretching frequencies shifted only slightly downward by 2 cm^{-1} in the γ position upon addition of solely AP molecules. Unlike Transcutol, ethanol stayed in the interface and the solubilized additives were localized more in the water-rich core than in the interface. As a result, a stronger effect was detected in the water-rich core (water behavior) than in the interface that was neutralized by the presence of ethanol (3/1 or 0.5 mole ratio of ethanol/AA or AP). Additionally, based on similar effects in the β and γ positions, it was assumed that the ethanol contributed to the location of the AP molecule and attracted it closer to the hydrophilic region.

The changes in the carbonyl bands upon addition of AA or AP are shown in Figure 4.18 (third block). The incorporation of AA molecules did not affect the peaks' positions or their intensities. However, addition of AP increased the intensity of the hydrogen-bonded carbonyl (lower $\nu_{C=O}$) at the expense of the free carbonyl (by 5%), demonstrating interactions with the carbonyl group and confirming that AP was localized closer to the hydrophobic region than the AA molecule.

In the presence of Transcutol, both additives affected the peaks' intensities (increase of 5–10%) and increased the peaks' width at half-height but only the AP molecule slightly increased (by $2\,cm^{-1}$) the hydrogen-bonded carbonyl vibration wave number (Figure 4.18, third block). The authors concluded that the accommodation of both additives in the interfacial region perturbed the C=O stretching mode, particularly the amphiphilic AP that broke a fraction of the intramolecular hydrogen bindings (in the GMO molecule) due to its presence.

In the presence of the interface-located ethanol, which was assumed to neutralize the effects of AA and AP, indeed only a slight increase in the intensity of the hydrogen-bonded carbonyl at the expense of the free carbonyl (by 2–3%), was observed (Figure 4.18, third block). It confirmed that the polar moieties of the AA and AP were localized in the same position within the mesophase.

Information on the conformational order of the GMO acyl chains was obtained from the stretching modes of the CH_2 segments (ν_{CH_2} at ~2853 and ~$2918\,cm^{-1}$). However, in all three systems the GMO methylene peaks were not affected by the incorporation of either AA or AP guest molecules (Figure 4.18, fourth block). To reconfirm the effect of AP that was seen in DSC measurements on the mesostructure, the CH_2 wagging mode was followed in the IR spectral region of 1330–$1400\,cm^{-1}$. By normalizing the intensities of the CH_2 wagging bands to the CH_3 bending vibrations, only the gg and the *kink* bands (at ~1354 and ~$1367\,cm^{-1}$, respectively) showed dependence on the solubilized AP (data not shown). In the typical H_{II} phase, where the presence of AP slightly decreased the hydrophobic melting temperature by ~1 °C, only the double-*gauche* band (gg) was shifted. However, in the fluid Tra and Eth H_{II} phases, the addition of the amphiphilic AP increased the fractions of both double-*gauche* and *gauche-trans-gauche* conformers in the GMO tails. The effect was more distinct in the mixture containing ethanol, which is in line with lower melting temperatures of the GMO tails observed by DSC.

The increase in *gauche* conformations resulted in increased chain disorder and looser acyl chain packing, hence the incorporation of AP in the presence of Transcutol or ethanol was found to increase chain disorder and lead to looser acyl chain packing even at very low temperatures.

4.4 Summary

The impact of surfactant geometry, mainly its tail volume and area per headgroup, and the corresponding mesophase formation have been demonstrated and analyzed.

It was shown that replacing the traditional surfactants by environmentally friendly molecules, such as poly(oxyethylene) cholesteryl ethers, to overcome the biodegradation process and aquatic toxicity resulted in the formation of diverse unique liquid-crystalline phases. The ability to obtain novel intermediate phases was attributed to the bulky and nonflexible hydrophobic part of the surfactants and intricate hydrophobic–hydrophilic balance controlled by the different EO chain length.

Shorter ethylene oxide chain length (trioxyethylene) led to the production of a thermally stable L_α phase that consisted of rigid bilayers and was attributed to the rigid multiring skeleton cholesterol group. However, as the number of EO units increased, the flexibility of the surfactant increased and the melting temperature of the liquid-crystalline phase was found to decrease. It was further concluded that in the case of short chain lengths, behavior was dominated by the rigidity of the cholesterol group imposing its properties on the surfactants; however, above ten units of ethylene oxide behavior was dominated by the EO chain.

It was demonstrated that systems based on either $ChEO_{10}$ or $ChEO_{15}$ exhibited a number of novel and intermediate phases. In the water/$ChEO_{15}$ system, I_1, H_1, R_1, and L_α phases were formed, whereas in the water/$ChEO_{10}$ mixture, W_m, R_1, L_α^H, and L_α phases were produced.

In terms of structural parameters, the surfactant with the shorter EO chain length exhibited smaller a_S that resulted in tighter packing than with the $ChEO_{15}$ molecule. Additionally, d_L was smaller in the $ChEO_{15}$-based system only when the L_α structure was formed, compared to the water/$ChEO_{10}$ system. 40–45 and 55–65 molecules were found to hydrate the $ChEO_{10}$ and $ChEO_{15}$ surfactants, respectively. On this basis, it was deduced that transformation into the intermediate R_1 structure took place when the hydration of the oxyethylene chains was incomplete compared to the fully hydrated H_1 or I_1 structures.

Other nonionic surfactants, that is, GMO, formed unique structures upon addition of ethanol or Transcutol. The unique isotropic fluid Q_L phase formed in the GMO/ethanol/water mixture and is surmised to be a transition phase between the cubic bicontinuous phase and the sponge phase.

In the GMO/TAG/water system above 35 °C or upon addition of Transcutol or ethanol (at room temperature), a low-viscosity H_{II} was formed. The fluidity of the hexagonal phase was explained by the significant reduction in domain size and cylinder length. This phenomenon was characterized by greater mobility of the GMO, lesser mobility of the water, and a significant dehydration process. The dehydration process enhanced the mobility of the hydrocarbon chains, resulting in increased curvature and concomitant reduction of the lattice parameter.

References

1 Svensson, B., Olsson, U., and Alexandridis, P. (2000) *Langmuir*, **16**, 6839.

2 Lodge, P.T., Pudil, B., and Hanley, K.J. (2002) *Macromolecules*, **35**, 4707.

3 Lindblom, G., and Rilfors, L. (1989) *Biochim. Biophys. Acta*, **988**, 221.

4 Huang, K.L., Shigeta, K., and Kunieda, H. (1998) *Prog. Colloid Polym. Sci.*, **110**, 171.

5 Antonelli, D.M., and Ying, J.Y. (1996) *Curr. Opin. Colloid Interface Sci.*, **1**, 523.

6 Braun, P.V., Osenar, P., and Stupp, S.I. (1996) *Nature*, **380**, 325.

7 Desai, S.D., Gordon, R.D., Gronda, A.M., and Cussler, E.L. (1996) *Curr. Opin. Colloid Interface Sci.*, **1**, 519.

8 Alexandridis, P., Olsson, U., and Lindman, B. (1998) *Langmuir*, **14**, 2627.

9 Floudas, G., Vazaiou, B., Schipper, F., Ulrich, R., Wiesner, U., Iatrou, H., and Hadjichristidis, N. (2001) *Macromolecules*, **34**, 2947.

10 Kunieda, H., Shigeta, K., Ozawa, K., and Suzuki, M. (1997) *J. Phys. Chem. B*, **101**, 7952.
11 Griffin, W.C. (1954) *J. Soc. Cosmet. Chem.*, **5**, 249.
12 Israelachvili, J.N., Mitchell, D.J., and Ninhan, B.W. (1976) *J. Chem. Soc.*, **72**, 1525.
13 Mitchell, D.J., and Ninham, B.W. (1981) *J. Chem. Soc. Faraday Trans. 2*, **77**, 601.
14 Seddon, J.M. (1990) *Biochim. Biophys. Acta*, **1031**, 1.
15 Hyde, S.T. (2001) Identification of lyotropic liquid-crystalline mesophases, in *Handbook of Applied Surface and Colloid Chemistry* (ed. K. Holmberg), John Wiley & Sons, Inc., New York, pp. 299–327.
16 Larsson, K. (1989) *J. Phys. Chem.*, **93**, 7304.
17 Larsson, K., Fontell, K., and Krog, N. (1980) *Chem. Phys. Lipids*, **27**, 321.
18 Uddin, Md.H., Morales, D., and Kunieda, H. (2005) *J. Colloid Interface Sci.*, **285**, 373.
19 Qiu, H., and Caffrey, M. (2000) *Biomaterials*, **21**, 223.
20 Misquitta, Y., and Caffrey, M. (2001) *Biophys. J.*, **81**, 1047.
21 Caffrey, M. (1987) *Biochemistry*, **26**, 6349.
22 Folmer, B.M., Svensson, M., Holmberg, K., and Brown, W. (1999) *J. Colloid Interface Sci.*, **213**, 112.
23 Kunieda, H., Horii, M., Koyama, M., and Sakamoto, K. (2001) *J. Colloid Interface Sci.*, **236**, 78.
24 Kunieda, H., Ozawa, K., and Huang, K.L. (1998) *J. Phys. Chem. B*, **102**, 831.
25 Hossain, Md.K., Acharya, D.P., Sakai, T., and Kunieda, H. (2004) *J. Colloid Interface Sci.*, **277**, 235.
26 Rodriguez, C., Naito, N., and Kunieda, H. (2001) *Colloids Surf. A*, **181**, 237.
27 Kanei, N., Watanabe, K., and Kunieda, H. (2003) *J. Oleo Sci.*, **52**, 607.
28 Efrat, R., Aserin, A., Kesselman, E., Danino, D., Wachtel, E.J., and Garti, N. (2007) *Colloids Surf. A*, **299**, 133.
29 Amar-Yuli, I., Wachtel, E., Shalev, D.E., Aserin, A., and Garti, N. (2008) *J. Phys. Chem. B*, **112**, 3971.
30 Anderson, B., and Olofsson, G. (1987) *Colloid Polym. Sci.*, **265**, 318.
31 Adam, C., Durrant, J., Lowry, M., and Tiddy, G. (1984) *J. Chem. Soc. Faraday Trans.*, **180**, 789.
32 López-Quintela, M.A., Akahane, A., Rodriguez, C., and Kunieda, H. (2002) *J. Colloid Interface Sci.*, **247**, 186.
33 Sato, T., Hossain, Md.K., Acharya, D.P., Glatter, O., Chiba, A., and Kunieda, H. (2004) *J. Phys. Chem. B*, **108**, 12927.
34 Holmes, M.C. (1998) *Curr. Opin. Colloid Interface Sci.*, **3**, 485.
35 Engström, S., Alfons, K., Rasmusson, M., and Ljusberg-Wahren, H. (1998) *Progr. Colloid Polym. Sci.*, **108**, 93.
36 Hyde, S.T. (1997) *Langmuir*, **13**, 842.
37 Fritz, G., Scherf, G., and Glatter, O. (2000) *J. Phys. Chem. B*, **104**, 3463.
38 Burgoyne, J., Holmes, M.C., and Tiddy, G.J.T. (1995) *J. Phys. Chem.*, **99**, 6054.
39 Sato, T., and Buchner, R. (2003) *J. Chem. Phys.*, **119**, 10789.
40 Efrat, R., Aserin, A., and Garti, N. (2008) *J. Colloid Interface Sci.*, **321**, 166.
41 Rodríguez, C., Shigeta, K., and Kunieda, H. (2000) *J. Colloid Interface Sci.*, **223**, 197.
42 Kunieda, H., Umizu, G., and Aramaki, K. (2000) *J. Phys. Chem. B*, **104**, 2005.
43 Uddin, Md.H., Kanei, N., and Kunieda, H. (2000) *Langmuir*, **16**, 6891.
44 Kunieda, H., Uddin, Md.H., Horii, M., Furukawa, H., and Harashima, A. (2001) *J. Phys. Chem. B*, **105**, 5419.
45 Rodríguez, C., Uddin, Md.H., Watanabe, K., Furukawa, H., Harashima, A., and Kunieda, H. (2002) *J. Phys. Chem. B*, **106**, 22.
46 Rodriguez-Abreu, C., Acharya, D.P., Aramaki, K., and Kunieda, H. (2005) *Colloids Surf. A*, **269**, 59.
47 Amar-Yuli, I., Wachtel, E., Ben-Shoshan, E., Danino, D., Aserin, A., and Garti, N. (2007) *Langmuir*, **23**, 3637.
48 Amar-Yuli, I., Wachtel, E., Shalev, D.E., Moshe, H., Aserin, A., and Garti, N. (2007) *J. Phys. Chem. B*, **111** (**13**), 544.
49 Soni, S., Brotons, G., Bellour, M., Narayanan, T., and Gibaud, A. (2006) *J. Phys. Chem. B*, **110**, 15157.

50 Ivanova, R., Lindman, B., and Alexandridis, P. (2000) *Langmuir*, **16**, 3660.

51 Wadsten-Hindrichsen, P., Bender, J., Unga, J., and Engstrom, S. (2007) *Colloid Interface Sci*, **315**, 701.

52 Amar-Yuli, I., Aserin, A., and Garti, N. (2008) *J. Phys. Chem. B*, **112**, 10171.

53 Sharp, K.A., Madan, B., Manas, E., and Vanderkooi, J.M. (2001) *J. Chem. Phys.*, **114**, 1791.

5
Swelling of Vesicle Precipitates from Alkyldimethylaminoxide and a Perfluoroalcohol by Refractive-Index Matching with Glycerol

Yun Yan, Yuwen Shen, Ying Zhao, and Heinz Hoffmann

5.1
Introduction

Th effect of glycerol on surfactant systems is gaining increasing attention owing to the inevitable combined use of them in many practical formulations [1, 2]. For example, pharmaceutical formulation for drug delivery uses cosolvents such as glycerol in order to improve the solubility of the active compounds or to aid the sensory perception. There is enough experimental evidence that the surfactant self-aggregation in glycerol occurs as that in water [3–6], but the Krafft point tends to increase, whereas the stability of aggregates tends to decrease [7–9], which was attributed to changes in the hydration of surfactant headgroups and in the solvophobic effect. It has been known that addition of glycerol increases the CMC of surfactants and decreases the micellar aggregation number [10–12]. Recent studies indicate that with increasing glycerol concentration, the average area of the surfactant headgroups is increased, which is accompanied by the decrease of critical packing parameter and therefore of smaller micelles [13]. It was found, however, in the water/C12EO8 system, that micellar aggregates tend to grow with increasing glycerol content, which is attributed to the dehydration of the ethylene oxide chain [14].

The aforementioned changes brought about by glycerol mainly focused on the microscopic inspection, whereas what occurs to the macroscopic phases is rarely reported. Actually, upon increasing the content of glycerol in the aqueous systems of surfactants, considerable phase and macroscopic changes are often observed [15–18], indicating that interaction on the level of the self-associated structures has occurred. Therefore, it is very necessary to probe the physical origins of the phase and other macroscopic changes introduced by glycerol.

Recently, we found that the influence of glycerol on surfactant Lα-phases can be related to the matching of the refractive index of water–glycerol mixed solvent and that of surfactant bilayers [16]. Experimentally, we observed swelling of block copolymer $EO_{15}[Si(CH_3)_2]_3EO_{15}$ Lα-phases and lowering of turbidity for vesicular suspensions upon replacing solvent water gradually with glycerol [16]. At 60% glycerol, the L_1/L_α two-phase system is transformed into single L_α-phase owing to

Self-Organized Surfactant Structures. Edited by Tharwat F. Tadros

ISBN: 978-3-527-31990-9

the increase of the interlayer spacing between bilayers, and the turbid vesicle phase becomes completely transparent. These changes are in line with the matching of the refractive index of 60% glycerol–water mixed solvent and that of the bilayers. In a theoretical inspection using DLVO theory, the matching of refractive index lowers the Hamaker constant of the system, which therefore results in a decrease of the attraction between the bilayers, and allows them to swell to a larger separation [16, 19]. We thus anticipate that by using the principle of index matching, multilayered vesicle precipitates can also be swollen owing to the decrease or disappearance of the attraction between the bilayers. Lately, we do get successful cases in surfactant vesicle precipitate systems.

In this chapter, we describe the swelling of vesicle precipitates in tetradecylaminoxide (C_{14}DMAO) and perfluoro cosurfactant 1, 1-H-dihydroperfluorooctanol (PFC) mixed surfactant system. This system was selected for two reasons: the first one is to have a lower refractive index of the bilayers. The refractive-index matching point in the silicone surfactant system is 60% glycerol. We therefore expect that a lower index matching point can be reached by lowering the refractive index of the surfactants. It is well known that the fluorocarbon compounds have very low refractive index close to water. Therefore, fluorocarbons were chosen to bring down the refractive index of the surfactant bilayers. The second reason is the phase behaviors of the selected system have been previously systematically investigated by Hoffmann *et al.* [20] It was known that a two-phase situation occurs in this system with a vesicle phase on top of a dense vesicle precipitate. The vesicle suspension contains relatively small vesicles, whereas the vesicle precipitates are composed of huge multilamellar vesicles (MLV) larger than 1 μm. The distance between the shells of the MLVs in the vesicle precipitates are so narrow that only a small fraction of water can be entrapped in the MLVs. Thus, the MLVs precipitate from water due to their larger density together with some small vesicles entrapped in the corners of the huge MLVs.

Our purpose in this work is to swell the MLV precipitates by replacing the solvent water with glycerol. Two compositions of the MLV precipitates were selected: (1) 100 mM C14DMAO / 100 mM PFC mixed water system; (2) 100 mM C14DMAO / 50 mM PFC / 10 mM NaCl mixed water system. Both systems are in the L_α/precipitate region in the phase diagram, except that the former has a larger volume of precipitates than the latter. Since the PFC amount is lowered in the second system, 10 mM NaCl is added to shield the charges. The experimental results indeed confirm that the vesicle precipitates can also be swollen by index matching.

5.2 Experimental

Materials. 1,1-H-dihydroperfluorooctanol (PFC), and tetradecyldimethylaminoxide, C_{14}DMAO, are gifts from Clariant/Gendorf. The purity of PFC is 97.8%, and the substance had a melting point at 45.4 °C. C_{14}DMAO was recrystallized twice from acetone, and its purity was characterized by the melting points and CMC values. Glycerol is 99.5% (wt%) with a density of 1.26. Bidistilled water was used throughout the experiments.

Sample Preparation. Stock solutions of 100 mM C_{14}DMAO and 100 mM C_{14}DMAO / 10 mM NaCl in water and glycerol were first prepared by weighing certain amounts of corresponding materials to a 200-ml flask followed by adding desired amount of solvent, respectively. At room temperature, all the solutions are transparent and colorless, which are L_1 phase. Then, a certain amount of PFC (solid) is weighed to the 100 mM C_{14}DMAO aqueous solution to prepare 100 mM C_{14}DMAO / 100 mM PFC, 100 mM C_{14}DMAO / 50 mM PFC in water and glycerol, respectively. Then, the mixed solutions with the same composition but in different solvent were mixed at different volume ratio to obtain C_{14}DMAO / 100 mM PFC, 100 mM C_{14}DMAO / 50 mM PFC / 10 mM NaCl in different water–glycerol mixed solvents. All the mixed systems were permitted for equilibrium at room temperature.

Measurements of Refractive Index. The refractive indices were measured by using a Julabo Abe refractometry with temperature controlled by a F30-C water bath system (Zeiss, Germany). To obtain the refractive index of C_{14}DMAO, a series of C_{14}DMAO aqueous solution of different concentrations were measured. By extrapolation of the refractive-index–concentration curve to 100%, we get the refractive-index value for the pure C_{14}DMAO. Since PFC is not soluble in water, we measured its refractive index at a temperature of 50 °C, which is beyond is melting point (45.5 °C). The measured refractive index for pure PFC at 50 °C was below the lowest limit of the instrument, but the value could be read with approximation. A refractive-index–composition curve of C_{14}DMAO/PFC was first constructed, then a linear-fitted function for the system was obtained. The refractive indices for the C_{14}DMAO/PFC mixtures were therefore estimated based on the standard curves. The refractive index for the swollen vesicles of C_{14}DMAO/PFC in different water–glycerol mixed system was measured directly.

5.3 Results and Discussion

5.3.1 Swelling of the Precipitates in the 100 mM C14DMAO / 100 mM PFC System by Replacing Water by Glycerol

As shown in Figure 5.1, in water the system has separated into two phases, with the upper phase being bluish, whereas the lower phase is crystalline precipitate. Upon replacing the solvent water by glycerol gradually, the bluish hue in the upper phase becomes lighter and lighter, until it completely turns transparent at about 30–40 glycerol. As discussed in our previous work [16], this increase of transparency is an indication of the matching of refractive index of the bilayers of C_{14}DMAO/PFC with the refractive index of the mixed solvent. Meanwhile, the volume of the

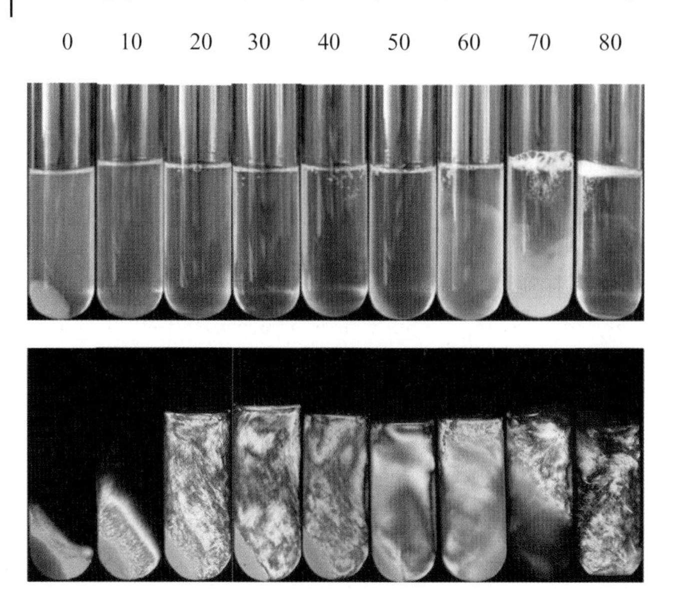

Figure 5.1 Photos of 100 mM C14DMAO / 100 mM PFC in water–glycerol mixed solvents. The numbers on the photo are the volume fraction of glycerol (v%). Upper: daylight; lower, in between polarizers.

precipitate becomes less on increasing the volume of glycerol. At glycerol contents higher than 40%, the system becomes turbid again. This is obviously due to the mismatch of the refractive index of the bilayers and that of the mixed solvent as the glycerol content increases. Between polarizers, it becomes evident that the precipitate of the samples at the bottom of the test tubes consists of two layers: a solid layer that does not change with the glycerol content and a birefringent layer that swells with glycerol. The solid layer consists of perfluoroalcohol that did not dissolve in the surfactant solution. On the verge of the precipitates are colorful birefringence patterns; upon addition of 10% glycerol, the volume of the birefringent area is obviously increased. It is noteworthy that the birefreingent area is obviously a new phase that differs from both the upper bluish phase and the crystalline precipitates, but it cannot be recognized without using polarizers. This means that the birefringent area has almost the same refractive index as that of the solvent. This result shows that most of the constituents in the birefringent area are PFC, since the refractive index of the fluorinate compounds is very close to that of water. Then, we can infer that the crystalline precipitate on the bottom of the phase is the unsolubilized PFC. It seems that the solubility of PFC increases with increasing glycerol content. The solubilized PFC forms birefringent Lα-phase with C_{14}DMAO in the system. At 20% glycerol, Lα-phase with colorful birefringence domains has filled the whole volume of the sample, but some PFC precipitates are still at the bottom. Upon careful inspection we found that the precipitates do not disappear even when 99.5% glycerol was used. The coexistence of swollen

Lα-phase and the PFC precipitates in the 100 mM $C_{14}DMAO$ / 100 mM PFC mixed systems at 99.5% glycerol indicates that 100 mM of PFC cannot be dissolved in the system. We therefore lower the PFC concentration to 50 mM to decrease the amount of precipitates.

5.3.2
Phase Behavior of 100 mM TDMAO / 50 mM $C_7F_{15}CH_2OH$ / 10 mM NaCl

As we lowered the content of PFC to 50 mM in the system, 10 mM NaCl is added to shield the charges on the TDMAO produced by its hydrolysis in water. Figure 5.2 shows the phase photos with increasing glycerol content. One first observed a turbid phase in water; after centrifugation, precipitates were observed, but the supernatant is still turbid, indicating the presence of vesicles. Upon replacing the

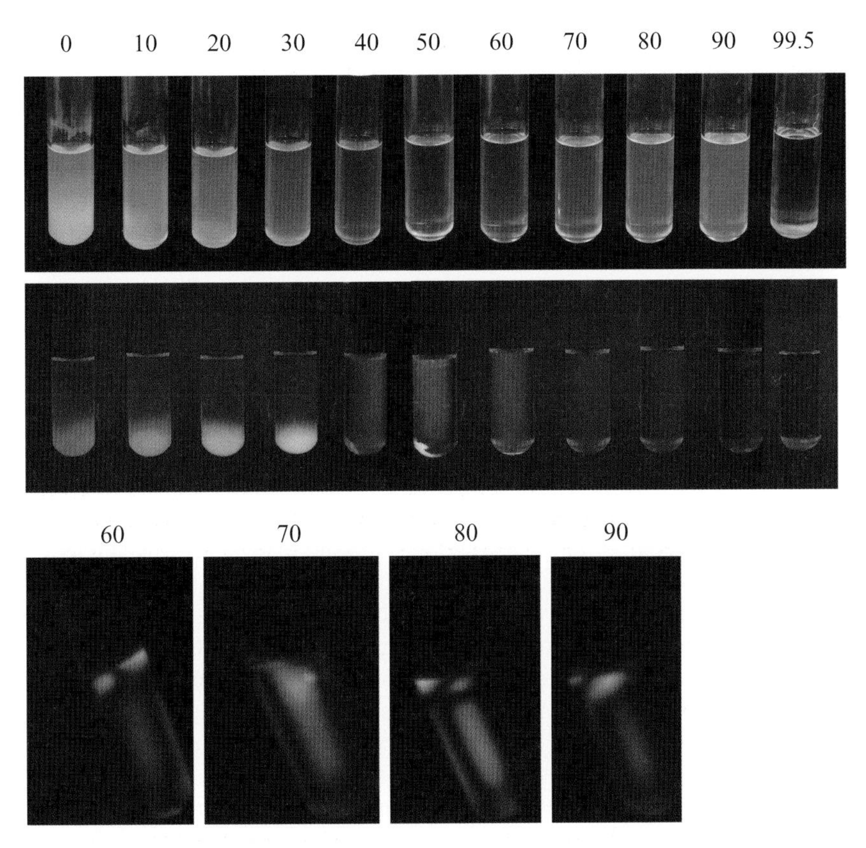

Figure 5.2 Photos of 100 mM C14DMAO / 50 mM PFC / 10 mM NaCl in water–glycerol mixed solvents. The numbers on the photo are the volume fraction of glycerol (v%). Upper: daylight; middle and lower, in between polarizers. The lower samples show flow birefringence for the Lα-phase in the 60–90% glycerol system. Note that in 50% glycerol, precipitate is still at the bottom of the tube.

solvent water by glycerol gradually, the system becomes clearer. At 60% glycerol, it becomes completely transparent. Further increase of the glycerol composition leads to an increasing bluish hue for the system. This demonstrates that refractive-index matching between 100 mM C_{14}DMAO / 50 mM PFC bilayers and the mixed solvent occurs at 60% glycerol; higher glycerol content leads to mismatch of the refractive index again. In between two polarizers, one can find that the amount of precipitate decreases with increasing glycerol; at 40–50% glycerol, there are still tiny amounts of precipitates in the bottom of the phase, but they completely disappear at 60% glycerol. This phase sequence indicates that the precipitates in 100 mM C_{14}DMAO / 50 mM PFC system can be fully swollen by glycerol. The phases with 60–90% glycerol are all single phase without any stationary birefringence, but they all show flow birefringence upon slight shaking. This means the phases are all single Lα-phases.

5.3.3
Microstructures in the System with Various Glycerol Content

The microstructures at various glycerol contents were examined by FF-TEM technique. As reported by Hoffmann *et al.*, both the upper and lower phases in the 100 mM C_{14}DMAO / 50 mM PFC / 10 mM NaCl system in water are composed of vesicles. The diameters for the vesicles in the upper phases are in the range of 80 nm to 1 μm, with some nonspherical vesicles present. Due to the fracture plane following the midplane of the bilayer the interior of the vesicles is not revealed. This way of fracturing is typical for vesicles of perfluoro systems that has been discussed in detail in ref. 19. The occurrence of unspherical vesicles indicates a demixing of C_{14}DMAO and PFC in the vesicle membrane, because areas with a different composition of the bilayers should have a different curvature. Compared with those in the upper phase, the vesicles in the lower phase are much more polydisperse: both huge vesicles as large as 5 μm and smaller ones similar to those in the upper phase exist. The small vesicles in the lower phase are constrained in the corners between the large ones. A significant feature revealed by the cross-fractured micrograph is the very dense shells of the MLVs. There is almost no extra space for the presence of solvent layers in between two vesicular shells. It is possible that the huge vesicles in the lower phase are perfluoro-rich, so that they separate from the upper phase due to the higher density. Upon addition of glycerol, the volume of the lower phase decreases gradually, indicating a decrease of the amount of huge vesicles. Up to 60% glycerol, the lower phase is diminished, and the system is dominated by a phase that shows flow birefringence. The gain of flow birefringence demonstrates that the number of vesicles in the phase is increased when compared with the vesicle population in the original upper phase where no glycerol is added. The FF-TEM micrograph indicates that at 40% glycerol, there are some large vesicles of 2–4 μm in the upper phase together with many smaller nonspherical vesicles (Figure 5.3c). It is striking that the population of nonspherical vesicles is much larger than that in the original upper phase. According to previous analysis by Hoffmann *et al.*, these nonspherical vesicles are

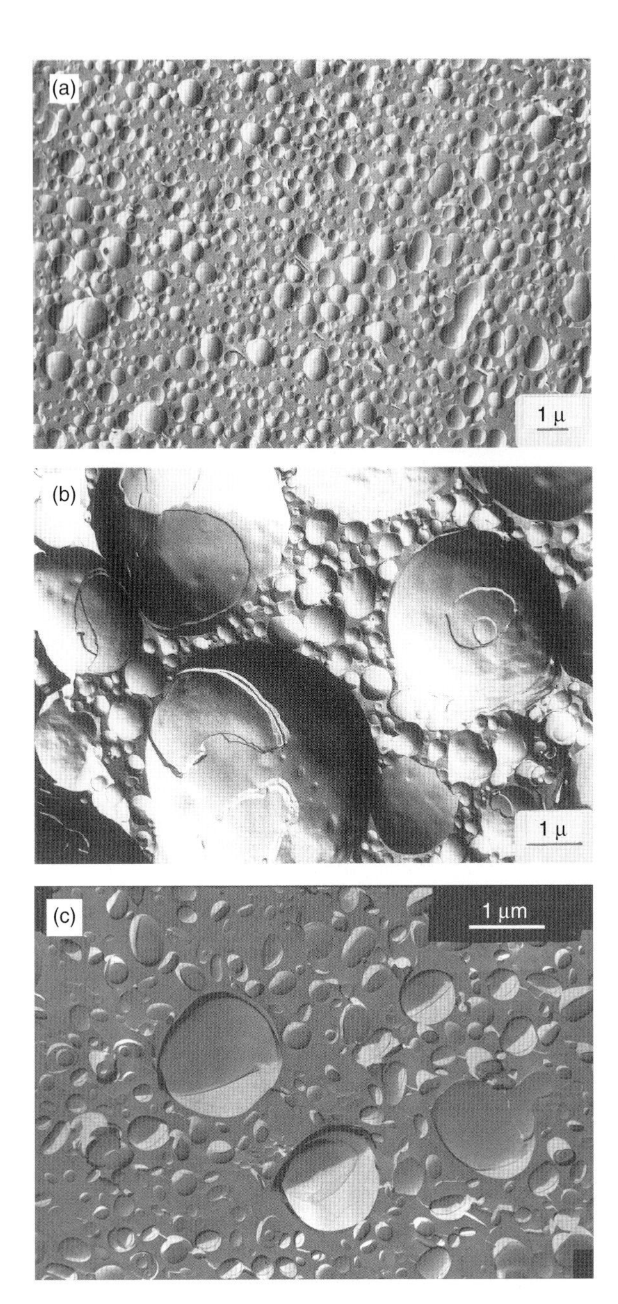

Figure 5.3 FF-TEM micrographs for the 100 mM C14DMAO / 50 mM PFC / 10 mM NaCl in water–glycerol mixed solvents. (a), no glycerol, upper phase; (b), no glycerol, lower phase; (c), 40% glycerol, upper phase; (d), enlargement of part of (c); (e), a MLV with broken outer shell in 40% glycerol; (f) 60% glycerol; 8), 80% glycerol.

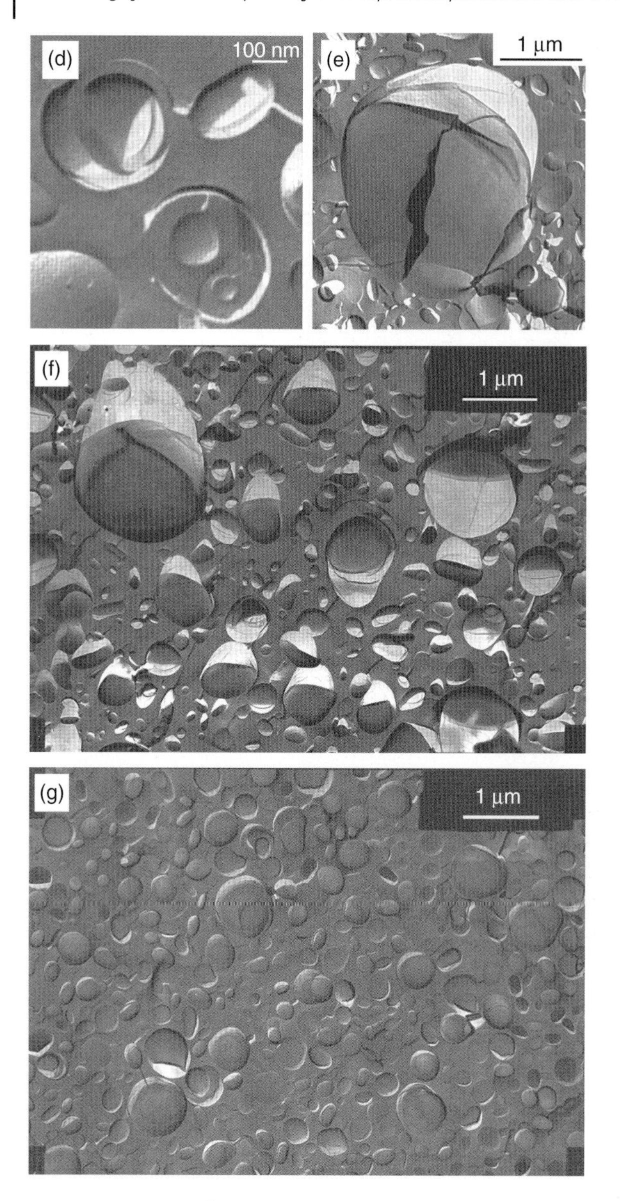

Figure 5.3 *Continued*

formed due to a demixing of perfluoro and hydrocarbon components in the vesicle membrane. The increased population of nonspherical vesicles is probably an indication that the huge vesicles in the lower phase that are rich in perfluoro cosurfactants have rearranged upon addition of glycerol and entered into the upper phase in the form of smaller unspherical vesicles. This rearrangement may arise from the swelling of the shells for the huge MLVs. As revealed in an enlarged micrograph of Figure 5.3c, we did see increased interlamellar spacings in the 40% glycerol system (Figure 5.3d). Although not all the interiors can be revealed in FF-TEM, we believe that at least a considerable portion of the MLVs have swollen in the presence of glycerol. However, for closed bilayer structures, the swelling of the shell may result in breaking of the bilayer. Actually, we did observe such broken membrane structures as increasing glycerol concentration in the 5% EO_{15}-$PDMS_{30}$-EO_{15} vesicular system. In addition, discernible cracks on the outer shell of the huge MLVs and membrane fragments in the bulk are indeed observed in the C14DMAO/PFC system (Figure 5.3e), which suggest that the outer shells are broken due to swelling. Consequently, it results in a decrease of the average size for the vesicles. This is in line with the decrease of the amount of precipitates that are formed due to presence of dense MLVs. As the glycerol content is increased to 60%, where the system has entered into one single phase, the number of vesicles larger than 1 μm is further decreased, while the population of medium-sized vesicles around 800 nm increases (Figure 5.3f). These medium-sized vesicles are probable from the broken MLVs. At even higher glycerol content of 80%, vesicles larger than 1 μm can hardly be seen, and the size of the vesicles in the system is very close to that found in the upper phase when no glycerol is added, except that the population of nonspherical vesicles is much larger (Figure 5.3g). It is clear that at this glycerol content, the huge perfluoro-rich MLVs with very tight shells have all transformed into smaller swollen vesicles that have smaller density. The decreased density of the swollen vesicles facilitate the system enters into single-phase region. It should be noticed that although the average size of the vesicles in the 80% glycerol system is obviously smaller than that in the 60% glycerol system, the turbidity of the former is larger than the latter. This demonstrates that the transparency of the system is indeed determined by the index matching between the vesicular membranes and that of the mixed solvent.

5.3.4 Rheological Results

The increase of the vesicle population with increasing glycerol is also reflected in the rheological results. In Figure 5.4 the plot of viscosity versus shear rate after subtraction of the solvent contribution is shown. All systems show non-Newtonian behavior, which is in line with the FF-TEM observations of the existence of large vesicles that are deformable under shear. The average viscosities increase with increasing glycerol content, and a remarkable increase occurs between 60% and 80% glycerol. This change can be correlated to the decreased vesicle size and increased vesicle population. Theoretical results have shown that the smaller the vesicles and the larger the vesicle population, the higher shear moduli and viscosity a system can have. This tendency is in good agreement with the FF-TEM results. As revealed in Figures 5.3c and f, the number of huge vesicles with diameter larger than 2 μm decreases with increasing glycerol from 40% to 60%, which favors an increase of the viscosity. However, the population of medium-sized vesicles around 800 nm is larger in the latter system, which underlines the increase in viscosity.

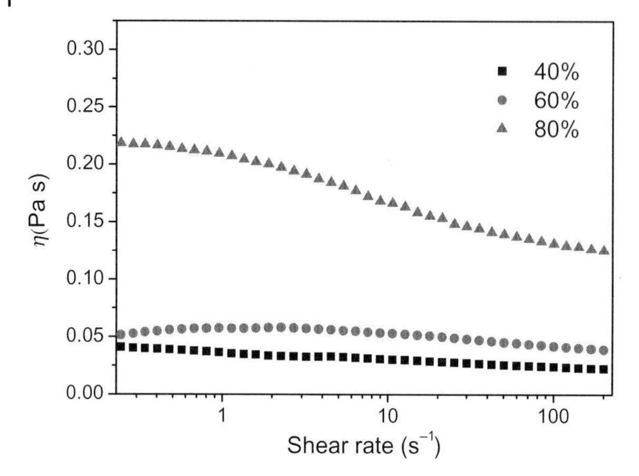

Figure 5.4 Viscosity of the systems with different glycerol content. The background contribution of the corresponding water–glycerol mixed solvent was subtracted. The viscosity in the 40% glycerol system was measured without disturbing the tiny amount of precipitates in the bottom.

As a result, the overall viscosity in the 60% glycerol system does not increase dramatically compared with that in the 40% glycerol system. As for in the 80% glycerol system the FF-TEM clearly reveals that vesicles larger than 1 μm can hardly be seen, and most of the vesicles are in the range of 60–800 nm. The microstructure change between 80% and 60% glycerol systems is therefore much larger than that between 60% and 40% glycerol systems. Taking into account the vanished precipitates that are obviously transformed into smaller vesicles in the system, it is clear that the occurrence of smaller vesicles is accompanied by the increase of the vesicle population, so that the viscosity in the 80% glycerol system increases markedly.

5.3.5 Discussion

The change of the system can be explained by matching the refractive index of the bilayers and that of the solvent. One may recall the phase behaviors in the 100 mM C_{14}DMAO with 100 mM and 50 mM PFC systems, respectively. The transparent phases occur at different glycerol–water ratios for the two PFC concentrations. For the 100 mM PFC system, 30% glycerol is required to produce a transparent phase, while 60% glycerol is required for the 50 mM PFC system. The transparency for a colloidal system is determined by the difference between the refractive index of the solvent and that of the colloidal particles. The refractive index of the solvent increases on increasing the glycerol content in water–glycerol mixtures; while the refractive index for the vesicles composed of C_{14}DMAO and PFC can be calculated by the weight fraction of the two materials in the system. Since the refractive index for C14DMAO and PFC is about 1.5285 and 1.3000, respectively, the mixtures of 100 mM C14DMAO / 100 mM PFC and 100 mM C14DMAO / 50 mM PFC are calculated to have a refractive index about 1.388 and 1.434, respectively. Then, the refractive-index matching solvent for the two vesicle systems has to be different. The value of 1.388 is close to that of 30% glycerol (1.380, Figure 5.4). Therefore, the system with 100 mM PFC arrives at index matching at this glycerol content. It should be kept in mind that the estimated refractive index of the bilayers only represents the solid C_{14}DMAO and PFC; water molecules that are involved in the interlayers are not considered. Therefore, the estimated (calculated) refractive index is always higher than that of vesicle bilayers. Actually, the directly measured value of the bilayers in 30% glycerol is 1.3777, which is almost the same as the 30% glycerol. For the 50 mM PFC system, the refractive index of the bilayers is higher than that for the 100 mM PFC system, since the low refractive index component PFC is less. At this composition, the mixed bilayers are estimated to have a refractive index of about 1.434, which is close to that of the 60% glycerol (1.428), so that the transparent phase occurs at 60% glycerol (Figure 5.5).

The effect of glycerol on planar Lα-phase has been discussed in detail in our previous paper. For nonclosed planar structures, the bilayers have a well-defined spacing that is the result of equilibrium between attractive van der Waals forces and repulsive undulation forces. Upon substitution of the solvent water gradually by glycerol–water mixed solvent, the attractive van der Waals forces approach zero. This is because the refractive index of the solvent and that of the bilayers becomes closer, so that the Hamaker constant in the expression of van der Waals forces gets close to zero:

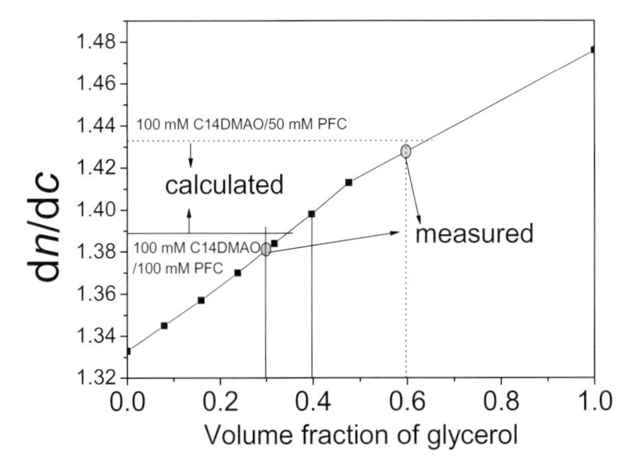

Figure 5.5 Comparison of the refractive index of the bilayers and that of the water–glycerol mixed solvent. Gray circles are the measured values for the swollen vesicles; the horizontal solid line is the calculated value for the 100 mM C14DMAO / 100 mM PFC mixture; the vertical solid line indicates the refractive index for 30% glycerol; the dotted lines are the same parameters for the 100 mM C_{14}DMAO / 50 mM PFC system.

$$P_A = -\frac{A}{6\pi}\left[\frac{1}{d^3} - \frac{2}{(d+\delta)^3} + \frac{1}{(d+2\delta)^3}\right] \tag{5.1}$$

$$A = \frac{3}{4}kT\left[\frac{\varepsilon_d - \varepsilon_m}{\varepsilon_d + \varepsilon_m}\right]^2 + \frac{3}{16\sqrt{2}}\hbar\omega\frac{\left(n_d^2 - n_m^2\right)^2}{\left(n_d^2 + n_m^2\right)^{3/2}} \tag{5.2}$$

where P_A is the attractive van der Waals pressure per unit area of the bilayer, A the Hamaker constant of the bilayer in medium, d the spacings between neighboring stacked bilayers, δ the thickness of the bilayer, k is the Boltzmann constant, T the absolute temperature, n_d and n_m are the refractive indices of the particles and the medium, respectively, for visible light, ω is the frequency of the dominating UV absorption (about 1.7–2.4 × 10^{16} rad/s), and $h = 2\pi/\hbar$ is Planck's constant. The quantities ε_d and ε_m are the dielectric constants of particles and the medium at zero frequency (static field). Since the second term in Equation 5.2 is very much larger than the first one, which is smaller than kT, the Hamaker constant A is mainly determined by the second term. As n_m gets closer and closer to n_d, the value of A thus approaches to zero. Therefore, the van der Waals attractive pressure, as expressed in Equation 5.1, tends to vanish. Meanwhile, the repulsive Henfrich undulation pressure P_R, expressed as Equation 5.3, is not affect by the change in refractive index:

$$P_R = \frac{3\pi^2}{64}\frac{(kT)^2}{\kappa}\frac{1}{(d-\delta)^3} \tag{5.3}$$

where κ is the bending modulus of the bilayer. As a result, the system has to swell, driven by the repulsive forces. For planar bilayer structures in a L_1/L_α two phase system, this swelling leads to an increase of the volume of lamellar phases, while for vesicle dispersions, the swelling may result in breaking of the vesicle shells of the MLVs. Thus, the mechanism of swelling of vesicle precipitates can be illustrated as Scheme 5.1. For comparison, the swelling process of the planar lamellae is also shown:

5.4 Conclusion

The vesicle precipitates in the 100 mMC14DMAO / 100 mM PFC/water and 100 mMC14DMAO / 50 mM PFC/water system can be swollen by matching the refractive index of it with the solvent. The bluish hue of the vesicle suspensions on the top of the precipitates becomes lighter and the amount of precipitates diminishes with increasing glycerol content due to the decrease of the refractive-index difference between the solvent and that of the vesicle bilayers. For the 100 mM C14DMAO / 100 mMm PFC system, the precipitates are too numerous so that they cannot be completely swollen even if all the solvent water is replaced by glycerol; decrease the PFC concentration to 50 mM allows production of fewer precipitates in the system, which can thus be completely swollen by 60% glycerol. Different from swelling in a planar lamellar system, swelling of MLVs also results in the breaking

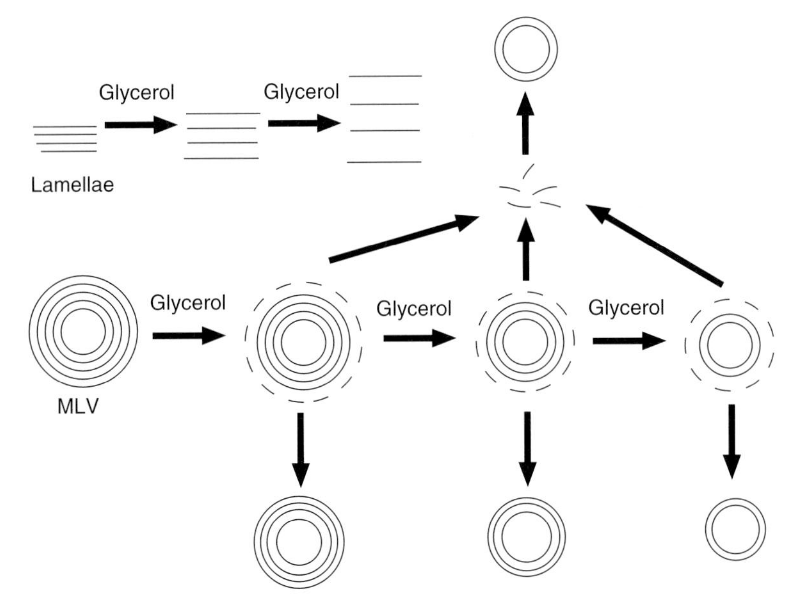

Scheme 5.1 Schematic representation of the swelling of lamellar phase and vesicle precipitates with increasing glycerol. It demonstrates swelling of lamellar phase simply leads to an increase of the interlayer spacing, whereas swelling of MLV results in swelling and breaking of the outer shell and consequently gives birth to smaller vesicles. Meanwhile, the broken membrane fragments can arrange into new vesicles. The shell number of the MLVs is just for illustration.

of vesicle shells, which therefore produces smaller vesicles. Meanwhile, the broken membrane fragments can rearrange to form new vesicles.

Acknowledgment

Dr. Yan Y greatly thanks the National Nature Science Foundation of China (NSFC) for financial support.

References

1 Martin, M., and Swaebrick, J. (1983) *Physical Pharmacy* (ed. K.P. Ananthapadamanabhan), Lea & Febiger Press, Philadelphia, PA.

2 Sjöberg, M., and Warnheim, T. (1997) *Surfactant Sci. Ser.*, **67**, 179.

3 Auvray, X., Petipas, C., Anthore, R., Rico, I., and Lattes, A. (1989) *J. Phys. Chem.*, **93**, 7458.

4 Lin, Z., Davis, H.T., and Scriven, L.E. (1996) *Langmuir*, **12**, 5489.

5 Alexandridis, P., and Yang, L. (2000) *Macromolecules*, **33**, 5574.

6 Ivanova, R., Lindman, B., and Alexandridis, P. (2000) *Langmuir*, **16**, 3660–3675.

7 Kumar, A., Kunieda, H., Vazquez, C., and Lopez-Quintela, M.A. (2001) *Langmuir*, **17**, 7245.

8 Martino, A., and Kaler, E.W. (1995) *Colloids Surf. A*, **99**, 91.

9 Rodríguez, C., Shigeta, K., and Kunieda, H. (2000) *J. Colloid Interface Sci.*, **223**, 197.

10 Cantú, L., Corti, M., Degiorgio, V., Hoffmann, H., and Ulbricht, W. (1987) *J. Colloid Interface Sci.*, **116**, 384.

11 Ionescu, L.G., Trindade, V.L., and de Douza, E.F. (2000) *Langmuir*, **16**, 988.

12 D'Erricol, G., Ciccarelli, D., and Ortona, O. (2005) *J. Colloid Interface Sci.*, **286**, 747.

13 Carnero Ruiz, C., Diaz-Lopez, L., and Aguiar, J. (2007) *J. Colloid Interface Sci.*, **305** (2), 293.

14 Aramaki, K., Olsson, U., Yamaguchi, Y., and Kunieda, H. (1999) *Langmuir*, **15**, 6226.

15 Alam, M.M., Shrestha, L.K., and Aramaki, K. (2009) *J. Colloid Interface Sci.*, **329**, 366.

16 Yan, Y., Hoffmann, H., Makarsky, A., Richter, W., and Talmon, Y. (2007) *J. Phys. Chem. B*, **111**, 6374.

17 Zou, A.H., Hoffmann, H., Freiberger, N., and Glatter, O. (2007) *Langmuir*, **23**, 2977.

18 Yan, Y., Hoffmann, H., Leson, A., and Meyer, C. (2007) *J. Phys. Chem. B*, **111**, 66161.

19 Leneveu, V.A., Rand, R.P., Parsegian, V.A., and Gingell, D. (1997) *Biophys. J.*, **18**, 209.

20 Bergmeier, M., Hoffmann, H., Witte, F., and Zourab, S. (1998) *J. Colloid Interface Sci.*, **203**, 1.

6
Si QDots: Where Does Photoluminescence Come From?

Xuejun Duan, Javier Calvo-Fuentes, and M. Arturo López-Quintela

6.1
Introduction

Silicon nanoparticles (Si NPs) with sizes in the order of bulk exciton Bohr radius [1, 2] present interesting optical properties for fluorescent labeling in biological imaging applications with their potential nontoxicity [3–6]. However, the origin of their photoluminescence has been subjected to intense debate for almost two decades. This debate has been focused on whether quantum-confinement effects or atomic-scale defects at the surface of the nanocrystals are responsible for the light emission [7].

There have been reports of Si NPs produced by electrochemical etching of Si wafer, laser pyrolysis, plasma deposition and solution-phase synthesis [8]. The last approach seems to be the most promising method because the size of nanoparticles can be controlled by the use of inverse micelles as nanoreactors in solution phase [3, 4]. Moreover, these synthesis methods allow chemical functionalization of Si NPs surfaces with relative ease, to offer numerous stabilization, solubility and bioconjugation options [4, 8–13]. In recent years, Wilcoxon's and Tilley's groups have synthesized water-dispersible Si NPs by using $LiAlH_4$ reduction of silicon halide [3], and then modified the nanoparticles' surface with allylamine by using a platinum catalyst [4]. However, we have found that reproducibility and scalability of these reported methods of synthesis have not been completely controlled yet due to many factors, such as kinetics, oxidation, contamination, etc. For example, since $LiAlH_4$ is a very strong reducing agent, it is difficult to control the speed of the reducing reaction required for good reproducibility. It has also been previously reported that H_2PtCl_6 can be reduced to platinum nanoparticles by methanol or other reducing agent [14]. The presence of other types of nanoparticles as impurities could introduce uncertainties in the origin of the fluorescence. A possible solution for this problem could be the use of UV irradiation instead of H_2PtCl_6 to functionalize the nanoparticle surface. Since all these methods involve the evaporation of the initial nonpolar solvent before redispersing in water, the possibility of contamination due to noneliminated toluene must be taken into account, as the optical properties of this solvent have been previously mistaken in the literature with those of Si NPs [15].

Self-Organized Surfactant Structures. Edited by Tharwat F. Tadros

ISBN: 978-3-527-31990-9

6.2
Experimental

6.2.1
Materials

Silicon tetrachloride (Aldrich, 99.998%), allylamine (Aldrich, 99.8%), lithium aluminum hydride solution (1.0 M in tetrahydrofuran, Aldrich), toluene (Aldrich, 99.8%, dehydrated), methanol (Aldrich, 99%), hydrogen hexachloroplatinate (Aldrich, 99.995%), isopropyl alcohol (Aldrich, 99.5%, anhydrous), sodium borohydride (Fluka) and tetraoctylammonium bromide (TOAB) (Sigma, 99%) were purchased from Sigma-Aldrich and used without further purification.

6.2.2
Methods

In order to prevent the oxidation of the silicon to silica, all the synthesis was performed in an argon-filled glove box with low oxygen and all the reagents were completely dehumidified.

$SiCl_4$ (25 µl) was dissolved in TOAB (0.375 g) and dehydrated toluene (25 ml) and left for three hours under magnetic stirring. Si NPs were formed by adding 8 mg of $NaBH_4$ (or slowly adding 0.25 ml $LiAlH_4$ 1 M in THF) and leaving it to react, stirring vigorously, for three hours. Then, any remaining reducing agent was eliminated by adding 5 ml of dehydrated methanol. After quenching the reducing agent, 0.5 ml of allylamine was added, stirring under UV irradiation (254 nm Pen-Ray lamp, UVP Light Sources) for 3 h (or catalyzing by addition of 10 µl of H_2PtCl_6 0.05 M in isopropanol). After modification of the surface, the sample was taken out of the glove box and the solvent removed in a rotary evaporator until only the TOAB and Si NPs were left. For purifying the hydrophilic Si NPs, first, 3 ml of distilled water was added. Then TOAB was removed by filtration through a 0.22 µm filter. Finally, 25 ml of acetone was added as a cleansing solvent. Extraction in the rotary evaporator was repeated four times by removing the acetone from the round-bottomed flask. After extraction, the product was recovered by dissolving it in 3 ml of water.

6.2.3
Results

Figure 6.1 shows a TEM image of Si NPs obtained by reduction with $NaBH_4$. The size histogram was determined by surveying 850 nanoparticles in several images. It shows a homogeneous distribution of almost spherical nanoparticles with an average size of 1.9 ± 0.6 nm. In Figure 6.2, AFM analysis confirms that the sizes of most of the nanoparticles are about 2.0 nm.

TEM image and size histogram in Figure 6.3 correspond to allylamine-capped Si NPs synthesized by reduction with $LiAlH_4$. A broader size distribution with an average size of 4.4 ± 2.0 can be observed.

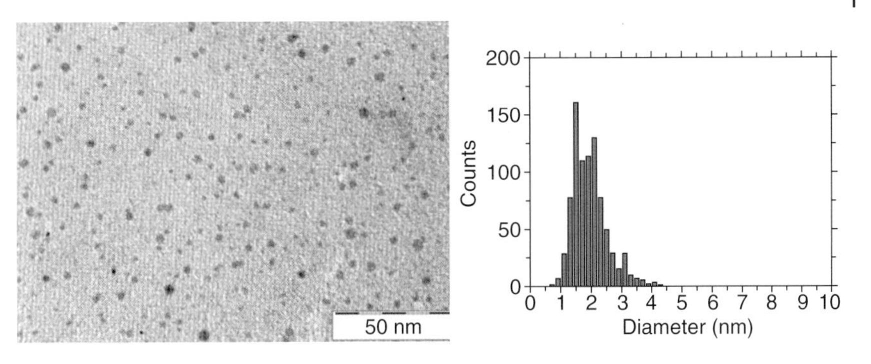

Figure 6.1 TEM image and size histogram (measured over 850 particles) of allylamine-grafted Si NPs reduced by $NaBH_4$.

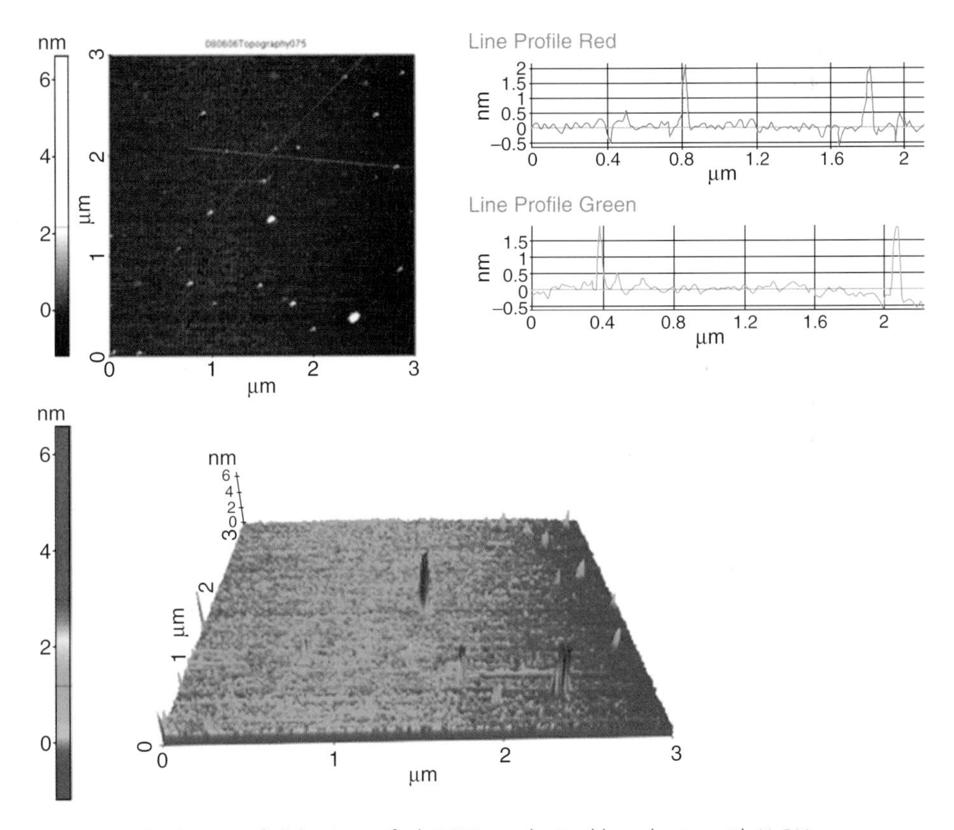

Figure 6.2 AFM image of allylamine-grafted Si NPs synthesized by reduction with $NaBH_4$.

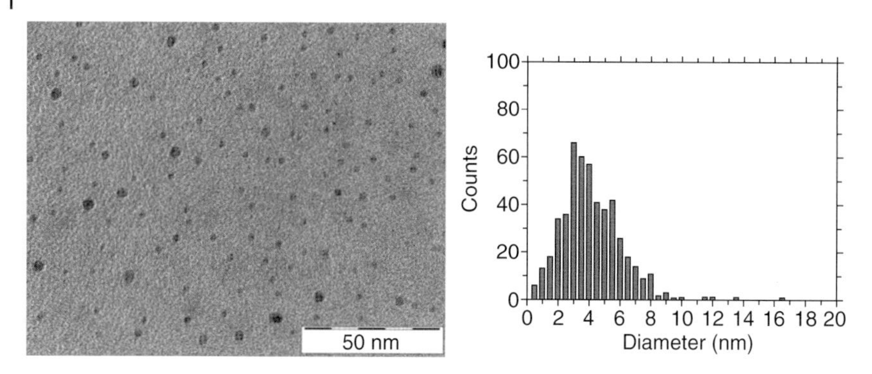

Figure 6.3 TEM image and size histogram (measured over 500 particles) of allylamine-grafted Si NPs reduced by $LiAlH_4$.

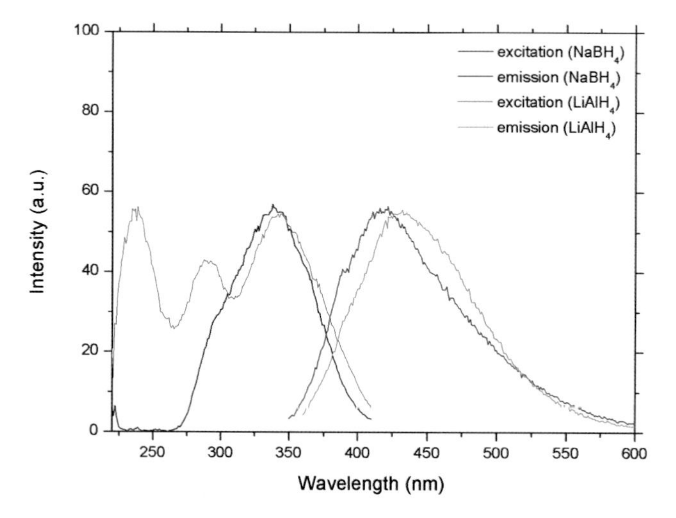

Figure 6.4 Photoluminescence spectra of allylamine-grafted Si NPs in water solution synthesized by reduction with $NaBH_4$ and $LiAlH_4$.

Photoluminescence spectra of as-synthesized nanoparticles in water solution, shown in Figure 6.4, display a maximum emission band centered at 430 nm while the excitation band is at 340 nm, independently of the reducing agent used in the synthesis. However, the excitation bands at 240 nm and 280 nm only appear in the spectrum of Si NPs synthesized by reduction with $LiAlH_4$.

Fluorescence lifetimes were very similar for both reduction approaches. In both cases, fluorescence decays (Figure 6.5) required a three-exponential fit with average lifetimes of 6 ns when reducing with $LiAlH_4$ and 4.7 ns when with $NaBH_4$. These short lifetimes reveal a rapid recombination associated with direct bandgap transitions in silicon nanocrystals.

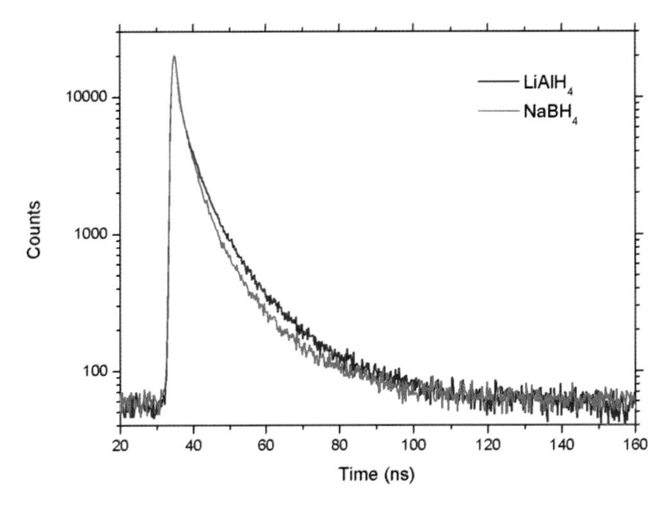

Figure 6.5 Time-resolved fluorescence decays for an excitation wavelength of 360 nm (recorded at maximum emission) of allylamine-grafted Si NPs synthesized by reduction with $NaBH_4$ and $LiAlH_4$.

FTIR spectroscopy of Si NPs (Figure 6.6) shows that the main peaks are in the same positions for both reducing agents with only minor changes in the peak intensities. It is especially noticeable in the peak attributed to grafted allylamine, at 3413 cm^{-1}, which is more intense when $NaBH_4$ is used.

After the UV-induced grafting of allylamine to the surface of Si NPs, the obtained sample was a clear water solution with a light green color. Photoluminescence quantum yields of the resulting solutions were up to 5.4% relative to the standard (quinine sulfate in H_2SO_4 0.1 M aqueous solution) under an excitation wavelength of 350 nm. This fluorescence was stable for at least 9 months in air at room temperature.

However, a significant change in the fluorescence was observed when the final solution was subjected to an annealing process at temperatures in the range of 50 °C to 80 °C. The emission band at 430 nm corresponding to an excitation wavelength of 340 nm significantly increased its intensity after only some hours at these temperatures (Figure 6.7).

Besides, after annealing, a new band appeared at 535 nm when exciting at 460 nm (Figure 6.8). Time-resolved measurements revealed that this fluorescence also had a fast decay, with an average lifetime of 4.7 ns for excitation at 460 nm and emission at 530 nm.

A control sample was prepared in order to discard the possibility of spectral changes coming from the "natural" evolution of the sample and not from the annealing. The fluorescence and absorption of this control sample was determined at the beginning and end of the experiment, not finding any significant change in the spectra. Surprisingly, these changes in the fluorescence were only noticeable in Si NPs synthesized with $LiAlH_4$. There was not any change observed in the

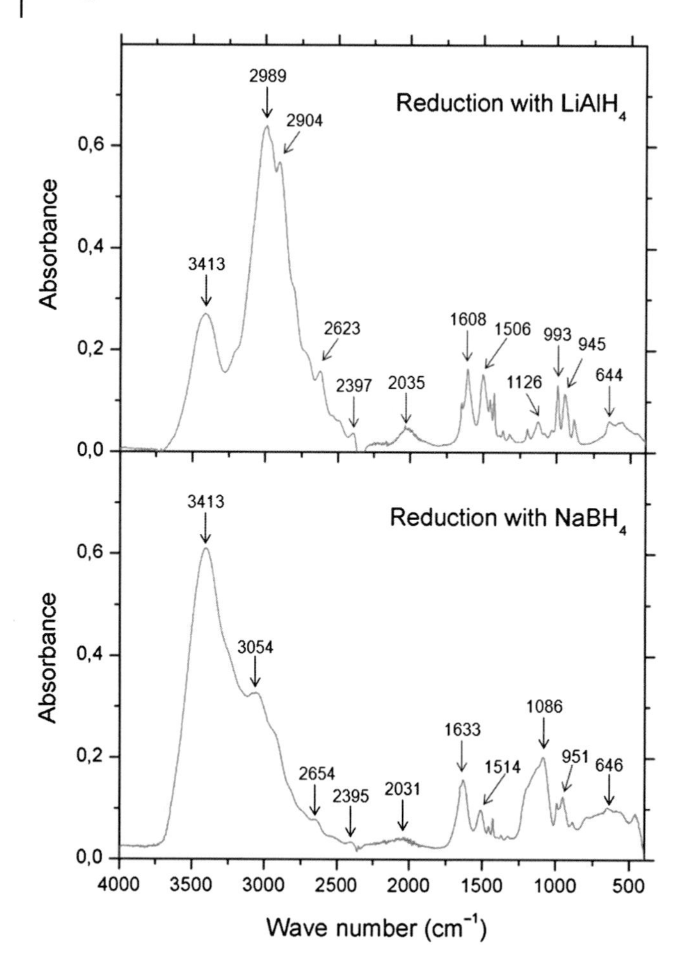

Figure 6.6 FTIR spectra of allylamine-grafted Si NPs.

fluorescence of annealed Si NPs synthesized by reduction with $NaBH_4$, where intensities and wavelengths of the fluorescence bands remained stable before or after annealing.

6.2.4
Discussion

We report here that silicon tetrachloride can be reduced by sodium borohydride to Si NPs and these nanoparticles can be further modified with allylamine under UV irradiation to avoid the possibility of contamination by platinum nanoparticles (Scheme 6.1).

According to our experiments, by using $LiAlH_4$ as reducing agent to synthesize Si NPs, reproducibility of the resulting fluorescence was poor, indicating that the

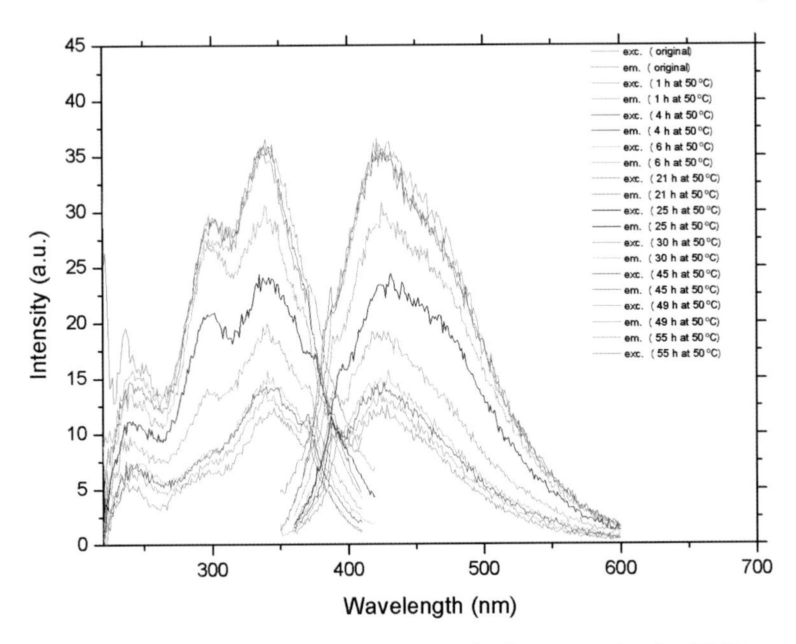

Figure 6.7 Influence of annealing temperature on the fluorescence bands of Si NPs (excitation at 345 nm).

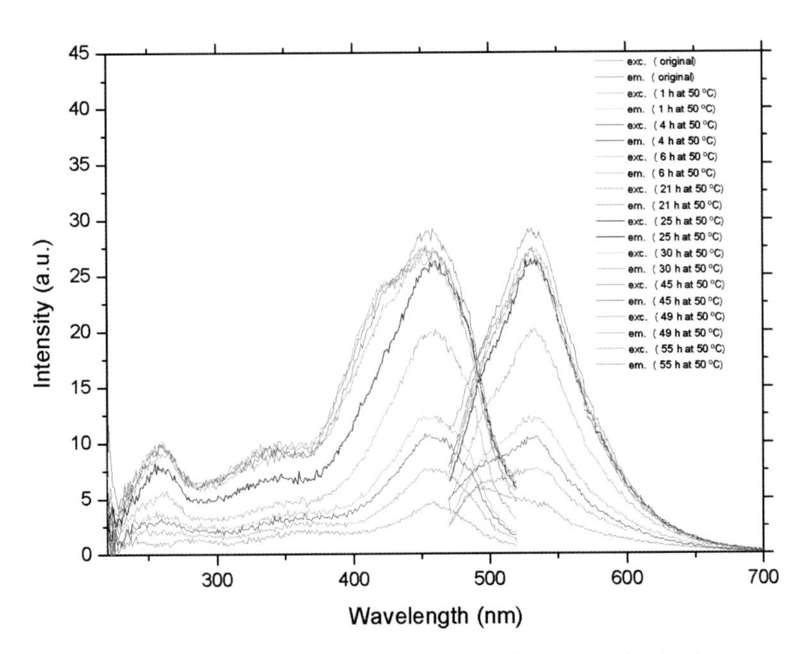

Figure 6.8 Influence of annealing temperature on the fluorescence bands of Si NPs (excitation at 460 nm).

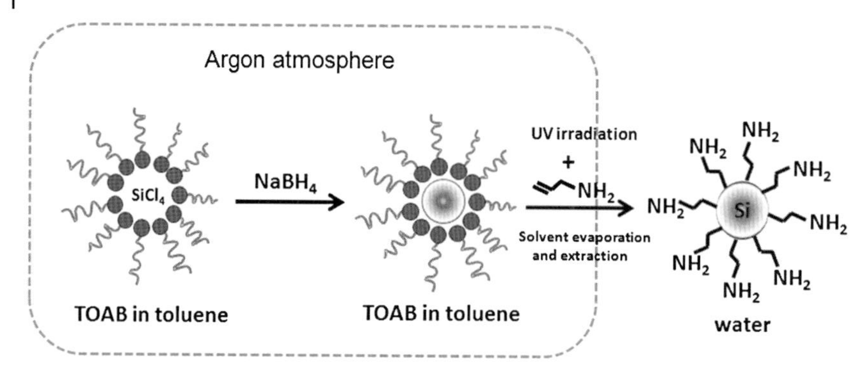

Scheme 6.1 Reduction of $SiCl_4$ with sodium borohydride followed by surface modification under UV irradiation.

experimental addition procedures of the $LiAlH_4$ have a large influence on the final size and crystallization of the obtained nanoparticles. It has been reported that a rapid addition of $LiAlH_4$ leads to polydisperse size distributions, whereas a slow addition of reducing agent leads to more monodisperse size distributions [16]. This can be attributed to its strong reductive properties, making it difficult to reproduce the results because the adding rate plays an important role on the final size and size distribution of resulting Si NPs. Unlike $LiAlH_4$, sodium borohydride is a less strong reducing agent and can be used more easily to control the speed of the reaction. In this case, the size of the obtained NPs is smaller, more monodisperse and reproducible than by using $LiAlH_4$ because of its smaller dependence on the addition speed of reducing reagent. The smaller particle size obtained with $NaBH_4$ is in agreement with the results obtained from FTIR spectra, where the peak at 3413 cm^{-1}, attributed to conjugated allylamine is much more intense in Si NPs synthesized with $NaBH_4$ than in those synthesized with $LiAlH_4$. Smaller particles have a larger surface/volume ratio and, therefore, would have more attached allylamine molecules than bigger particles.

Crystallization of the Si NPs is also affected by the rate of reduction. High reduction rates would produce a fast disordered growth of Si NPs and a poor crystallization with abundance of amorphous nanoparticles, while a low rate of reduction would give better crystalline Si NPs.

Annealing results could indicate that as-synthesized Si NPs are almost perfectly crystalline when they have been reduced with $NaBH_4$, while most of them are amorphous when reduced with $LiAlH_4$, as it is explained in Scheme 6.2.

A mixture of different sizes of Si NPs would be produced when $LiAlH_4$ is used, including small nanocrystals and nanoparticles, as well as some bigger nonfluorescent amorphous nanoparticles. For small Si NPs fluorescence intensity will increase by annealing due to crystallization. For bigger NPs new fluorescence emission (about 530 nm when excited at 460 nm) is observed by crystallization due to annealing. In contrast, Si NPs synthesized by $NaBH_4$ mainly consists of small

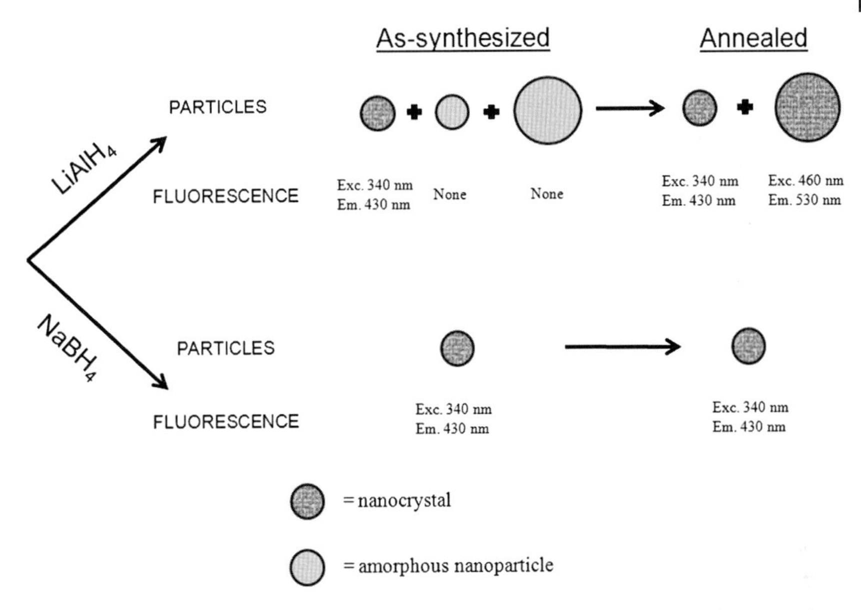

Scheme 6.2 Sizes, crystallization and fluorescence of Si NPs synthesized by reduction with LiAlH4 and $NaBH_4$, before and after annealing.

monodisperse silicon nanocrystals with a crystalline structure. Their fluorescence emission is at 430 nm wavelength when excited at 340 nm and is not influenced by annealing.

6.3 Conclusion

In this work, we have reported for the first time that sodium borohydride can be used to reduce silicon tetrachloride to silicon nanocrystals and that these nanocrystals can be further modified under UV irradiation with unsaturated long-chain organics, such as allylamine or hexenoic acid, through formation of a Si–C bond.

In this way, some of the problems that arose in previously reported synthesis, such as polydispersity or contamination by Pt nanoparticles, are resolved. Special care must also be taken in purification of final solutions, as any remaining trace of the solvents used during the synthesis could mislead the interpretation of observed properties.

It has been observed that the choice of the reducing agent is of extreme importance as different reagents lead to different final nanoparticle sizes. Crystallization of the nanoparticles has been found to play an important role in observed fluorescence, which can be influenced by a simple annealing process.

Acknowledgments

This work has received financial support from MCI, Spain (projects MAT2008-06503/NAN and NanoBioMed CONSOLIDER-INGENIO 2010-0012) and from the European Commission (Contract No. 037465-FLUOROMAG).

References

1 Stucky, G., and Mac Dougall, J.E. (1990) *Science*, **247**, 669–678.
2 Alivisatos, A.P. (1996) *Science*, **271**, 933–937.
3 Wilcoxon, J.P., Provencio, P.P., and Samara, G.A. (1999) *Phys. Rev. B*, **60**, 2704–2714.
4 Warner, J.H., Hoshino, A., Yamamoto, K., and Tilley, R.D. (2005) *Angew. Chem. Int. Ed.*, **44**, 4550–4554.
5 Kumar, V. (2007) *Nanosilicon*, Elsevier, London.
6 Michalet, X., Pinaud, F.F., Bentolila, L.A., Tsay, J.M., Doose, S., Li, J.J., Sundaresan, G., Wu, A.M., Gambhir, S.S., and Weiss, S. (2005) *Science*, **307**, 538–544.
7 Heitmann, J. (2005) *Adv. Mater.*, **17**, 795–803.
8 Veinot, J.C. (2006) *Chem. Commun.*, 4160–4168.
9 Wang, L., Reipa, V., and Blasic, J. (2004) *Bioconjug. Chem.*, **15**, 409–412.
10 Yang, C., Bley, R.A., Kauzlarich, S.M., Lee, H.W.H., and Delgado, G.R. (1999) *J. Am. Chem. Soc.*, **121**, 5191–5195.
11 Sato, S., and Swihart, M. (2006) *Chem. Mater.*, **18**, 4083–4088.
12 Li, Z.F., and Ruckenstein, E. (2004) *Nano Lett.*, **4**, 1463–1467.
13 Tilley, R.D., Warner, J.H., Yamamoto, K., Matsui, I., and Fujimori, H. (2005) *Chem. Commun.*, 1833–1835.
14 Rivadulla, J.F., Vergara, M.C., Blanco, M.C., López-Quintela, M.A., and Rivas, J. (1997) *J. Phys. Chem. B*, **101**, 8997–9004.
15 Rosso-Vasic, M., Spruijt, E., van Lagen, B., Cola, L.D., and Zuilhof, H. (2008) *Small*, **4**, 1835–1841.
16 Warner, J.H., and Tilley, R.D. (2005) *Proc. SPIE*, **6038**, 603815.

7
Worm-Like Micelles in a Binary Solution of Nonionic Surfactant $C_{16}E_7$ and Water

Tadashi Kato, Yuka Shimada, Daisuke Nozu, and Youhei Kawabata

Static and dynamic light-scattering and rheological measurements have been made in semidilute and concentrated region of the heptaethylene glycol n-hexadecyl ether ($C_{16}E_7$)/D_2O system. Light-scattering data as well as zero-shear viscosity strongly suggest an entanglement of worm-like micelles. The concentration dependence of the zero-shear viscosity and the stress-relaxation time obtained from the viscoelastic spectra, shear-rate dependence of the viscosity, and the slow mode of dynamic light scattering at 35 °C can be partly explained by the living polymer model of Cates. However, the behaviors above 40 °C are completely different from his prediction. On the other hand, modification of Hoffmann's simple theory can explain these results semiquantitatively. It is demonstrated that combination of the stress relaxation time and the surfactant self-diffusion coefficient is useful to discuss which model is better between the ghost-like crossing proposed by Shikata *et al.* and the sliding of connections proposed by Appel *et al.*

7.1
Introduction

It is now well recognized that surfactant micelles grow into rod-like or worm-like micelles under specific conditions and that worm-like micelles become entangled when the surfactant concentration exceeds a certain value (c^*) called the overlap concentration, like semidilute solutions of flexible polymers. On the other hand, dynamical properties of worm-like micelles are very different from those of polymers. During 10 years from the mid-1980s, experimental and theoretical methods have been developed to clarify the dynamics of entanglement based mainly on rheological properties [1–4]. In the past 10 years, rheological properties for many kinds of surfactant systems have been studied based on these methods [5–7].

Worm-like micelles are formed in solutions of both ionic and nonionic surfactants. For aqueous solutions of ionic surfactants, inorganic or organic salt or cosurfactants are necessary except for special system like dimeric surfactants. On the other hand, nonionic surfactants can form worm-like micelles in water without any additional components. Surfactants of polyoxyethylene type $C_nH_{2n+1}(OC_2H_4)_mOH$

Self-Organized Surfactant Structures. Edited by Tharwat F. Tadros

ISBN: 978-3-527-31990-9

(abbreviated as C_nE_m) are typical nonionic surfactants and the static properties of micelles have been studied in detail [8–10]. Kato *et al.* [11–13] have shown for the first time that entanglement of worm-like micelles occurs in the polyoxyethylene surfactant systems based on the scaling analysis of static and dynamic light-scattering data. However, the relaxation time for polyoxyethylene micelles is generally much faster than for ionic micelles. Constantin *et al.* [14, 15] have performed high-frequency rheological measurements on a binary system of $C_{12}E_6$ and water by using the piezorheometer. In spite of the fact that this special equipment can cover the frequency range up to 6×10^4 rad s^{-1}, only the low-frequency part of the viscoelastic relaxation spectrum is observed except for the particular concentration and temperature where the crossing of the storage modulus (G') and the loss modulus (G'') is observed at about 10^4 rad s^{-1}, which is about three-to-five orders of magnitude higher than for so-called viscoelastic worm-like micelles. Such fast relaxation time makes it difficult to study dynamical properties of worm-like micelles for polyoxyethylene surfactants by using a conventional rheometer.

More than 10 years ago, Kato *et al.* measured dynamic light scattering in a binary system of $C_{16}E_7$ and D_2O paying attention to the slow mode that gives stress relaxation time [16–18]. The observed relaxation time is 10^{-4}–10^{-1} s, depending on the concentration and temperature. These values are still shorter than in ionic systems, but much longer than in the $C_{12}E_6$ system. It is well known that the rate constant for the monomer/micelle exchange processes decreases by about one order of magnitude as the number of carbon atoms increases by two [19]. So, the longer relaxation time for the $C_{16}E_7$ system than the $C_{12}E_6$ system may be due to the increment of the carbon number from 12 to 16. Acharya and Kunieda have shown that addition of short oxyethylene chain polyoxyethylene dodecyl ether ($C_{12}E_m$, $m = 1$–4) surfactants to a dilute solution of polyoxyethylenecholesteryl ether ($ChE_{m'}$, $m' = 10$ and 15) induces formation of worm-like micelles with relaxation times longer than 1 s [20]. They explain these results in terms of the average section area per surfactant in a micelle that decreases upon the addition of short-EO-chain $C_{12}EO_m$. However, such a long relaxation time is probably due to the large number of carbon atoms in the cholesteric group.

In the present study, we have performed rheological measurements on the $C_{16}E_7/D_2O$ system in addition to static and dynamic light scattering. Combining these results with the relaxation time obtained from the slow mode of dynamic light scattering [16–18] and the surfactant self-diffusion coefficient [21–23] reported previously, dynamics of worm-like micelles is discussed in more detail than in our previous study.

7.2
Experimental

$C_{16}E_7$ was purchased from Nikko Chemicals, Inc. in crystalline form (>98%) and used without further purification. Deuterium oxide purchased from ISOTEC, Inc.

(99.9%) was used as a solvent after being degassed by bubbling of nitrogen to avoid oxidation of the ethylene oxide group of surfactants.

Static and dynamic light scattering was measured with a monomode-fiber compact goniometer system (ALV) using the 488-nm line of an argon-ion laser (Spectra Physics, Stabilite 2016) for the scattering angle 30–130°. The solutions were filtered through Syrfil-MF (0.22 μm pore size, Nuclepore), directly into a cylindrical cell of 10 mm diameter made of quartz.

Rheological measurements were performed by using a rheometer MCR 300 (Anton Paar, Germany) with a Couette Cell.

7.3 Results

7.3.1 Light Scattering and Surfactant Self-Diffusion Coefficients

Figure 7.1 shows the concentration dependence of the apparent aggregation number of micelles at different temperatures obtained from the following equations

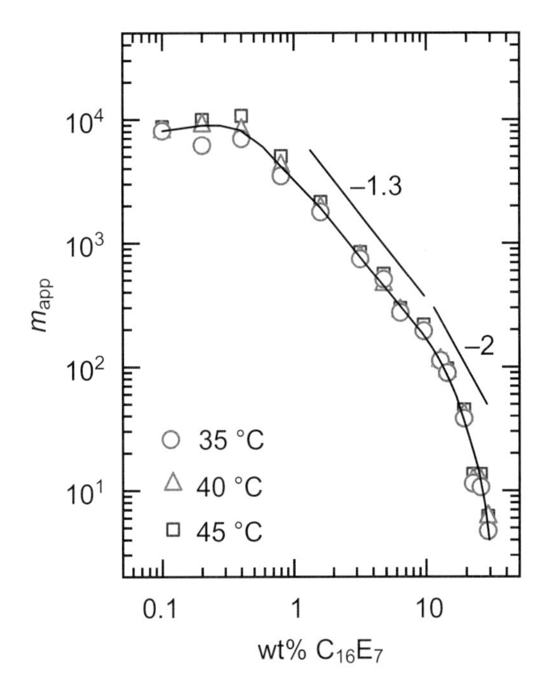

Figure 7.1 Concentration dependence of the apparent aggregation number at different temperatures obtained from light-scattering intensity (see Equation 7.1) for $C_{16}E_7/D_2O$ system. The solid lines indicate the slopes predicted by the scaling theory for semidilute solutions of flexible polymers in good solvent (−1.3) and θ solvent (−2).

$$m_{\mathrm{app}} = \frac{R(0) - R_0(0)}{M_1 H(c - c_0)} \tag{7.1}$$

$$H \equiv \frac{4\pi^2 n^2}{\lambda^4 N_{\mathrm{A}}} \left(\frac{\partial n}{\partial c} \right)_{T,P}^2 \tag{7.2}$$

where c is the weight concentration (g cm^{-3}) of surfactant, c_0 is the CMC in g cm^{-3}, $R(0)$ and $R_0(0)$ are the Rayleigh ratios at zero-scattering vector at c and c_0, respectively, λ is the wave length in vacuum, n is the refractive index of solution, and N_{A} is the Avogadro's number. In polyoxyethylene surfactant/water systems, micelles grow with increasing temperature when the temperature exceeds a certain value, T_{S}, depending on the concentration. At around 1 wt%, for example, $T_{\mathrm{S}} \cong T_{\mathrm{c}} - 30\,°\mathrm{C}$, where T_{c} is the lower critical solution temperature. For the $C_{16}E_7/D_2O$ system, T_{c} is 51.0 ± 0.5 °C [13] and so micellar growth occurs above about 20 °C at around 1 wt%. In the temperature range studied (above 35 °C), therefore, micelles are already long enough to entangle.

When intermicellar interactions can be neglected, the apparent aggregation number is equal to the weight averaged aggregation number and so m_{app} increases with increasing concentration. When worm-like micelles begin to entangle, m_{app} takes a maximum and then decreases with further increase in the concentration. During 10 years after 1985, applications of the scaling theory for entangled polymers to semidilute solutions of surfactants had been reported [1]. According to the scaling theory, the osmotic compressibility, $\partial\pi/\partial c$, and the correlation length of concentration fluctuations, ξ, follow the following power laws above the overlap concentration c^*

$$\frac{\partial \pi}{\partial c} \propto c^{1/(3\nu - 1)} \tag{7.3}$$

$$\xi \propto c^{\nu/(1-3\nu)} \tag{7.4}$$

where ν is 0.5 and 0.588 for θ solvent and good solvent, respectively. The osmotic compressibility is related to the $R(0)$ and m_{app} through as

$$\frac{\partial \pi}{\partial c} \propto \frac{RTH(c - c_0)}{R(0) - R_0(0)} = \frac{RT}{m_{\mathrm{app}}} \tag{7.5}$$

Combining Equations 7.3 and 7.5, we obtain

$$m_{\mathrm{app}} \propto c^{1/(1-3\nu)} \tag{7.6}$$

Figure 7.1 includes the slopes predicted from Equation 7.6 for θ solvent (−2.0) and good solvent (−1.3). The observed slope in the concentration range of 1–10 wt% is close to the theoretical prediction for good solvent ($c^* \cong 0.3$ wt%). When the concentration exceeds about 10 wt%, the slope of the log–log plots becomes steeper. Similar results have been obtained in our previous study on a $C_{12}E_5$ system [12].

Figure 7.2 shows concentration dependences of the collective diffusion coefficient (D_{C}) and the self-diffusion coefficient (D_{S}) [21–23] obtained from the dynamic

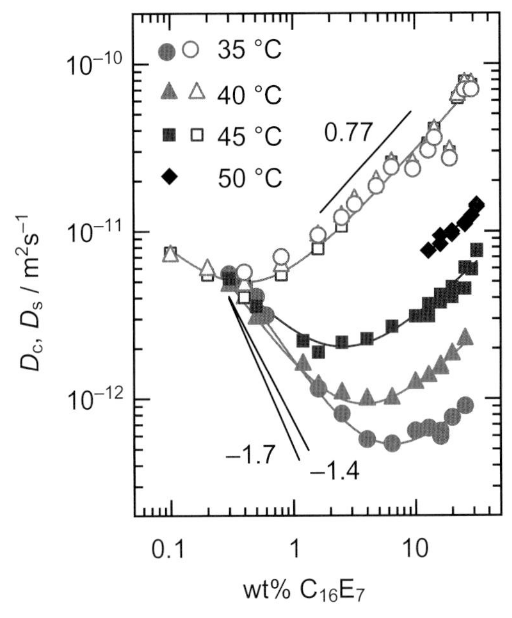

Figure 7.2 Concentration dependences of the collective diffusion coefficient (D_c) obtained from dynamic light scattering (open symbols) and surfactant self-diffusion coefficient (D_S) obtained from the pulsed-gradient spin-echo (closed symbols, from Ref. [21]) for $C_{16}E_7/D_2O$ system at different temperatures. The upper solid line indicates the slope predicted by the scaling theory for semidilute solutions of flexible polymers in good solvent. The lower solid lines indicate the slope predicted by Cates [24] for the reversible scission (−1.7) and bond-interchange (−1.4) mechanisms.

light scattering and the pulsed-gradient spin echo, respectively. It can be seen from the figure that D_C takes a minimum at c^* and increases with increasing concentration. The collective diffusion coefficient can be expressed in terms of the correlation length ξ as

$$D_C = \frac{kT}{6\pi\eta_s\xi} \tag{7.7}$$

where k is the Boltzmann Constant and η_s is the viscosity of the solvent. Substitution of Equation 7.4 into Equation 7.7 gives the following power law

$$D_C \propto c^{\nu/(3\nu-1)} \tag{7.8}$$

The straight line in Figure 7.2 indicates the slope 0.77 for good solvent. Again good agreement is obtained between observed and theoretical slope in the concentration range 1–10 wt%.

On the other hand, the self-diffusion coefficient takes a minimum at the concentration much higher than c^*. Scaling argument of the self-diffusion coefficient in the entangled micelles was made by Cates [24] for the first time. He proposed a theoretical model ("living polymers") taking into account chain breakage and recombination [1, 24, 25]. In this theory, the survival time of a chain before it breaks into two pieces, τ_{break}, is introduced in addition to the classical reptation time, τ_{rep}. In the case where $\tau_{break} \ll \tau_{rep}$, the monomer diffusion coefficient follows the power law

$$D_S \propto c^{-2\alpha/3-1/(3\nu-1)} = c^{-1.7} \quad (\text{for } \nu = 0.588 \text{ and } \alpha = 0.6) \tag{7.9}$$

In addition to this "reversible scission" mechanism, Turner and Cates [26] proposed "end-interchange" and "bond-interchange" mechanisms. They have calculated the exponent for D_S to be −1.6 and −1.4 for end-interchange and bond-interchange, respectively. In Figure 7.2, the slopes for these models are indicated by the straight lines. The observed self-diffusion coefficient decreases with concentration as expected, but the slope is gentler than these predictions. Moreover, as the concentration increases further, the diffusion coefficient increases after taking a minimum. Such a self-diffusion behavior is completely different from the prediction of Cates. Similar results have been obtained for $C_{14}E_6$ and $C_{14}E_7$ systems [21].

7.3.2 Rheological Properties

Figures 7.3a and b show shear-rate dependences of the viscosity at 35 °C and 40 °C, respectively, for different concentrations. The viscosity is almost constant in the shear rate range up to 10–100 s^{-1}, depending on the concentration and temperature. So, the viscosity at 1 s^{-1} is regarded as the zero-shear viscosity and plotted against the concentration in Figures 7.4 and 7.5. The data in the lower concentration range in Figure 7.4 have been already reported before [21].

According to the Cates's theory, the terminal time for stress relaxation (τ_R) and the zero-shear viscosity (η_0) follows the scaling relation [24]

$$\tau_R \propto c^{\alpha+(3\nu-3)/[2(1-3\nu)]} = c^{1.4} \tag{7.10}$$

$$\eta_0 \propto c^{\alpha-(3+3\nu)/[2(1-3\nu)]} = c^{3.7} \tag{7.11}$$

where the numerical values for the exponents are calculated for ν = 0.588 and α = 0.6. It should be noted that the plateau modulus G'_∞ is assumed to be proportional to the density of the blobs and so follow the power law

$$G'_\infty \propto \xi^{-3} \propto c^{3\nu/(3\nu-1)} = c^{2.3} \tag{7.12}$$

The predicted slopes are shown in Figures 7.4 and 7.5. It can be seen from these figures that the observed slopes at 25 and 35 °C are close to the prediction of Equation 7.11. At higher temperatures, however, the observed slope decreases with increasing temperature and reaches about unity at 50 °C.

Figure 7.6 shows viscoelastic spectra at 35 °C for different concentrations. This figure demonstrates that the crossing of the storage modulus (G') and the loss modulus (G'') is observed at the frequencies 20–100 rad s^{-1}, which is much lower

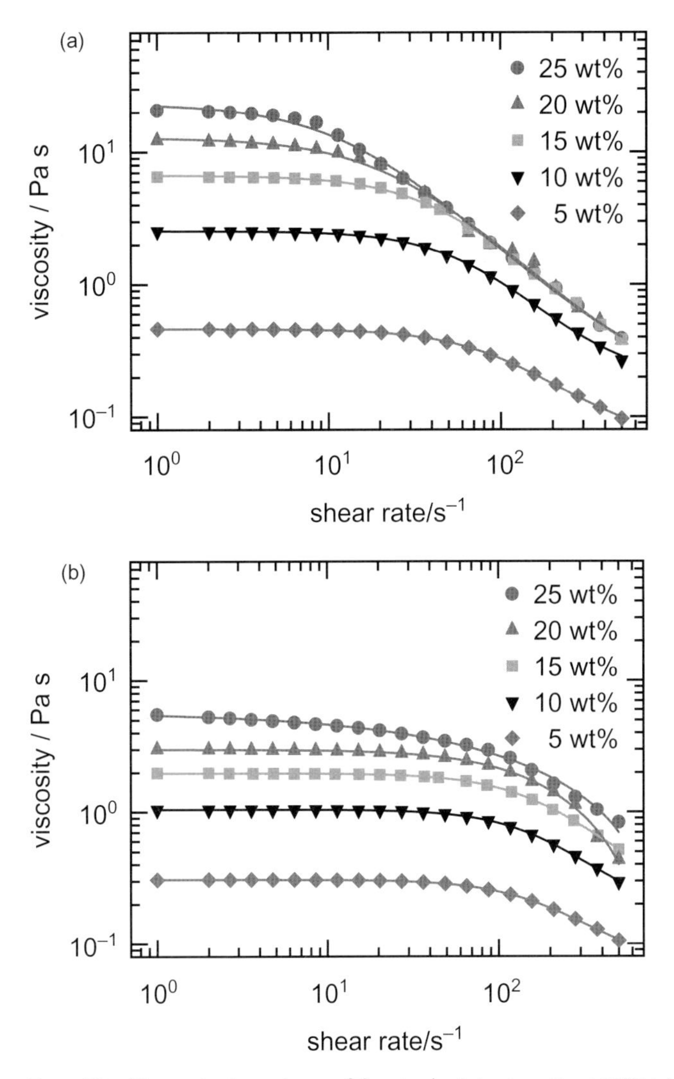

Figure 7.3 Shear-rate dependence of the steady-state viscosity at 35 °C (a) and 40 °C (b) at different concentrations for $C_{16}E_7/D_2O$ system.

than those reported for the $C_{12}E_6$ system [14, 15]. The solid lines in these figures indicate fits to the Maxwell model

$$G' = \frac{G'_{\infty}\omega^2\tau_R^2}{1+\omega^2\tau_R^2} \tag{7.13}$$

$$G'' = \frac{G'_{\infty}\omega\tau_R}{1+\omega^2\tau_R^2} \tag{7.14}$$

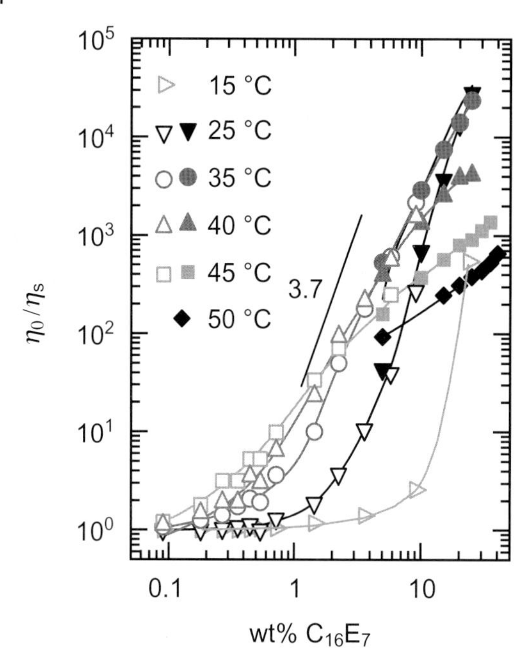

Figure 7.4 Concentration dependence of the relative viscosity at different temperatures for $C_{16}E_7/D_2O$ system. The open symbols indicate the data reported previously [21]. The solid line indicates the slope predicted by Cates [24] for the reversible scission mechanism.

We have also fitted the shear-rate dependence of the viscosity to the following equation [27, 28]

$$\eta(\dot{\gamma}) = \eta_\infty + \frac{\eta_0 - \eta_\infty}{1 + (\dot{\gamma}\tau_v)^m} \tag{7.15}$$

where η_0 and η_∞ are the limiting viscosities at the zero and infinite shear rates, respectively, m is a constant around unity, and τ_v is the relaxation time. The solid lines in Figure 7.3 indicate the fitting results.

Figure 7.7a shows concentration dependences of the relaxation times τ_R and τ_ν thus obtained (open symbols). At 35 °C, τ_R and τ_ν increase with increasing concentration. The observed slope is close to or slightly smaller than 1.4 predicted by Cates (see Equation 7.10). At 40 °C, however, both τ_R and τ_υ decrease with increasing concentration. In this figure, we compare these data with the relaxation time obtained from the slow mode of dynamic light scattering (τ_{LS}) reported previously [16–18]. The present rheological data (τ_R and τ_v) are in agreement with the light-scattering data (τ_{LS}) within the experimental error except for the region between 10–20 wt% at 35 °C where τ_R and τ_υ monotonously increases, whereas τ_{LS} levels off. As the concentration exceeds 20 wt%, τ_{LS} increases abruptly and again good agreement is observed between rheological and light-scattering data.

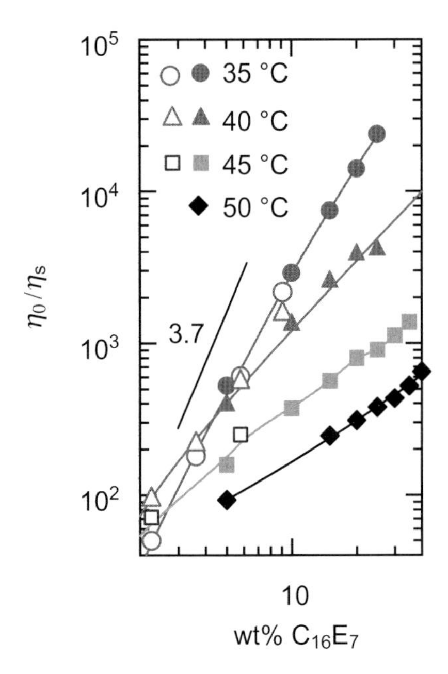

Figure 7.5 Concentration dependence of the relative viscosity at different temperatures for $C_{16}E_7/D_2O$ system (the same as closed symbols in Figure 7.4). The solid line indicates the slope predicted by Cates [24] for the reversible scission mechanism.

Figure 7.7b shows the concentration dependence of the plateau modulus obtained from the fitting of the viscoelastic spectra. We have also calculated G'_∞ from τ_{LS} and η_0 by using the relation

$$\eta_0 = G'_\infty \tau_R \tag{7.16}$$

and an assumption $\tau_R = \tau_{LS}$ (closed symbols in Figure 7.7b). Figure 7.7b demonstrates that G'_∞ does not depend on temperature very much and that good agreement is obtained between the data obtained from the viscoelastic spectra and those obtained from τ_{LS} and η_0. The slope of the plot is slightly smaller than the predicted slope 2.3 (see Equation 7.12).

7.4 Discussion

Figure 7.8 shows the phase diagram of the $C_{16}E_7/D_2O$ system reported previously [22, 29] where experimental points for the steady-state viscosity and viscoelastic spectra are shown by crosses and circles, respectively. The experimental points for the light-scattering and self-diffusion measurements are almost the same as those for the steady-state viscosity.

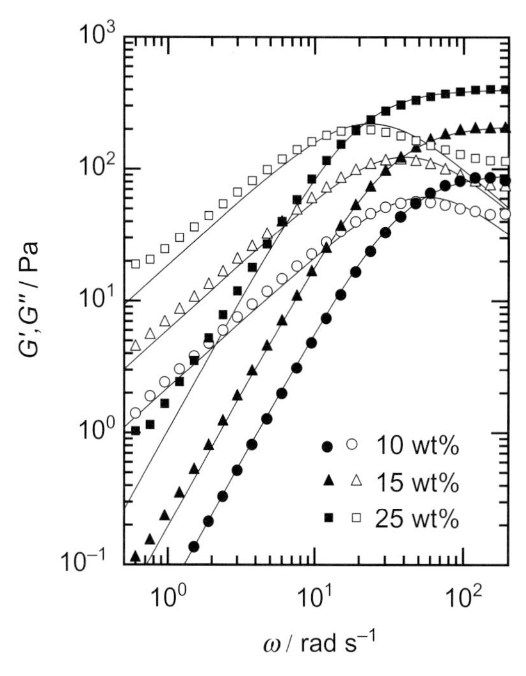

Figure 7.6 Viscoelastic spectra at 35 °C and different concentrations for $C_{16}E_7/D_2O$ system. The closed and open symbols indicate the storage modulus (G') and the loss modulus (G''), respectively. The solid lines indicate a least-square fitting to the Maxwell model (see Equations 7.13 and 7.14).

The concentration dependences of the apparent aggregation number (Figure 7.1), the cooperative diffusion coefficient (Figure 7.2), and the zero-shear viscosity (Figure 7.4) strongly suggest entanglement of worm-like micelles. The overlap concentrations are obtained to be about 0.7, 0.4, and 0.2 wt% at 35, 40, 45 °C, respectively, from the concentrations above which the zero-shear viscosity increases rapidly.

7.4.1
Rheological Properties

The slopes of the double logarithmic plots of the zero-shear viscosity *vs.* concentration at 25 °C and 35 °C may be explained by the Cates's theory if a particular concentration range is chosen. As the temperature increases up to 50 °C, however, the slope decreases to about unity, which is much smaller than his prediction (3.7). Imae *et al.* have measured the viscosity on several homologs of polyoxyethylene surfactants ($C_{12}E_7$, $C_{14}E_7$, and $C_{16}E_7$) in water with and without NaCl in the concentration range where the double-logarithmic plot gives a straight line [30]. They show that the exponent for the viscosity decreases from ~4 to ~1 with the elevation

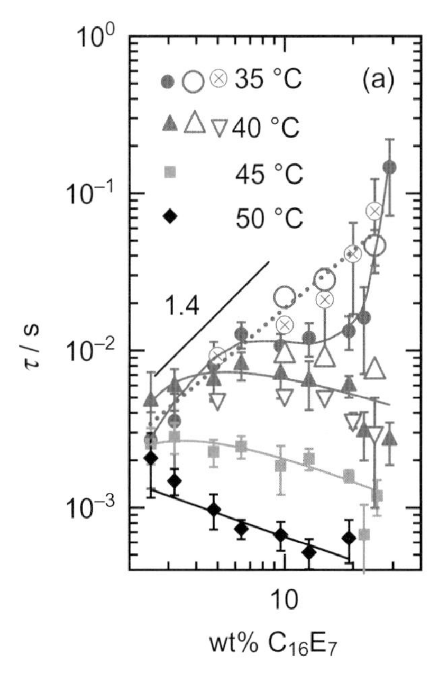

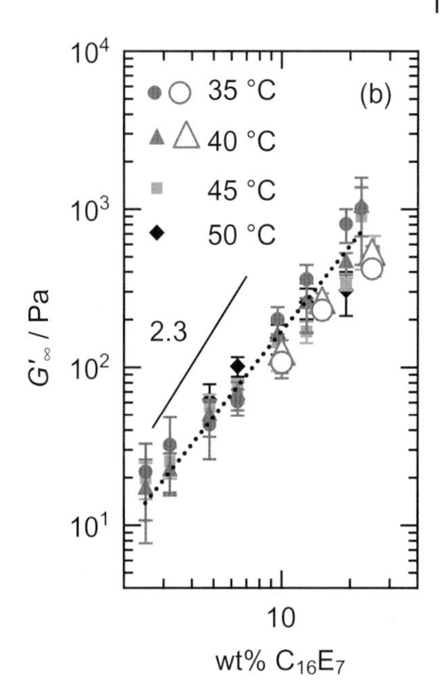

Figure 7.7 (a) Concentration dependences of the relaxation time for $C_{16}E_7/D_2O$ system obtained from the fitting of the viscoelastic spectra to the Maxwell model (τ_R, open circles and triangles), fitting of the shear rate dependence of the steady-state viscosity to Equation 7.15 (τ_ν, open circles with crosses and inverse triangles), and the slow mode of dynamic light scattering (τ_{LS}, closed symbols) reported before [16–18]. The solid line indicates the slope predicted by Cates [24] for the reversible scission mechanism. (b) Concentration dependences of the high-frequency plateau modulus for $C_{16}E_7/D_2O$ system obtained from the fitting of the viscoelastic spectra to the Maxwell model (open symbols) and from τ_{LS} and η_0 by using Equation 7.16 (closed symbols). The solid line indicates the slope for entangled polymers.

of temperature. For the $C_{16}E_7/H_2O$ system, the exponent decreases from 4.0 to 1.3 as the temperature increases from 20 °C to 45 °C.

The large deviation in the exponent for the viscosity from the prediction of the living polymer model has also been reported for other systems [2, 3]. For example, Khatory *et al.* have performed rheological measurements on cetypyridinium chlorate micelles in $NaClO_3$ brine [31]. The obtained exponents are 2 and 1 for 0.1 M and 1 M $NaClO_3$, respectively. Hoffmann has measured the viscosity on tetradecyldimethylaminoxide (C_{14}DMAO)/ decanol/water system for different ratios of C_{14}DMAO:decanol [2, 3]. He has shown that the exponent decreases from about 5.4 to 1.3 as the ratio of decanol increases from 0 to C_{14}DMAO : decanol = 5 : 1.

Roughly speaking, the plateau modulus G'_∞ depends on the number density of the entanglement points, and so the large deviation in the exponent for the viscosity from the prediction may originate from the deviation in the exponent for the stress relaxation time. It can be seen from Figure 7.7a that the slope of the double-logarithmic plot of the relaxation times τ_R and τ_υ *vs.* concentration at 35 °C is close

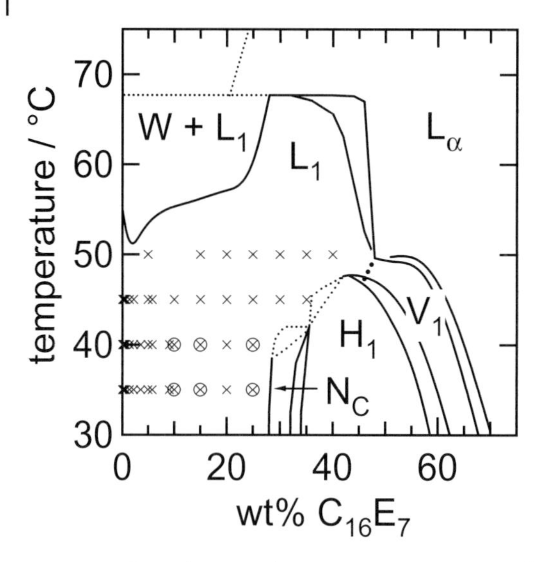

Figure 7.8 Phase diagram of $C_{16}E_7/D_2O$ system [22, 29]. The circles and crosses indicate experimental points for the viscoelastic spectra and steady-state viscosity, respectively.

to the prediction of the living polymer model (1.4). Above 40 °C, however, the slope becomes negative, which again cannot be explained by this model. According to the data of Hoffmann on the C_{14}DMAO system (C_{14}DMAO : decanol = 5 : 1), the relaxation time increases with increasing concentration up to about 50 mM [2, 3]. As the concentration increases further, the relaxation time decreases after reaching a maximum. Although it is difficult to find the concentration range where the plot gives a straight line, the exponent is about −1 in the range 200–400 mM (5–10 wt%). A similar concentration dependence in the relaxation time has been reported for Candau and Oda [32] and Koehler *et al.* [33]. In their systems, however, electrostatic interactions play an important role.

To explain the decrease in the relaxation time with increasing concentration, we need different models other than that of Cates. Shikata *et al.* measured viscoelastic spectra for CTAB micelles in the presence of sodium salicylate [34]. To explain the dependence of the relaxation time on the concentration of salicylate ions, they proposed a ghost-like crossing model. According to this model, micelles can go through like a ghost to relax the stress applied to the entangled network. On the other hand, Appel *et al.* considered that connections can slide freely to relax the stress [35]. Construction of a theory taking into account the sliding of connections was performed by Lequeux [36, 37]. He shows that the relaxation time and the viscosity are reduced by the connections. However, quantitative comparison between the theory and experiments cannot be achieved because the concentration dependence of the parameters in the theory is not known.

To the authors' knowledge, there has been reported no theory that gives a negative exponent for the relaxation time except for a rather simple theory of Hoffmann [2].

He assumes that the relaxation time is proportional to the time necessary for two arms of worm-like micelles to diffuse a distance equal to the mesh size x, that is,

$$\tau_R \propto \frac{x^2}{D} \tag{7.17}$$

where D is the diffusion coefficient of the arms. By using the relation $x \propto c^{-1/2}$, we obtain

$$\tau_R \propto \frac{1}{D \cdot c} \propto c^{-1} \tag{7.18}$$

$$\eta_0 = G'_\infty \tau_R \propto c^{2.3} c^{-1} \propto c^{1.3} \tag{7.19}$$

Hoffman also proposed another model giving the same exponent as in Equations 7.18 and 7.19 taking into account the sliding of connection. He assumes that relaxation time is determined by the time necessary for two network points to meet on their random walk along the contour length. For a saturated network the mean distance between the network points is proportional to $c^{-1/2}$. If x in Equation 7.17 is replaced by the mean distance between the network points, we obtain the same equations as Equations 7.18 and 7.19.

Equations 7.18 and 7.19 can explain the observed exponent for his system and some other systems mentioned above. Concerning our data, however, the slopes of the plots in Figure 7.7a above 40 °C is much gentler than −1. It should be noted, however, that D is assumed to be constant in Equations 7.18 and 7.19. As can be seen from our experimental results in Figure 7.2, D may decrease with increasing concentration (note that D is not always equal to D_S, see the next section). Then the exponent for the relaxation time becomes less negative than −1.

Thus, our results above 40 °C can be explained semiquantitatively by modification of Hoffmann's theory. However, it is still difficult to determine which model is more appropriate between the ghost-like crossing proposed by Shikata and the sliding of connections proposed by Appel from the stress-relaxation time alone.

The difference in the concentration dependence of the relaxation time between 35 °C and 40 °C is remarkable. This may result from the fact that a nematic phase exists above 27 wt% at 35 °C, whereas it melts at about 40 °C (see Figure 7.8).

7.4.2 Surfactant Self-Diffusion

The concentration dependence of the surfactant self-diffusion coefficient in Figure 7.2 is completely different from the prediction of Cates. Based on these data together with the dependence of the self-diffusion coefficient on diffusion time, one of the authors has proposed a diffusion model for entangled worm-like micelles [13, 21]. In this model, it is assumed that (1) a surfactant molecule diffuses in a micelle along its contour during the time τ_m and then migrates to adjacent micelles at one of the entanglement points and (2) τ_m satisfies the condition $R_g^2/D_L \ll \tau_m \ll \Delta$, where R_g is the radius of gyration of worm-like micelles,

D_L is the intramicellar (lateral) diffusion coefficient, and Δ is the diffusion time. Then, the mean-square displacement of surfactant molecules for the time much longer than τ_m can be estimated by random walks where a jump of length d is performed at each time τ_m, which gives the surfactant self-diffusion coefficient as

$$D_S = D_M + \langle d \rangle^2 / (6\tau_m) \tag{7.20}$$

where D_M is the self-diffusion coefficient of micelles, $\langle d \rangle$ is the mean distance between the centers of mass of adjacent micelles.

Figures 7.2 and 7.7 demonstrate strong correlation between D_S and τ_{LS} in the concentration range of 2–20 wt%, except for 35 °C. In fact, the product $D_S \tau_{LS}$ is almost constant in this concentration and temperature range, indicating that self-diffusion and structural relaxation are dominated by the same kinetic process. If we consider that surfactant molecules migrate to another micelle through a transient connection at an entanglement point, the formation rate of such a transient connection should dominate both self-diffusion and structural relaxation.

In other words, τ_m should be nearly equal to τ_{LS}. In our previous study, we have calculated the average spacing $\langle d \rangle$ from observed D_S and τ_{LS} by using Equation 7.20 and then compared it with that obtained from static light scattering [18]. The results indicate that the relation $\tau_m \cong \tau_{LS}$ holds at least at 40 °C and 45 °C.

Then, the concentration dependence of the self-diffusion coefficient can be explained as follows. In dilute solutions (but still much higher than the cmc where contribution from the monomer diffusion can be neglected), the self-diffusion coefficient is dominated by the first term of Equation 7.20, D_M. As the concentration increases, D_M decreases rapidly due to the entanglement of worm-like micelles, whereas the second term increases due to the decrease of τ_m ($\langle d \rangle$ depends on the concentration only slightly [13, 18, 21]), leading to the minimum in the plot of D_S *vs.* c.

From these results, we can infer that entangled micelles form a transient connection whose lifetime is of the order of 10^{-3} s, depending on concentration and temperature. In other words, our results at 40 °C and 45 °C can be explained by the ghost-like crossing proposed by Shikata, rather than the sliding of connections proposed by Appel. It should be emphasized that this conclusion is drawn by the measurements of both stress-relaxation time and surfactant self-diffusion coefficient. According to our previous self-diffusion results, however, the number of connections increases with increasing temperature above 45 °C, and so the sliding of connections may play an important role [22, 23].

At 35 °C, the concentration dependence of D_S cannot be explained in terms of τ_{LS} alone. This may be correlated with the fact that the concentration dependence of τ_{LS} is different from those of τ_R and τ_υ, which cannot be explained at present.

7.5 Summary

We have performed rheological measurements in addition to static and dynamic light scattering on the $C_{16}E_7/D_2O$ system. To the authors' knowledge, this is the first report showing the crossing of G' and G'' in the binary system of polyoxyethylene surfactant and water by using a conventional rheometer. Light-scattering data as well as zero-shear viscosity strongly suggest an entanglement of worm-like micelles. The concentration dependence of the zero-shear viscosity and the stress-relaxation time obtained from the viscoelastic spectra, shear-rate dependence of the viscosity, and the slow mode of dynamic light scattering at 35 °C can be partly explained by the living polymer model of Cates. However, the behaviors above 40 °C are completely different from his prediction. On the other hand, modification of Hoffmann's simple theory can explain these results semiquantitatively. It is demonstrated that combination of the stress relaxation time and the surfactant self-diffusion coefficient is useful to discuss which model is better between the ghost-like crossing proposed by Shikata *et al.* and the sliding of connections proposed by Appel *et al.*

References

1. Cates, M.E., and Candau, S.J. (1990) *J. Phys. Condens. Matter*, **2**, 6869 and references therein.
2. Hoffmann, H. (1994) ACS Symposium Series 578, American Chemical Society, Washington DC, p. 2.
3. Hoffmann, H. (2003) *Structure-Performance Relationships in Surfactants*, 2nd edn (eds K. Esumi and M. Ueno), Marcel Dekker, p. 433.
4. Lequeux, F., and Candau, S.J. (1994) ACS Symposium Series 578, American Chemical Society, Washington DC, p. 51.
5. Acharya, D.P., and Kunieda, H. (2006) *Adv. Colloid Interface. Sci.*, **123–126**, 401.
6. Erzahi, S., Tuval, E., and Aserin, A. (2006) *Adv. Colloid Interface. Sci.*, **128–130**, 77.
7. Dreiss, C.A. (2007) *Softmatter*, **3**, 956.
8. Degiorgio, V. (1985) *Physics of Amphiphiles: Micelles, Vesicles, and Microemulsions* (eds V. Degiorgio and M. Corti), North-Holland, Amsterdam, p. 303.
9. Meguro, K., Ueno, M., and Esumi, K. (1987) *Nonionic Surfactants. Physical Chemistry* (ed. M.J. Schick), Marcell Dekker, New York, p. 109.
10. Magid, L.J. (1987) *Nonionic Surfactans. Physical Chemistry* (ed. M.J. Schick), Marcel Dekker, New York, p. 677.
11. Kato, T., Anzai, S., and Seimiya, T. (1987) *J. Phys. Chem.*, **91**, 4655.
12. Kato, T., Anzai, S., and Seimiya, T. (1990) *J. Phys. Chem.*, **94**, 7255.
13. Kato, T., Terao, T., Tsukada, M., and Seimiya, T. (1993) *J. Phys. Chem.*, **97**, 3910.
14. Constantin, D., Palierne, J.-F., Freyssingeas, E., and Oswald, P. (2002) *Europhys. Lett.*, **58**, 236.
15. Constantin, D., Freyssingeas, E., Palierne, J.-F., and Oswald, P. (2003) *Langmuir*, **19**, 2554.
16. Kato, T. (1997) *Prog. Colloid Polym. Sci.*, **106**, 57.
17. Kato, T., Nozu, D., and Shikata, T. (1999) Slow dynamics, in *Complex Systems* (eds M. Tokuyama and I. Oppenheim), American Institute of Physics, p. 150.
18. Kato, T., and Nozu, D. (2001) *J. Mol. Liq.*, **90**, 167.
19. Aniansson, E.A.G., Wall, S.N., Almgren, M., Hoffmann, H., Klelmann, I., Ulbricht, W., Zana, R., Lang, J., and Tondre, C. (1976) *J. Phys. Chem.*, **80**, 905.
20. Acharya, D.P., and Kunieda, H. (2003) *J. Phys. Chem. B*, **107**, 10168.
21. Kato, T., Terao, T., and Seimiya, T. (1994) *Langmuir*, **10**, 4468.
22. Kato, T., Taguchi, N., Terao, T., and Seimiya, T. (1995) *Langmuir*, **11**, 4661.
23. Kato, T. (1996) *Prog. Colloid Polym. Sci.*, **100**, 15.
24. Cates, M.E. (1988) *J. Phys. France*, **49**, 1593.
25. Cates, M.E. (1987) *Macromolecules*, **20**, 2289.

26 Turner, M.S., Marques, C., and Cates, M.E. (1993) *Langmuir*, **9**, 695.
27 Cross, M.M. (1965) *J. Colloid Interface Sci.*, **20**, 417.
28 Raghavan, S.R., Kaler, E.W., and Wagner, N.J. (2003) *Langmuir*, **19**, 4079.
29 Minewaki, K., Kato, T., Yoshida, H., Imai, M., and Ito, K. (2001) *Langmuir*, **17**, 1864.
30 Imae, T., Sasaki, M., and Ikeda, S. (1989) *J. Colloid Interface Sci.*, **127**, 511.
31 Khatory, A., Kern, F., Lequeux, F., Appell, J., Porte, G., Morie, N., Ott, A., and Urbach, W. (1993) *Langmuir*, **9**, 933.
32 Candau, S.J., and Oda, R. (2001) *Colloids Surf. A*, **183–185**, 5.
33 Koehler, R.D., Raghavan, S.R., and Kaler, E.W. (2000) *J. Phys. Chem. B*, **104**, 11035.
34 Shikata, T., Hirata, H., and Kotaka, T. (1988) *Langmuir*, **4**, 354.
35 Appell, J.A., Porte, G., Khatory, A., Kern, F., and Candau, S.J. (1992) *J. Phys. France II*, **2**, 1045.
36 Lequeux, F. (1992) *Europhys. Lett.*, **19**, 675.
37 Elleuch, K., Lequeux, F., and Pfeuty, P. (1995) *J. Phys. I France*, **5**, 465.

8
Mesophase Morphologies of Silicone Block Copolymers in a Selective Solvent Studied by SAXS

Dietrich Leisner, Md. Hemayet Uddin, M. Arturo López-Quintela, Toyoko Imae, and Hironobu Kunieda

8.1
Introduction

In previous papers, short-chained linear poly(dimethylsiloxane)-*b*-poly(ethyleneoxide) diblock copolymers (PDMS-*b*-PEO) have been evaluated as nonionic surfactants suitable for the stabilization of reverse (w/o) micelles [1–7]. The main advantages of these diblock copolymers are a high segregation between the different blocks [7, 8], providing predominant amphiphilicity and the liquid-like behavior of even rather long PDMS chains at ambient temperature. The phase behavior of their mixtures with water and silicone oil, tetra(dimethylsiloxane) (D_4), has been investigated systematically by polarization microscopy and small-angle X-ray scattering (SAXS) [1–3, 6]. The whole range of classical mesophases has been found: cubic micellar I_1 and hexagonal columnar H_1 phases with water, and lamellar, and reverse micellar hexagonal H_2 and reverse cubic I_2 phases with oil.

In this chapter, the study is extended to homologues with somewhat high molecular weight and is designed to probe the influence of silicone oil (D_4) volume fraction and of the PDMS chain length at constant PEO block length and temperature. The main question to be answered is how mesophase stability responds to increased PDMS chain length, where according to Flory–Huggins theory, not only higher segregation but also slightly higher incompatibility between PDMS and D_4 can be expected [7, 9]. The geometric parameters of the morphologies (micellar radius, core–core distance and interfacial area per molecule) are evaluated by SAXS study. Another question is how sensitive the morphologies respond to the history of their generation. Shear orientation of lamellar, cylindrical, and even cubic liquid crystals (LCs) is well known but complicated [10]. Important variables are the shear rate, strain, mode (oscillatory or unidirectional), and temperature T with respect to the glass temperature T_g and the melting temperature T_m of the blocks. Generally, for nanostructure formation, copolymers with high incompatibility of blocks and not too high molecular weight (M_w) are desirable in order to achieve high segregation at fast diffusion (low glass temperature T_g). The present situation could match this requirement, since T_g(PDMS) $\ll$ T_m(PEO) ~

Self-Organized Surfactant Structures. Edited by Tharwat F. Tadros

ISBN: 978-3-527-31990-9

40 °C, enabling the crystallization of the more cohesive PEO blocks and the arrangement of the PEO core structures through movement in the liquid PDMS/oil shells.

A large-scale oriented nonequilibrium morphology, generated by the shear, can be a template for the formation of an oriented equilibrium morphology. Here, by shearing the undercooled melt, nonequilibrium PEO domains could crystallize just by the shear activation, allowing only crystals compatible with gliding layers. After cessation of the shear in the undercooled system, the nucleation rate would substantially decrease (no further mechanical activation), and the growth of the well-oriented PEO crystallites would prevail. Even after melting the PEO crystallites at 58 °C, the LC would maintain the recent orientation, although it is disturbed more and more, when T comes closer to the order–disorder transition temperature (ODT), for example, at high oil volume fraction (φ_{D4}) where the ODT decreases rapidly.

8.2 Experimental Section

Materials. PDMS-*b*-PEOs with general formula $Me_3Si\text{-}(OSiMe_2)_{m-1}\text{-}(CH_2)_3\text{-}(OCH_2CH_2)_nOH$ (abbreviated $Si_mC_3EO_n$) were delivered from Dow Corning Toray Silicone Co. Ltd., Japan. Me denotes a methyl group attached to Si, m is the total number of silicone, and n is the average number of ethylene oxide (EO) unit. The trimethylene linking unit contributes less than 2% to the copolymer volume and is compatible with the PDMS block at the investigated temperatures; therefore we classify the copolymers as AB-type block copolymers with respect to segregation behavior. Table 8.1 shows data corresponding to the copolymers as received. The polydispersity indices, M_w/M_n of poly(oxyethylene) and poly(dimethylsiloxane) blocks of copolymers were measured by gel permeation chromatography. The main impurity is the unreacted polyether, except for $Si_{25}C_3EO_{51.6}$, which contains the unreacted poly(dimethylsiloxane). The silicone oil was removed by washing the copolymer with hexane at least three times. The copolymers were then washed

Table 8.1 Data for $Si_mC_3EO_n$ copolymers as received.

m	n	Purity/wt%[a]	P. I. m[b]	P. I. n[b]	Water/wt%	M_w/Da	f_{EO}
14	51.6		1.19	1.13	0.67	3369	0.64
25	51.6	94.6	1.20	1.13	0.34	4185	0.50
52	51.6		1.07	1.13	0.06	6187	0.33

a) Excluding water content.
b) P. I. m and P. I. n stand for the polydispersity index of poly(dimethylsiloxane) and poly(oxyethylene) chains, respectively. P. I. = M_w/M_n, where M_w is the weight average molecular weight and M_n the number average molecular weight.

with methanol to remove water and unreacted polyether. After evaporation of the solvent, they were dried over P_2O_5. The remaining water content was measured by Karl-Fischer titration with a Mitsubishi CA-06 moisturemeter.

Samples for SAXS were prepared in sealed vials as reported elsewhere [2]. Two Kapton™ (polyimide) films (7.5 μm thick) were buffed in the y-direction with an ethanol-wetted tissue paper. About 100 mg of sample were molten at ~55 °C to a liquid crystal and transferred to a tense, fixed Kapton film. Before cooling to below 35 °C, another Kapton film was put over the sample and both tense films were gently rubbed against each other at ~0.5 mm/s in the y-direction for ~10 s. Within three minutes the sandwiched sample was mounted in a hot copper block with a hole for the beam, and annealed at 58 ± 1 °C for 30 min prior to the start of a 22-h exposure, during which annealing was continued.

SAXS measurements were performed with a Rigaku 3-slit pinhole camera, equipped with a Rigaku ultraX18HB X-ray source operated at 45 kV, 80 mA, and a 3 × 3 mm focus on the rotating copper target, Osmic CMF25-77.5Cu6 confocal multilayer optics, 0.5 and 0.4 mm pinholes, and a 126 × 126 mm^2 Fuji image plate (IP) detector, placed at a distance of 1000 mm from the sample. The wavelength (CuK_α) was $\lambda = 1.541$ Å. The X-ray beam passed normal to the sandwiched sample (in the z-direction), and its divergence at the detector (full width at half-maximum, FWHM) was $\Delta q = 0.034\,nm^{-1}$. Scanned 2D patterns were corrected for absorption and for background from Kapton films prior to azimuthal averaging and evaluation of the azimuthal intensity distribution of the 1st-order Bragg reflections. The azimuthal angle $\beta = 0°$ corresponds to the "axial" y-direction ($x = 0$).

8.3 Results

Lyotropic LC "binary" samples of $Si_mC_3EO_{51.6}$ and D_4 (up to 50 mass%) were prepared and aliquots was examined by microscopy. The phase behavior is shown in Figure 8.1. Two-dimensional SAXS patterns (Figure 8.2) of the samples were also recorded to explore the ordering and the mesoscopic lattice parameters in the range of the lamellar L_α, hexagonal H_2, and cubic I_2 morphologies. By azimuthal averaging of the patterns the $I(q_{xy})$ functions were obtained. The patterns do not contain information of different spacings in the z-direction which is always the direction normal to the film support. Thus, $I(q_{xy})$ may lack some reflections of the powder averaged $I(q)$ function, if the anisotropy of the sample orientation is high. Keeping this in mind, we use it as well for indexing the morphology and especially to obtain the lattice parameters from Gaussian fits to the deconvoluted Bragg peaks. Finally, the relations derived from the geometric arrangement of the indexed morphology (Table 8.2), the known volume ratios of the segregated domains (Table 8.3), and the known molecular volume of the PEO domain (3.46 nm^3/molecule) [2] were applied to calculate the micellar cross section and the interfacial area per molecule in the supramolecular arrangement.

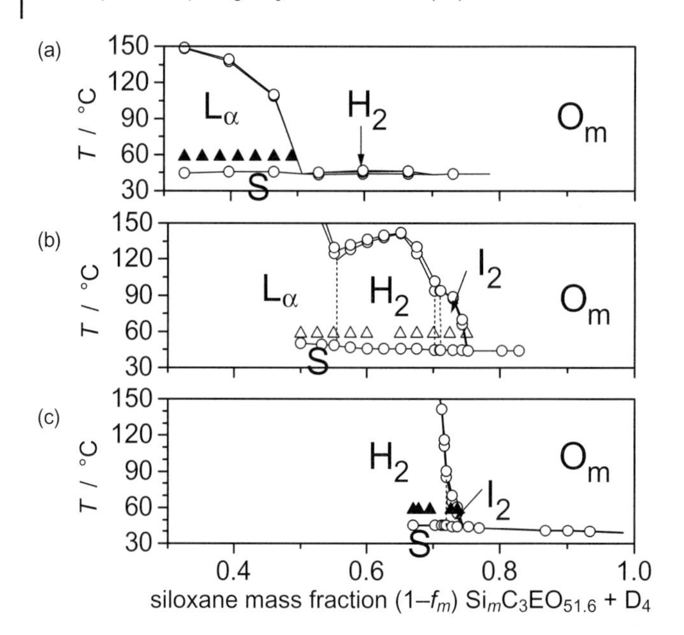

Figure 8.1 Phase diagram of $Si_mC_3E_{51.6}/D_4$ binary mixtures, as a function of the total mass fraction of siloxane (PDMS blocks and D_4), in terms of the PDMS block length (m). Triangles denote the compositions of SAXS samples measured here, whereas circles denote binodal points obtained by polarization microscopy.

In the following the patterns in Figure 8.2 are described and analyzed in order. With $m = 14$ ($f_{EO} = 0.64$), the only LC phase observed was L_α. The shown inner ring in the SAXS pattern, with higher intensities in the direction with a larger q_x component, corresponds to the [100] Bragg reflections of one-dimensional lamellar stacks, which are preferentially oriented with their normal in the x-direction. [200] and [300] (in some cases) reflections were observed but lay outside the q_{xy} range as displayed in Figure 8.2. The third image in Figure 8.2a is a very weak [100] Bragg reflection ring. Thus, an additional weak ring of diffuse scattering becomes apparent (with the denser gray scale). It is obvious that the ring is a maximum of the form factor of a single micelle. Figure 8.3a shows the azimuthally averaged scattering intensity $I(q_{xy})$ for the patterns of the lamellar samples. Gaussian fits of the Bragg peak yield the q_{100} values from which the repeat distances $d = 2\pi/q_{100}$ are calculated and displayed in Figure 8.3b. The anisotropy of $I(q_{100})$, shown in Figure 8.3c, has been used to calculate the nematic order parameter S [11],

$$S = \frac{\int I(\beta)\cdot\left[3\cos^2(\beta - \langle\beta\rangle) - 1\right]d\beta}{2\int I(\beta)\cdot d\beta} \tag{8.1}$$

where $\langle\beta\rangle$ is the average of the $I(\beta)$ weighted azimuthal angle β. As shown in Figure 8.3d, S decreases towards the edge of the stability range of the L_α phase, where the ODT also decreases towards the actual temperature (see Figure 8.1).

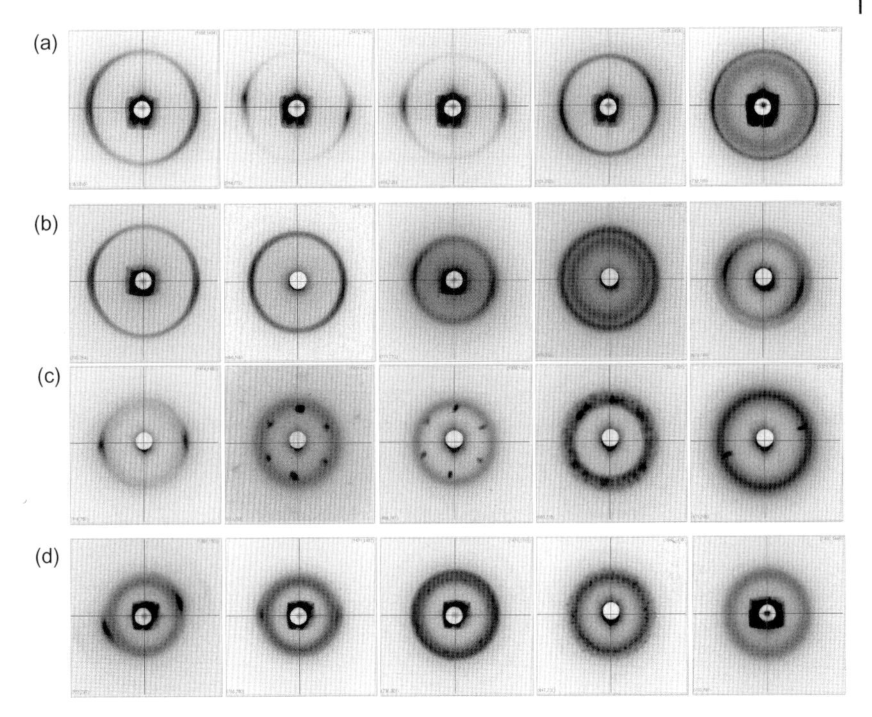

Figure 8.2 2D-patterns in the q_x, q_y range $\pm$ 0.71 nm^{-1} (2θ range $\pm$ 1°) of oriented morphologies in the system $Si_mC_3E_{51.6}/D_4$ at 58 °C, with the central structure featuring parasitic scattering (limited by the 3rd slit), beam-stop and attenuated direct beam image. Row (a), $m = 14$: 5 typical L_α samples with $\phi_{D4} = 0$, 0.04, 0.12, 0.20, and 0.24. Row (b), $m = 25$: 5 samples around the L_α–H_2 transition with $\phi_{D4} = 0$, 0.05, 0.10, 0.15, and 0.20. Row (c), $m = 25$: 5 samples around the H_2–I_2–O_m transition with $\phi_{D4} = 0.25$, 0.35, 0.40, 0.45, and 0.50. Row (d), $m = 51.6$: 5 samples around the H_2–I_2–O_m transition with $\phi_{D4} = 0$, 0.075, 0.17, 0.20, and 0.50.

For the balanced copolymer ($m = 25$, $f_{EO} = 0.5$), lamellar (L_α), hexagonal (H_2), and cubic (I_2) morphologies were distinguished. The results for the L_α phase are displayed in Figure 8.4. As with $m = 14$, [100], [200] and [300] (in some cases) Bragg reflections of one-dimensional lamellar stacks preferentially oriented with normal in the x-direction were observed. However, S appears to be much lower, indicating less perfect ordering, which is possibly due to the wider liquid-like PDMS/D_4 layer.

The additional broad peak is attributable to the form factor maximum of the lamellar stacks consisting of two wide layers with similar electron densities: 0.52 (PEO block) and 0.60 (PDMS block) mol electrons/mL, separated by one layer of the C_3 middle block (only 1.8% of volume but considerably less electron density: 0.37 mol e$^-$/mL). Thus, the middle block layer provides the highest X-ray scattering contrast and the highest contribution to the total scattering. The cylinder form factor maximum almost superimposes the 1st order Bragg reflections of the inter-micellar structure factor and thus lowers the order-parameter estimate obtained from the anisotropy of the total scattering in that q_{xy} range.

Table 8.2 Calculation of ratios R_{core}/d as a function of the PEO core volume fraction, $\phi_{EO} = (1 - \phi_{D4}) \times f_{EO}$, for the classical mesophases.

	a_x/d	a_y/d	a_z/d	V_{xyz}/d^3	n_{mic}/V_{xyz}	V_{mic}/d^3	R_{core}/d	$n_{neighbors}$
fcc	$2^{1/2}$	$2^{1/2}$	$2^{1/2}$	$8^{1/2}$	4	$(1/2)^{1/2}$	$(3\phi_{EO}/2\pi)^{1/3}/2^{1/2}$	12 (3D)
hcp	1	$3^{1/2}$	$(8/3)^{1/2}$	$8^{1/2}$	4	$(1/2)^{1/2}$	$(3\phi_{EO}/2\pi)^{1/3}/2^{1/2}$	12 (3D)
bcc	$(4/3)^{1/2}$	$(4/3)^{1/2}$	$(4/3)^{1/2}$	$(4/3)^{3/2}$	2	$(16/27)^{1/2}$	$(3\phi_{EO}/\pi)^{1/3}/3^{1/2}$	8 (3D)
P6mm	1	$3^{1/2}$	1	$3^{1/2}$	2	$(3/4)^{1/2}$	$(3^{1/2}\phi_{EO}/2\pi)^{1/2}$	6 (2D)
L_α	1	1	1	1	1	1	$\phi_{EO}/2$	2 (1D)

Table 8.3 Calculated ratios R_{core}/d as a function of the oil volume fraction and PEO volume fraction in the copolymer, f_{EO}, for the classical micellar LC mesophases.

f_{EO}	ϕ_{D4}	ϕ_{EO}	R_{core}/d			
			fcc, hcp	bcc	P6mm	L_α
0.5	0.2	0.40	0.41	0.42	0.33	0.20
0.5	0.3	0.35	0.39	0.40	0.31	0.18
0.5	0.35	0.33	0.38	0.39	0.30	0.16
0.5	0.4	0.30	0.37	0.38	0.29	0.15
0.5	0.45	0.28	0.36	0.37	0.28	0.14
0.5	0.5	0.25	0.35	0.36	0.26	0.13
0.33	0	0.33	0.38	0.39	0.30	0.17
0.33	0.025	0.32	0.38	0.39	0.30	0.16
0.33	0.075	0.31	0.37	0.38	0.29	0.15
0.33	0.17	0.27	0.36	0.37	0.27	0.14
0.33	0.2	0.26	0.35	0.36	0.27	0.13

The results for the H_2 phase are displayed in Figure 8.5 and in the first part of Table 8.4. The anisotropy of the 1st-order maximum shows high-order parameters ($S = 0.44$ resp. 0.56, not shown). Higher-order maxima were visible in the direction of the nematic director but less clear in the azimuthally averaged data.

The averaged profile is reminiscent of a form factor for a cylinder coated with shell contrast that arises from the thin C_3 layer with low electron density on top of the PEO cylinder core. Figure 8.6 shows a unique result: the observed Bragg spots from several samples containing 35 or 40% D_4 display a long-range ordered structure. Some reflections that usually are observed in a hexagonal primitive P_3/mm lattice are systematically absent, whereas other reflexes up to 3rd order are observed. However, for the electron-density distribution sketched in Figure 8.6b (−1/2 times the excess scattering length on Wyckoff positions 2c than on Wyckoff positions 1a of a hexagonal P_3/mm lattice), the Fourier transform (Figure 8.6c) yields the observed pattern. The long axis of the unit cell depicted in Figure 8.6b

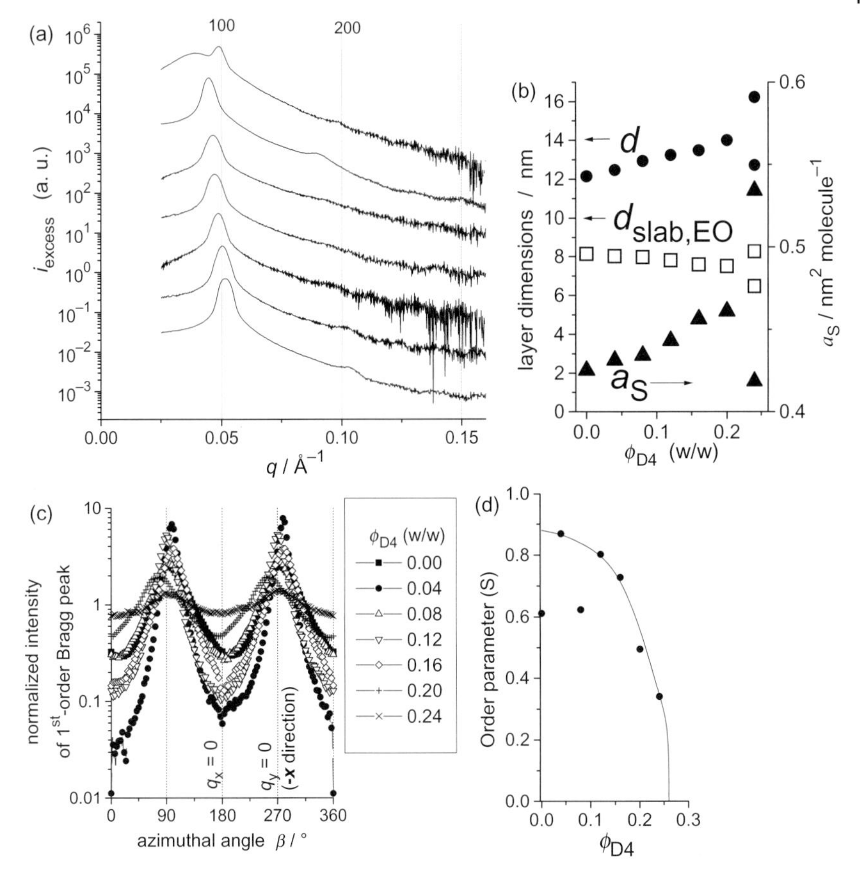

Figure 8.3 SAXS analysis of oriented lamellar morphologies in the system $Si_{52}C_3E_{51.6}/D_4$ at 58 °C: (a) Azimuthally averaged scattering intensity, after background correction: from bottom to top: $\phi_{D4} = 0$, 0.04, 0.08, 0.12, 0.16, 0.20, 0.24, scaling factor 10 with respect to former; (b) d spacings and derived parameters d_{slab} and a_S from Gauss fits to [100] and clear [200] peaks; (c) the anisotropy of the intensity distribution in the range of the 1st-order maximum; and (d) the derived nematic order parameter.

is reproducibly aligned with the shear direction, whereas the cylinders stand upright on the polyimide film support. The hexatic long-range order is obviously introduced by the unidirectional shear.

The scattering from the samples with cubic I_2 morphology is analyzed in Figure 8.7. Comparison of the observed Bragg peaks of the spotty patterns (Figure 8.2c4) with structure factors from Fourier synthesis (Figure 8.7b) [6] suggests that the structure is fcc, or eventually the $Fd\bar{3}m$ structure suggested by Charvolin *et al.* [4, 12] rather than bcc. The spottiness suggests a relative large grain size (of several micrometers).

Whereas only the results derived for the cylindrical and cubic micelles are entered in Table 8.4, Figure 8.8 summarizes all the quantitative SAXS results for

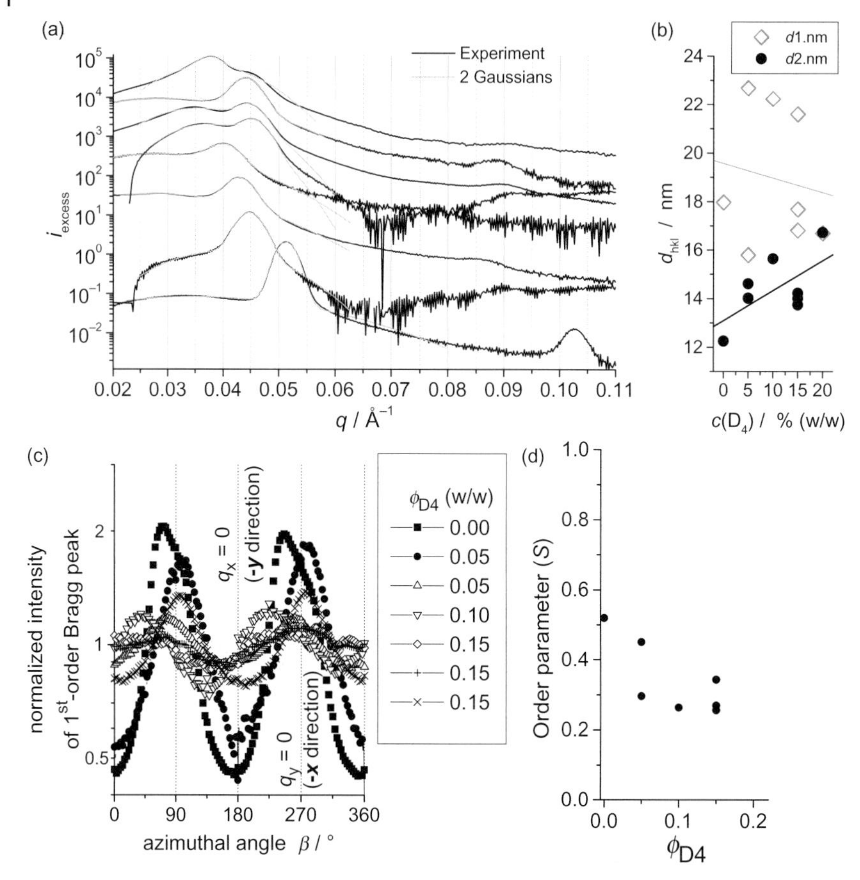

Figure 8.4 SAXS patterns of oriented lamellar morphologies in the system $Si_{52}C_3E_{51.6}/D_4$ at 58 °C: (a) Azimuthally averaged scattering intensity, after background correction: from bottom to top: $\phi_{D4} = 0$, 0.05, 0.05, 0.10, 0.16, 0.15, 0.15, 0.15, and 0.20, scaling factor ~5 with respect to former; (b) d spacings (and derived parameters d_{slab} and a_S) from Gauss fits to [100] and form-factor peaks; (c) the anisotropy of the intensity distribution in the range of the 1st-order maximum; and (d) the minimum estimate of the derived nematic order parameter.

the $Si_{25}C_3EO_{51.6}/D_4$ system. The distance between centers of the PEO cores remains at 15 ± 2 nm, independent of the D_4 concentration. Whereas this distance equals the repeat distance d in the L_α phase, the translational repeat distance in the H_2 and I_2 phases becomes 28 ± 4 nm. On the other hand, the half-thickness, resp., radius of the PEO core, d_{PEO}, increases from 6 nm in the L_α phase to 8.5 nm in the H_2 phase and to 10.5 nm in the cubic fcc phase. With increasing D_4 concentration within a phase, d_{PEO} decreases down to 25%. The parameter that changes almost linearly with the D_4 concentration is the interfacial area a_S. It increases from 0.5 nm^2/molecule in the lamellar phase to 0.8 nm^2/molecule in the H_2 phase and to 1.1 nm^2/molecule in the cubic fcc phase.

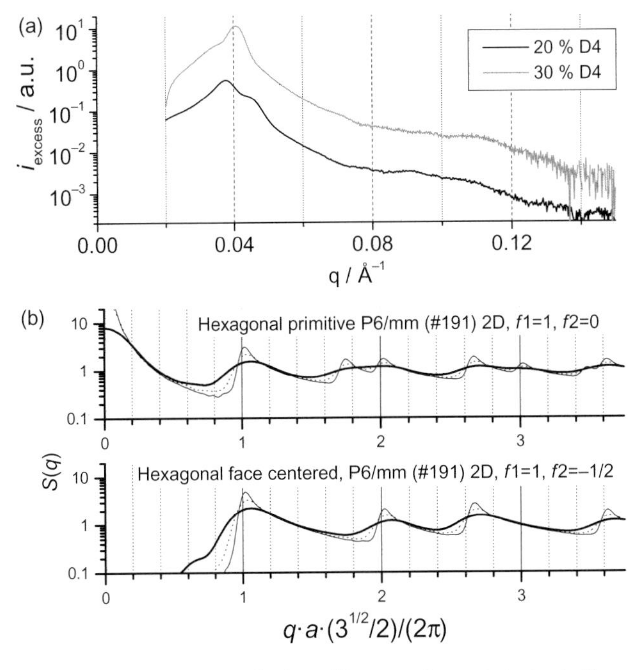

Figure 8.5 SAXS patterns of oriented hexagonal morphologies in the system $Si_{52}C_3E_{51.6}/D_4$ at 58 °C: (a) Azimuthally averaged scattering intensity, after background correction for $\phi_{D4} = 0.20$ and 0.30 (with factor 10); (b) modeled powder structure factors (2nd more likely).

Table 8.4 Calculated *d* spacings from measured and indexed Bragg peaks for hexagonal and cubic $Si_mC_3EO_{51.6}/D_4$ mixtures at 58 °C, and derived micellar core radius R_{core} and interfacial area per molecule, a_S.

f_{EO}	ϕ_{D4}	ϕ_{EO}	q_{hkl}/nm^{-1}	indexed	*hkl*	*d*/nm	R_{core}/nm	a_S/nm^2/ molecule
0.5	0.2	0.40	0.46	P6mm	100	27.4	9.1	0.76
(m = 25)	0.3	0.35	0.46	P6mm	100	27.1	8.4	0.82
	0.35	0.33	0.45	P6mm	100	27.8	8.3	0.83
	0.4	0.30	0.37	FCC	111	29.5	10.9	0.95
	0.45	0.28	0.38	FCC	111	29.0	10.4	1.00
	0.5	0.25	0.38	FCC	111	28.5	9.9	1.05
0.33	0	0.33	0.65	P6mm	200	19.3	5.8	1.19
(m = 52)	0.025	0.32	0.67	P6mm	200	18.9	5.6	1.23
	0.075	0.31	0.35	P6mm	100	18.2	5.3	1.31
	0.17	0.27	0.35	FCC	111	22.0	7.9	1.32
	0.2	0.26	0.35	FCC/BCC	111/110	22.0	7.8	1.33

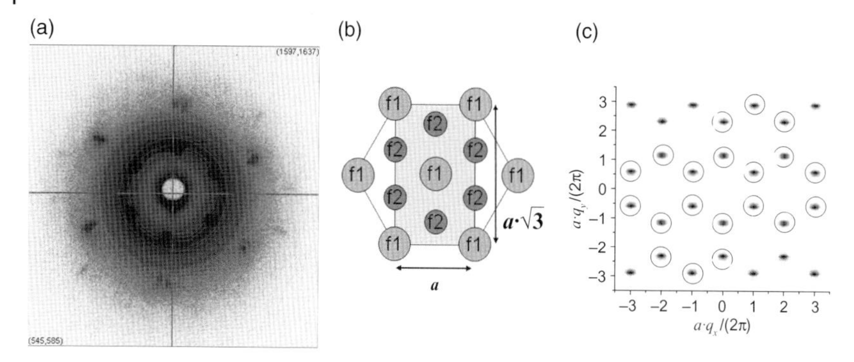

Figure 8.6 Interpretation of the hexagonal pattern observed in the H_2 range of $Si_{25}C_3E_{51.6}/D_4$ morphologies. (a) Observed 2D SAXS pattern for $\phi_{D4} = 0.35$ (inner structure contrast manipulated to show higher order Bragg spots as well); (b) scheme of possible honeycomb layer cross section where the electron density fluctuations located in centers of circles are $f_1 = -2f_2$; (c) Fourier transform of a stack of 5×3 unit cells of (b) with beam normal to the layer (along the cylinder axis), circled spots matching the observed Bragg reflections in (a).

The lyotropic LC phases of the copolymer with the longer PDMS block ($m = 52$, $f_{EO} = 0.33$) and D_4 exhibited only the H_2 and I_2 morphologies. The pattern (Figure 8.2d) analysis is shown in Figure 8.9. The cylinders in the H_2 phase show (again) a strong alignment with the shear axis (nematic order parameter $S \geq 0.44$ to 0.48) or stand upright on the support film at 7.5% D_4 concentration. The isotropic samples at higher D_4 concentrations can most likely be indexed as cubic fcc samples because of the positions of the first two structure factor minima and of the two rings of Bragg spots at a q ratio close to $(4/3)^{0.5}$ that make up the first broad peak in the azimuthally integrated scattering curves (compare with the calculated structure factors in Figure 8.7b). However, the structure could eventually be of the Charvolin $Fd\bar{3}m$ type [6], which would match perfectly, if the $q_{hkl}^{0.5}$ ratios of (weak) and intense Bragg peaks are (3) : (4) : 8 : 11 : (19) : 24 : 27 : 32 instead of 3 : 4 : 8 : 11 : (12) : (16) : 19 : 20 for the fcc structure.

8.4 Discussion

From Figures 8.3b and 8.8 and Table 8.4, the micelle radii, the core–core distance and the interfacial area per molecule can be compared for the three copolymers as a function of the total volume fraction of silicone and of the silicone block length. Molecular modeling shows that the contour length of the PEO block ($EO_{51.6}$) is about 19 nm, while the contour lengths of the PDMS blocks are about 5, 8, and 16 nm for the extended Si_{14}, Si_{25} and Si_{50} blocks, respectively. Thus, the cross section of the PDMS chain is more than three times larger than that of the PEO chain. With the molecular volumes 1.89 nm^3 for Si_{14}, resp., 3.29 nm^3 for Si_{25}

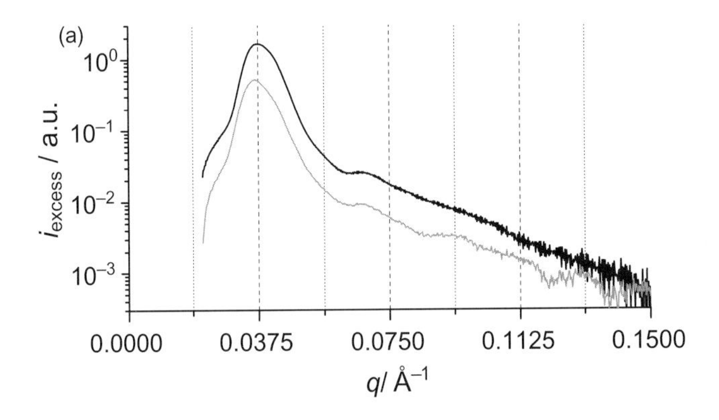

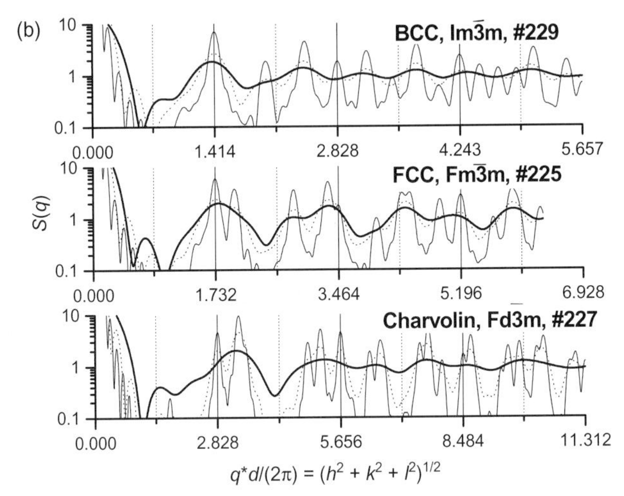

Figure 8.7 SAXS patterns of cubic or morphologies in the system $Si_{52}C_3E_{51.6}/D_4$ at 58 °C (a) Azimuthally averaged scattering intensity, after background correction: two samples with $\phi_{D4} = 0.40$; (b) modeled structure factors for possible structures (merit: bcc ≪ Charvolin < fcc).

[4] it can be calculated to be 0.40 nm^2. Indeed the measured lowest a_S in the lamellar phase of $Si_{14}EO_{51.6}$ was 0.42 nm^2, which would suggest an almost crystalline arrangement of the siloxane chains in the corresponding lamellae. However, as soon as D_4 is added a_S increases, indicating that D_4 enters into the palisade layer of PDMS chains. For the longer block homolog Si_{25}, a_S starts already at >0.5 nm^2 in the lamellar phase without D_4, indicating more liquid-like behavior of the chains. a_S does even increase to 1.33 nm^2 for the longer Si_{50} chain with oil close to the ODT from the cubic fcc phase to the oil-dissolved single micelles (O_m); obviously the longer chain still provides enough cohesion in spite of its liquid-like behavior. However, the probably partly crystallized state of the micellar PEO core should be the important factor here to support this cohesion.

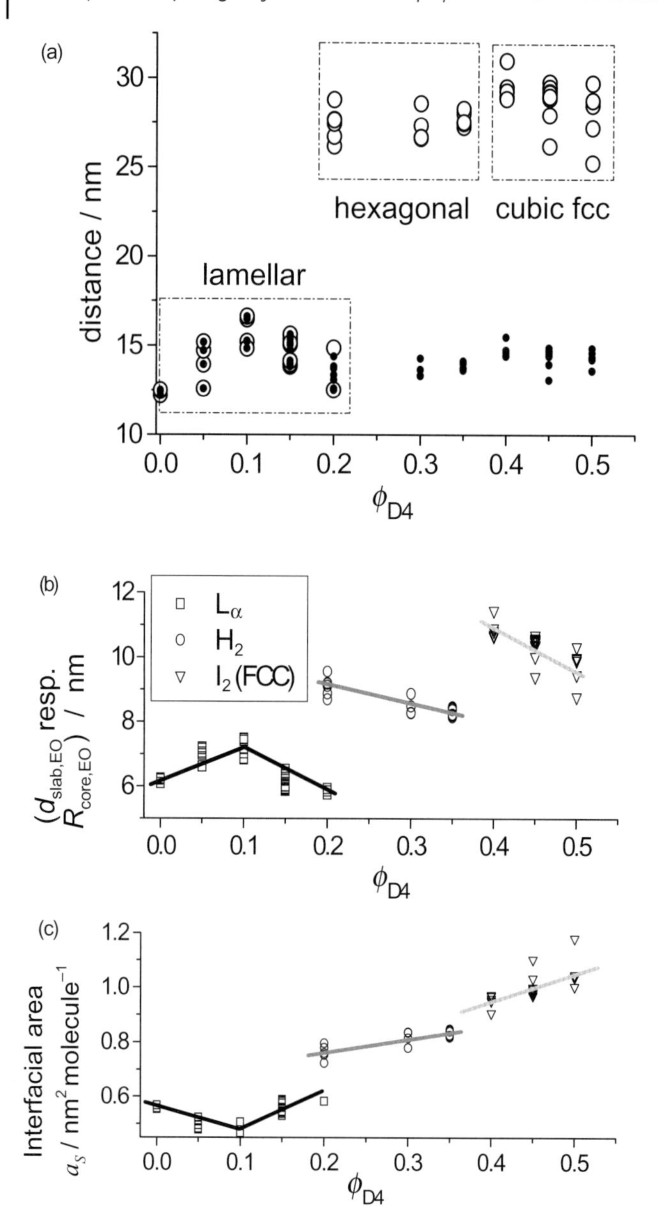

Figure 8.8 (a) Closest core–core distance d (solid circles) and translational repeat distance a of indexed unit cells (open circles), and derived parameters: (b) PEO slab thickness resp. core radius R_{core} and (c) interfacial area per molecule a_S, from Gaussian fits to indexed peaks for $Si_{25}C_3E_{51.6}$/D_4 at 58 °C.

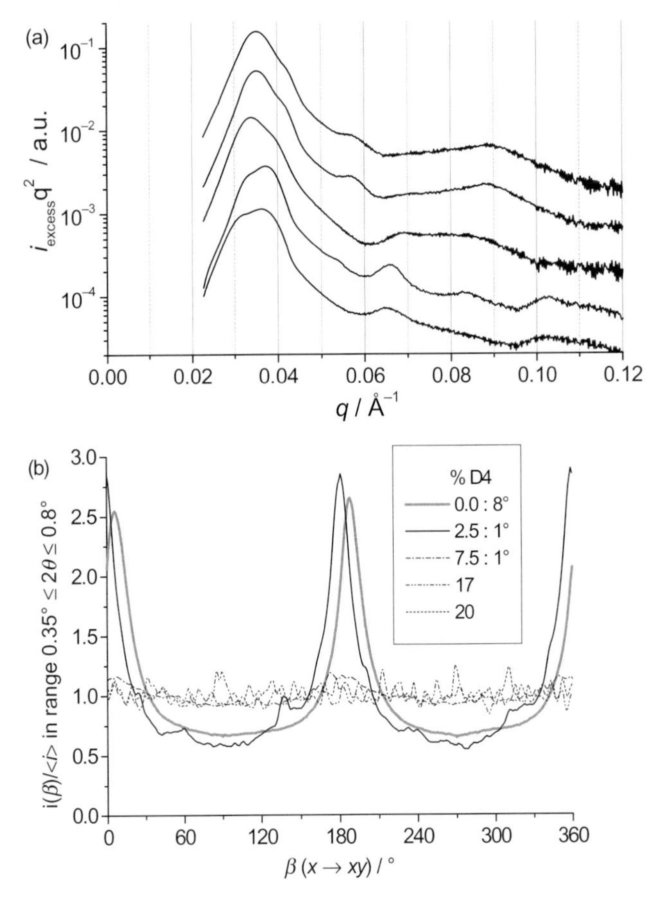

Figure 8.9 SAXS patterns of oriented hexagonal and of cubic morphologies in the system $Si_{52}C_3E_{51.6}/D_4$ at 58 °C: (a) Azimuthally averaged scattering intensity, after background correction: from bottom to top: $\phi_{D4} = 0$, 0.025, 0.075, 0.17, 0.20, scaling factor $10^{0.5}$ with respect to former; (b) the anisotropy of the intensity distribution in the range of the 1st-order maximum.

8.5 Conclusions

The phase behavior at 58 °C was correlated with the volume fraction of the more cohesive PEO blocks, which formed the cores of lamellar, cylindrical or spherical polymer micelles. Lamellar and hexagonal micelles were highly oriented by rubbing between two polyimide films. An epitaxially grown hexatic phase with cylinders standing on the buffed film was reproducibly observed in the vicinity of the stability optimum of the H_2 phase.

For the balanced copolymer in silicone oil, the PEO/PDMS interfacial area a_S increased from 0.5 nm^2/molecule in the lamellar phase to 1.1 nm^2/molecule in the cubic phase. This parameter correlates almost linearly with the silicone volume fraction and it also increases substantially with the molecular weight of the silicone block in the copolymer, obviously resulting in the fluid-like motions of the silicone block, which favor a Gaussian conformation of this block.

Acknowledgment

We thank A. Harashima and H. Furukawa at Dow Corning Toray Silicone Co. Ltd., Japan, for supplying the silicone copolymers. The experiments were partially funded by the Mitsubishi Foundation.

References

1 Uddin, Md.H., Rodriguez, C., Watanabe, K., López-Quintela, A., Kato, T., Furukawa, H., Harashima, A., and Kunieda, H. (2001) *Langmuir*, **17**, 5169–5175.

2 Kunieda, H., Uddin, Md.H., Furukawa, H., and Harashima, A. (2001) *Macromolecules*, **34**, 9093–9099.

3 Rodriguez, C., Uddin, Md.H., Watanabe, K., Furukawa, H., Harashima, A., and Kunieda, H. (2002) *J. Phys. Chem. B*, **106**, 22–29.

4 Kunieda, H., Uddin, Md.H., Horii, M., Furukawa, H., and Harashima, A. (2001) *J. Phys. Chem*, **105**, 5419–5426.

5 Kaneko, M., Matsuzawa, K., Uddin, Md.H., López-Quintela, M.A., and Kunieda, H. (2004) *J. Phys. Chem. B*, **108**, 12736–12743.

6 Uddin, Md.H., Rodriguez, C., López-Quintela, A., Leisner, D., Solans, C., Esquena, J., and Kunieda, H. (2003) *Macromolecules*, **36**, 1261–1271.

7 Kunieda, H., Uddin, Md.H., Yamashita, Y., Furukawa, H., and Harashima, A. (2002) *J. Oleo Sci.*, **51**, 113–122.

8 Galin, M., and Mathis, A. (1981) *Macromolecules*, **14**, 677–683.

9 Matsen, M.W. (1995) *Phys. Rev. Lett.*, **74**, 4225–4228.

10 Larson, R.G. (1999) *The Structure and Rheology of Complex Fluids*, Oxford University Press, New York, Chapters 12 and 13.

11 Roe, R.J. (2000) *Methods of X-Ray and Neutron Scattering in Polymer Science*, Oxford Univ. Press, New York, Section 3.6.

12 Charvolin, J., and Sadoc, J.F. (1988) *J. Phys.*, **49**, 521.

9
Molecular Dynamics Study of Isoprenoid-Chained Lipids: Salient Features of Isoprenoid Chains As Compared with Ordinary Alkyl Chains

Wataru Shinoda and Masakatsu Hato

9.1
Introduction

Synthetic isoprenoid-chained lipids are a new class of lipids, which have recently been attracted increasing attention [1–25]. A salient feature of the "isoprenoid-chained lipids" lies in their regularly branched chains consisting of poly-isoprenoid (C5) hydrophobic chains, which are remarkably different from those of ordinary lipids with linear hydrocarbon chains of variable length, different degree of unsaturation.

Apart from the synthetic isoprenoid-chained lipids, naturally occurring isoprenoid-chained lipids are found in plasma membranes of archaebacteria, which survive in extraordinary environments: halophiles at high salt concentration (e.g., saturated brine), thermoacidophiles at high temperature and acidic pH (e.g., 50–110 °C and pH = 2) [26–29]. In archaebacterial plasma membranes, the polar lipids are all derived from a common basic core structure, glycerol di-(or tetra-)*ether* of 3,7,11,15-tetramethylhexadecyl (phytanyl) group [26–29]. The phytanyl group of these lipids is remarkably different from the ordinary lipids of other organisms: glycerol di-*ester* lipids with fatty acid as the hydrophobic backbone (1,2-diacyl-sn-glycerol).

With the regularly branched chain structure, molecular volume, *trans*/*gauche* ratios, rotational and translational motions, and hence molecular packing could all be affected. The isoprenoid-chained lipids in fact display unique physical properties as compared with ordinary lipids; for example, very low values of hydrated solid–liquid-crystalline phase transition temperature, T_K [12, 13, 30], substantially lower solute permeability and higher salt tolerance of the bilayer membranes [31]. From the phase-behavior point of view, as molecular packing of lipid molecules in aqueous media is determined by competing interactions of the polar headgroups and the hydrophobic chains, the different chain structure in turn have significant effects on the phase behavior [32, 33]. Different phase preference of phytantriol to an L_3 phase from that of monoolein has recently been reported [34].

In addition to the scientific interests in the isoprenoid-chained lipids, they are of technical importance, because their physical properties mentioned above are preferable in such technical applications as liposomal drug-delivery systems [35].

Self-Organized Surfactant Structures. Edited by Tharwat F. Tadros

ISBN: 978-3-527-31990-9

Moreover, their low values of T_K make isoprenoid-chained lipids particularly useful in preparing sugar-based surfactants (or lipids) and a range of inverted liquid-crystalline phases that are stable over a wide span of temperature from 0 °C: a feature not readily realized by conventional sugar-based surfactants and inverted phases of ordinary lipid counterparts. It is well recognized that T_K values of conventional sugar-based surfactants are usually high [11, 36], imposing a severe limitation in exploiting sugar-based surfactants in many technical applications. The recent progress in the synthesis and understanding of the physical and biophysical properties of sugar-based isoprenoid-chained surfactants has been described [24]. The fact that the values of T_K are depressed below 0 °C has made the sugar-based isoprenoid-chained surfactants considerably easier to handle, opening up a new opportunity to prepare vesicles and/or black lipid membranes, which can successfully be applied in functional reconstitution of membrane proteins [24].

Essentially the same consideration applies to an inverted bicontinuous cubic phase (Q_{II}) and an inverted hexagonal (H_{II}) phase, which are usually formed with lipids with long hydrophobic chains. Q_{II} and H_{II} phases have recently received growing attention in the pharmaceutical or biological fields, for instance, as new carriers for drug-delivery systems, and matrices for membrane protein crystallization [37–42]. The conventional (e.g., monoolein) Q_{II} phase, however, often transforms into a solid phase at low temperatures around 4 °C [43–47], where temperature-sensitive proteins or actives are most preferably handled and preserved. It has recently been confirmed that isoprenoid-chained lipids can in fact give a range of Q_{II} phases that are stable at low temperatures [13].

In spite of the emerging widespread interests in isoprenoid-chained lipids, the molecular origin of their unique properties of isoprenoid-chained lipids is not yet well understood. Taking dipalmitoyl phosphatidylcholine (DPPC) and diphytanoyl phosphatidylcholine (DPhPC) as typical examples, we here discuss the molecular origin of the characteristic features of isoprenoid-chains as compared to those of the ordinary alkyl chains on the basis of a series of all-atomic MD simulations using the CHARMM PARAM27 force field [48]. DPPC is an ordinal, naturally occurring straight-chained lipid, and their bilayer is well studied experimentally and theoretically. DPhPC is the polyisoprenoid (i.e. phytanoyl)-chained counterpart of DPPC. The molecular structures of DPPC and DPhPC are depicted in Figure 9.1. The technical details are found in our previous reports [49].

9.2 Effect of Chain Branching on the Lipid Bilayer Properties

9.2.1 Structure

9.2.1.1 *Gauche/Trans* Ratio

Recent experimental and simulation studies have shown that a change in the bilayer thickness is minor due to the lipid chain branching, though DPhPC forms

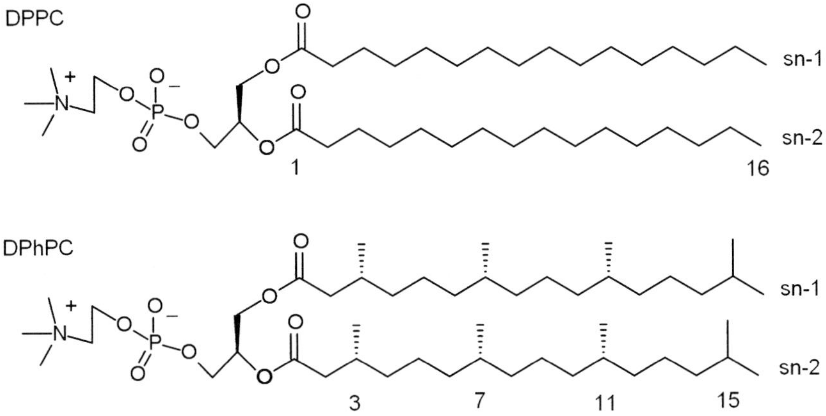

Figure 9.1 Molecular structure of dipalmitoyl phosphatidylcholine (DPPC) and diphytanoyl phosphatidylcholine (DPhPC).

a slightly thinner membrane than DPPC [49–51]. This does not mean a change in density of the bilayer core but in the cross-sectional area per lipid. Chain branching causes an area expansion. For example, the experimental molecular area is reported as 62.9 $Å^2$ and 76 $Å^2$ for DPPC and DPhPC bilayers, respectively [50, 51]. A similar trend is also observed in an MD study [49].

From the molecular viewpoint, a methyl branch of an alkyl chain changes a preferential dihedral state of the main chain in the thermal equilibrium due to the steric repulsion of a branched methyl group. In a straight hydrocarbon chain, the difference in energy between the *trans* and *gauche* states of the dihedral angle along the chain is measured to be about 0.7 kcal/mol [52]. However, the dihedral angle in the vicinity of *tert*-carbon shows almost equivalent energy for *trans* and *gauche* states (Figure 9.2b). Thus, in the liquid state of the chains, the probability of *gauche* state is quite large next to *tert*-carbon, compared with that in straight chains. Figure 9.2a shows a plot of *gauche* probability along the main chains in these two lipids obtained from MD simulations [49]. A high probability of *gauche* conformers is actually observed in the vicinity of the *tert*-carbons of DPhPC molecules. On the other hand, the two dihedrals located between the *tert*-carbons show a higher probability of *trans* conformers than do the straight-chained counterparts. Thus, the phythanoyl chain tends to bend selectively at the dihedrals in the vicinity of the *tert*-carbons. Figure 9.2c shows snapshots of palmitoyl and phytanoyl chains typically observed in DPPC and DPhPC bilayer membranes, respectively. Even though *gauche* conformers at the *tert*-carbons of DPhPC are preferred, the overall *gauche* probabilities averaged over the full chain lengths in DPPC and DPhPC show minor difference from each other (DPPC:27%, DPhPC:31%). Due to this, the change of chain length is minor, which explains the similar thickness of DPPC and DPhPC membranes.

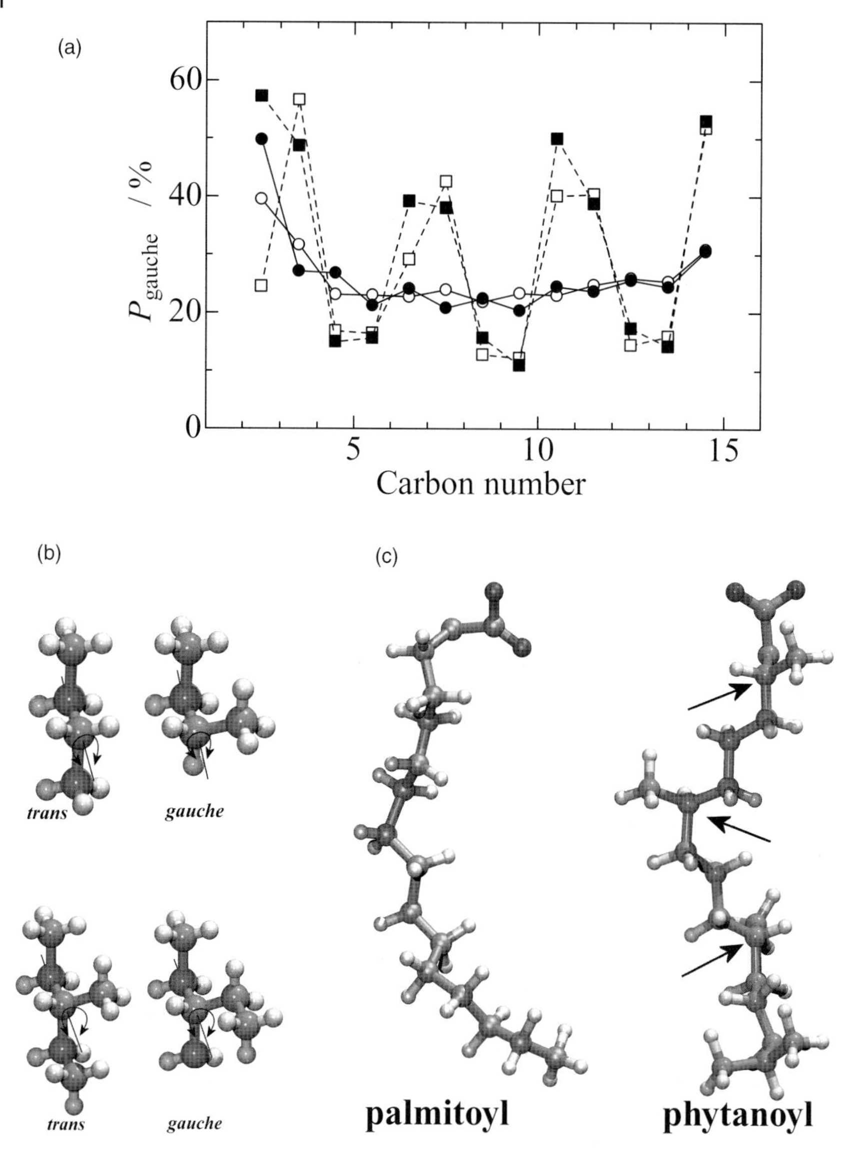

Figure 9.2 (a) Ratio of *gauche* conformers as a function of carbon number along chains. Closed circle: the *sn*-1 chain of DPPC, open circle: the *sn*-2 chain of DPPC, closed square: the *sn*-1 chain of DPhPC, and open square: the *sn*-2 chain of DPhPC. (b) trans and gauche conformers of straight and branched chains. (c) A likely found conformation of the straight palmitoyl chain and highly branched phytanoyl chain. Arrows point at the branching segments.

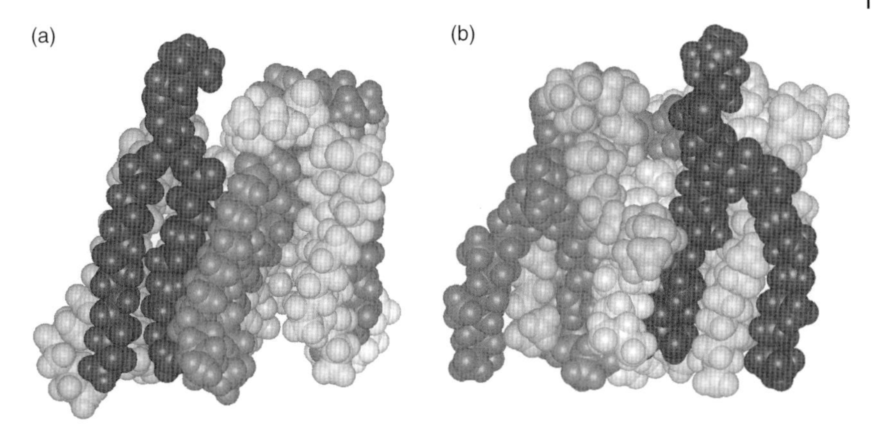

Figure 9.3 Typical snapshots of the lipid membrane patches. (a) DPPC and (b) DPhPC.

9.2.1.2 Chain Packing

The bend of the chain promotes the entanglement among the neighboring chains in the membrane layers, which is not so often observed in ordinal straight-chain lipid membranes. Figure 9.3 shows typical snapshots of DPPC and DPhPC lipid membrane patch taken from MD trajectories. In DPhPC, some phythanoyl chains are caught between the two chains of the neighboring lipid. An enhanced chain entanglement would change the packing of lipid chains in the membrane core.

In order to characterize the neighboring chain–chain correlation quantitatively, a measure of the order of the chain packing has been introduced [53]. We define the following function;

$$\Gamma^{\text{Neighbor}} = \langle \cos\theta^{\text{Neighbor}} \rangle = \langle \boldsymbol{\mu}_i \cdot \boldsymbol{\mu}_i^{\text{Neighbor}} \rangle \qquad (9.1)$$

$\hat{\boldsymbol{\mu}}_i$ and $\hat{\boldsymbol{\mu}}_i^{\text{Neighbor}}$ are the unit vectors of the ith chain and the neighboring chain, respectively. The neighboring chain is selected by the distance between centers of mass of the chain. Thus, the function Γ^{Neighbor} gives a measure for the orientational correlation among the neighboring chains; Γ^{Neighbor} should give 1 if all chains are completely aligned, like in the crystal phase. Estimated values of the function, Γ^{Neighbor}, are 0.806 and 0.680 for DPPC and DPhPC, respectively. The lower value of the correlation function, Γ^{Neighbor}, for DPhPC demonstrates that their neighboring chains are not well aligned in the same direction. This function, Γ^{Neighbor}, does not go to 0 as long as the lipids are packed in a bilayer. The lower bound of Γ^{Neighbor} is calculated using the probability distribution of the chain vector. The estimated lower bound values are 0.699 and 0.657 for DPPC and DPhPC, respectively [53]. In the DPPC bilayer, Γ^{Neighbor} is much larger than the lower bound, which indicates that the neighboring chains are prone to align parallel to each other, compared with the two arbitrarily chosen chains in the same layer. On the other hand, in the DPhPC bilayer, the orientational correlation between the neighboring chains, Γ^{Neighbor}, is almost comparable to that between a pair of arbitrary chains in the same lipid layer. This should be due to the characteristic bending conformation of the branched (DPhPC) chains, which would prohibit the parallel packing of the chains and enhance the chain crossing between the neighboring chains.

9.2.2 Dynamics

We can surmise a slowing down of the DPhPC lipid motion from the enhanced entanglements among the neighboring lipid chains. In this section, we will see how and which motion is actually decreased or not. One of the biggest advantages of the MD simulation is that the dynamical properties are straightforwardly evaluated from the MD trajectories using a variety of correlation functions [54].

9.2.2.1 Rate of *Trans–Gauche* Isomerization

The rate of *trans–gauche* isomerization is evaluated using the following state function of the dihedral angles [49];

$$S(t,\phi)=\begin{cases} 1 & (\phi \geq (2/3)\pi, \quad \textit{trans} \text{ state}) \\ -1 & (\phi < (2/3)\pi, \quad \textit{gauche} \text{ state}) \end{cases} \tag{9.2}$$

This function describes the state of the dihedral angle ϕ at time t. The relaxation time of the autocorrelation function, $C(t)$, of the state function gives a measure of the net isomerization rate.

$$C(t)=\frac{\langle \delta S(t)\cdot \delta S(0)\rangle}{\langle \delta S(0)^2 \rangle}, \tag{9.3}$$

where

$$\delta S(t)=S(t)-\langle S\rangle. \tag{9.4}$$

The relaxation time is estimated by the numerical integration of the correlation function, $C(t)$, and a correction for the larger time is made by fitting the tail region of the function to a single-exponential function. The relaxation time is plotted against the carbon number along the hydrophobic chain in Figure 9.4. For the straight-chained DPPC, the NMR relaxation time of segmental CD vectors is also presented in the figure [55]. The calculated profile of the relaxation time is similar to that obtained by NMR. The branched DPhPC shows a much longer relaxation time profile throughout the chain. It is clearly shown that the phytanoyl chain is less mobile than the palmitoyl one; for the middle part of the chains, the relaxation time of the correlation function of DPhPC is approximately 5 times that of DPPC. The peaks of DPhPC occur at the *tert*-carbon in the relaxation time profile. The same is also observed in the ^{13}C NMR relaxation time profile [56]. A relaxation time in the range of 200–300 ps observed for the middle part of the branched chain is comparable to that of dihedrals around glycerol groups in the DPPC molecule [54].

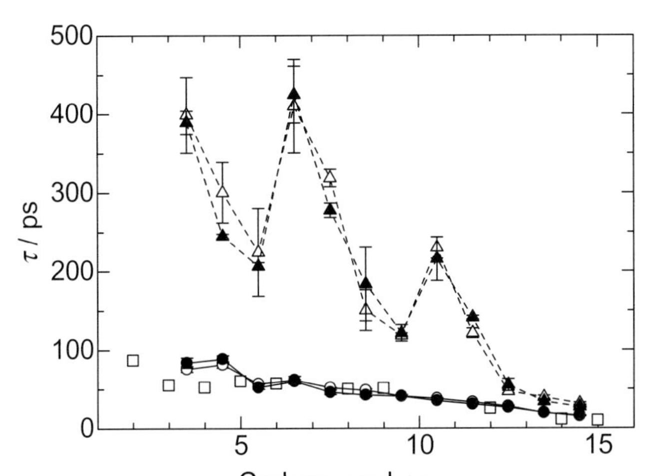

Figure 9.4 *Trans–gauche* isomerization rate calculated from 10ns-NVE-MD runs: The relaxation-time profile as a function of carbon number. Closed circle: the sn-1 chain of DPPC, open circle: the sn-2 chain of DPPC, closed triangle: the sn-1 chain of DPhPC, and open triangle: the sn-2 chain of DPhPC. The NMR relaxation time of segmental CD vectors of DPPC (open square) is also plotted.

9.2.2.2 Rotational Motion of the Chains (Wobbling Motion of the Chains) and of the Headgroup

To assess the rotational motions, in general, we use the following correlation function;

$$\chi(t) = \frac{P2(t) - P2(\infty)}{1 - P2(\infty)}, \tag{9.5}$$

where $P2(t)$ is the second term of the Legendre polynomials defined as,

$$P2(t) = \frac{1}{2}\langle 3\cos^2\theta(t) - 1\rangle = \frac{1}{2}\langle 3(\hat{\mu}_i(t)\cdot\hat{\mu}_i(0))^2 - 1\rangle, \tag{9.6}$$

where $\hat{\mu}_i(t)$ is the unit vector parallel to the target chain. $P2(t)$ is, in principle, the function that reaches zero after infinitely long time, $P2(\infty) = 0$, if and only if the overall relaxation of the chain vectors including the flip-flop motion of the lipid molecule occurs during the observation. However, the full relaxation is not expected in the MD simulation time. Thus, instead, we usually use the function $\chi(t)$ to evaluate the rotational relaxation. If we choose the unit vector as parallel to the hydrophobic chain, we can evaluate the wobbling motion of the lipid tail chain. If we choose the P–N vector, that is, the vector connecting phosphorus to nitrogen atom in the lipid headgroup, we can evaluate headgroup rotation (see Figure 9.5). Table 9.1 lists the calculated relaxation times of $\chi(t)$ for the wobbling motion of a lipid tail chain and the rotation of P–N vector, together with the relaxation time of the chain-wobbling motion determined by an analysis of nanosecond time-

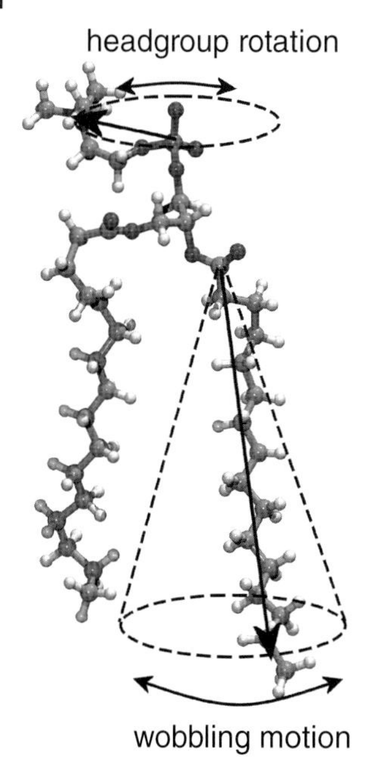

Figure 9.5 Schematic presentation of a wobbling motion of alkyl chain and the rotational dynamics of the headgroup of a lipid.

Table 9.1 A comparison of dynamical properties of lipids. Relaxation times of the wobbling motion of the hydrophobic chain and the rotational motion of the headgroup (PN vector) and lateral diffusion coefficient, *D*.

	DPPC		DPhPC	
	MD	Expt.	MD	Expt.
τ_{chain}(ns)	1.7 ± 0.3	1.79 ± 0.91[a),b)]	3.3 ± 0.3	2.27 ± 1.14[a)]
τ_{PN} (ns)	0.53 ± 0.03	–	0.45 ± 0.03	–
D ($\times 10^{-8}$ cm^2/s)	35.3 ± 0.6	20.7 ± 2.8[b),c)]	13.7 ± 0.5	18.1 ± 5.6[c)]

a) The experimental data at 298 K are taken from [57].
b) The data were obtained for egg yolk phosphatidylcholine bilayers.
c) The experimental data at 323 K are taken from [57].

resolved fluorescence spectroscopy data based on the wobbling-in-cone model [57]. The calculated relaxation times are qualitatively in good agreement with the experimental ones. The branch-chained DPhPC clearly shows a slower wobbling motion of their chains than does the straight-chained DPPC. On the contrary, interestingly, the headgroup rotation is slightly faster in DPhPC than in the DPPC membrane. A similar change of lipid dynamics is often observed, when the cholesterol is added to a lipid bilayer system [58, 59]. Because of the hydrophobic nature of the molecule, cholesterol molecules reside in the hydrophobic core of the membrane. In the cholesterol-rich environment, the lipids show a higher-order parameter of the hydrophobic chain, a slower mobility of lipid, and a faster rotation of the headgroup. In these systems, it is suggested that cholesterol molecules decrease the headgroup–headgroup interactions by increasing the spacing between the lipids [58, 59]. Similarly, in the DPhPC bilayer, the larger molecular area due to the phytanoyl chains should reduce the interactions between the lateral neighboring headgroups.

9.2.2.3 Lateral Diffusion Coefficient of Lipid Molecules

Lateral diffusion coefficient of lipid molecules, D, can be evaluated by mean square displacements (MSD) of the lipid center of mass (COM) in the membrane plane using the Einstein relation;

$$D = \frac{1}{4} \lim_{t \to \infty} \frac{1}{t} \left\langle |\mathbf{r}(t) - \mathbf{r}(0)|^2 \right\rangle \tag{9.7}$$

The term in the bracket on the right-hand side of the equation denotes the two-dimensional MSD of lipid COMs. Namely, the slope of the plot of MSD as a function of time gives a diffusion coefficient. From our MD analysis, the plots show linear slopes of the MSD in the time range from about 0.7 to 3 ns. Lateral diffusion coefficients of DPPC and DPhPC are estimated as $35.3 \pm 0.6 \times 10^{-8}\,\text{cm}^2\,\text{s}^{-1}$ and $13.7 \pm 0.5 \times 10^{-8}\,\text{cm}^2\,\text{s}^{-1}$, respectively (Table 9.1). Time-resolved fluorescence spectroscopy measurements of DPhPC [57] yielded a very similar D value, $18.1 \pm 5.6 \times 10^{-8}\,\text{cm}^2\,\text{s}^{-1}$, to our observation from MD. Thus, as a consequence of chain branching, the mobility of the lipids is reduced. In addition, lateral diffusion of the lipids depends on the hydrophobic chain flexibility and is less correlated with the headgroup rotation. It is often discussed that lateral diffusion of lipid molecules is limited by restrictions at the bilayer/water interface [60]. Our analyses clearly indicate that, in case of phytanoyl chained DPhPC, in contrast, highly viscous membrane core significantly contributes to reduce the lateral diffusion, in spite of the faster headgroup rotation.

9.2.3 Permeability

There have been several models for the permeation mechanism of ionic molecules. All models propose the ion permeation with a guide of water molecules. For example, the water wire is widely believed to help a proton transport across the membrane. Investigating the water wire is still a challenging subject for MD simulations, though a few studies have been reported [61, 62]. However, permeability of neutral small molecules across the bilayers can now be evaluated by MD simulations. Here, we demonstrate an evaluation of water permeability across DPPC and DPhPC bilayers.

9.2.3.1 Water Permeability through the Lipid Bilayer Membrane

Since spontaneous water permeation across the lipid membranes is a rare event in the simulation time range, the permeation rate cannot be straightforwardly measured via ordinal MD simulation. Therefore, we need an alternative approach to assess the water permeability across the lipid membranes. According to the inhomogeneous diffusion model proposed by Marrink and Berendsen [63], permeability coefficient, P, is derived by

$$1/P = \int_{z_1}^{z_2} \frac{\exp(\Delta G(z)/kT)}{D_z(z)} dz \tag{9.8}$$

where $D_z(z)$ is the diffusion coefficient of penetrant along the membrane normal, z, and $\Delta G(z)$ is the free-energy loss or gain of penetrant when moved to $z = z$ from the reference state $z = z_0$. k is the Boltzmann constant and T is the temperature. Integration has to be done across the membrane from z_1 to z_2, which are in the water layers separated by the membrane, respectively. By calculating the free-energy profile and local diffusion coefficients, we can estimate the permeation rate of the molecule across the membrane.

9.2.3.2 Free-Energy Profile of Water along the Bilayer Normal

Free-energy calculation is often computationally demanding so that many efforts have been devoted to develop an algorithm to obtain a free-energy profile precisely and efficiently. Thermodynamic integration [64], Widom insertion method [65], probability ratio method [66], overlapping distribution method [67], WHAM (weighted histogram analysis method) [68] and more recently, steered MD based on Jarzynski's equality [69], ABF (adapting biasing force) [70] and energy representation method [71] are all used to evaluate the free-energy profile of molecules going across the bilayer membranes. There is no method that always works best. Any of these methods under suitable conditions work better than any others. Therefore, a combination of several different methods is also possible and sometimes works more efficiently [63, 72–74]. Thus, we should choose an appropriate method depending on the target.

Figure 9.6 plots free-energy profiles of water across DPPC and DPhPC membranes. These two bilayers show quite similar free-energy barriers to water. A slight difference between these two profiles is the barrier height: about 6.3 and 6.0 kcal/mol for DPPC and DPhPC bilayers, respectively. The numerical convergence of the free energy (excess chemical potential) calculation in the middle part of the membrane is fairly good, with an estimated error of ~0.1 kcal/mol. Thus, assuming the same excess chemical potential of a water molecule in the bulk water region, the branched DPhPC bilayer can have a slightly smaller energy barrier for a water molecule, although the difference was much less than kT. These findings are consistent with the experimental observation that chain branching gives slightly higher solubility of water into the hydrocarbon liquids (n-hexadecane: 95.5 μg/cm^3 and 2, 6, 10, 15, 19, 23-hexamethyl tetracosane : 102.0 μg/cm^3 at 318 K) [75]. At the bilayer center, a small depression of about 0.2 kcal/mol in the excess free-energy profile is observed in both systems. The low local density of the system

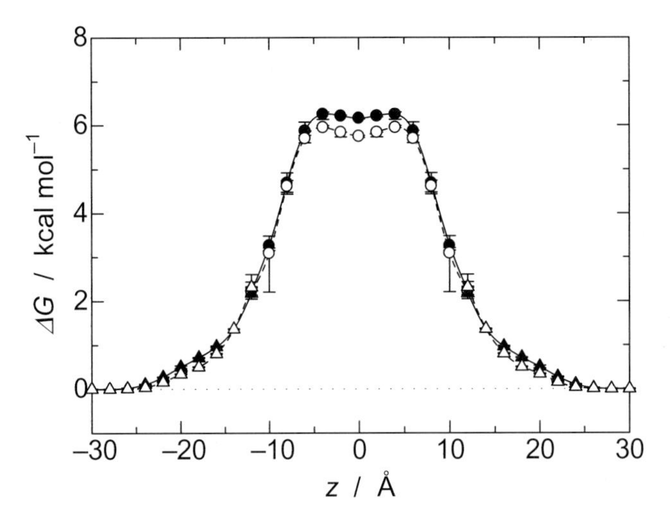

Figure 9.6 Free-energy profiles of water along the bilayer normal. The closed symbols denote the free-energy profiles across the DPPC bilayer. The open symbols denote those across the DPhPC bilayer. The circles and triangles are the free energy estimated by the cavity insertion and that estimated by the probability ratio method, respectively (the latter for the case of water only).

atoms in the vicinity of the slip plane between two leaflets of the bilayers would cause the small depression. A similar hollow is observed commonly in the free-energy profiles of small molecules [63, 72–74].

9.2.3.3 **Local Diffusion Coefficient of Water**

Although lipid chain branching does not alter the free-energy profile of the water molecule significantly, the phytanoyl chain's slower conformational motion, which is observed in the previous section, suggests that the dynamics of water (or of any penetrant) in the membrane interior will be changed. Figure 9.7 plots the local diffusion coefficients of water molecules along the bilayer normal, D_z, as a function of the distance from the bilayer center ($z = 0$). These coefficients are calculated by means of MSD ($|z| \geq 12$ Å) and force autocorrelation function (FACF) ($|z| \leq 12$ Å) [63].

The local diffusion coefficient profile of water in DPPC bilayer revealed the followings:

1) The diffusion coefficient in the middle of the water layer ($z \sim 33$ Å) is comparable to that observed in pure water ($\sim 0.77 \times 10^{-4}$ cm^2/s at 323 K; Note that modified TIP3P water [76] used here overestimates the experimental diffusion coefficient by a factor of ~2). If we construct the system with a larger concentration of water, it will be expected that the diffusion coefficient in the water layer completely coincides with that of bulk water. Although a slightly smaller diffusion coefficient observed in the middle of water layer is, in this sense, a finite

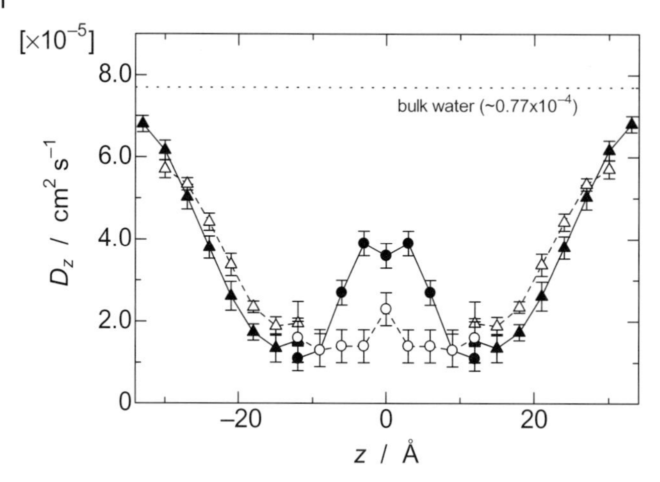

Figure 9.7 Local diffusion coefficients of a water molecule in the normal direction of the bilayer, D_z. The circles and triangles are the D_z values determined using the force autocorrelation function and mean square displacement, respectively. The closed and open symbols are D_z in the DPPC and DPhPC bilayers, respectively.

size effect of the simulated system, it should be noted that, in our simulations, the hydration level is fixed at 29 water/lipid, which is consistent with the saturated hydration for the DPPC bilayer [50].

2) From the bulk water to the interfacial region, $|z| > 15$ Å, the local diffusion coefficient of water decreases monotonously as water molecules approach the interfacial region.

3) The translational mobility of the water located in the region of $|z| = 9$–15 Å, where rather highly packed hydrophobic chains and ester groups are found, is significantly low.

4) In the hydrophobic bilayer center, $|z| < 9$ Å, water diffusion is promoted in the vicinity of the slip plane between the two leaflets of the bilayer.

A comparison of the diffusion coefficient profiles between DPPC and DPhPC bilayers demonstrates that chain branching reduces water diffusion significantly in the hydrophobic membrane interior. In the DPhPC bilayer, although an enhanced diffusion of water is still found in the vicinity of the slip plane, the low diffusion coefficient almost plateaus in the range of $|z| = \sim 3$–9 Å, demonstrating that the mobility of water is significantly affected by the slow conformational motion of phytanoyl chain. In the interfacial region, $|z| > 15$ Å, on the contrary, the water diffusion is relatively fast in the DPhPC bilayer. This enhanced diffusion can be explained by the higher exchange rate of the phosphate-bound water, which would be due to the larger molecular area of DPhPC.

Using Equation 9.8 with $z_1 = -30$ and $z_2 = 30$ Å, the permeability coefficients of water across DPPC and DPhPC bilayers are estimated to be about 1.6×10^{-2} cm/s and 1.1×10^{-2} cm/s, respectively. Since the components in the integrand of Equation 9.8 themselves show statistical errors, the calculated permeability may also contain a nontrivial error. In addition, the experimentally determined permeability showed wide variation, from 10^{-2} to 10^{-4} cm/s [77], because the values are sensitive to the experimental techniques and conditions, such as bilayer systems (liposomes or planar lipid bilayers), salt concentration, pH, *etc.* Therefore, we should restrict our discussion to the qualitative difference in the water permeability across the DPPC and DPhPC bilayers. However, as seen above, these MD simulations seem to give reasonable results qualitatively, and it is a significant advantage that we can measure permeability on the same standard. In our calculations, chain branching reduces the permeability of water by about 30%. One may think that the change of water permeability is not so large in the light of the structural and dynamical changes of lipid chain. Actually, the slower motion of the phytanoyl chains in an entangled packing significantly reduces the water diffusion in the bilayer core ($|z| < 10$ Å) as shown in Figure 9.7. However, the slightly lower free-energy barrier of the DPhPC bilayer (Figure 9.6) and slightly faster water diffusion near the interfacial region ($|z| \sim 20$ Å) cancel the effect to some extent. As a result, water permeability is moderately reduced in the DPhPC membrane compared with that in the DPPC membrane. A comparative experimental measurement of water permeability across DPhPC ($4.29 \pm 0.30 \times 10^{-3}$ cm/s) and across straight-chained *E. coli* lipid bilayers ($4.9 \pm 0.16 \times 10^{-3}$ cm/s) supports our observation qualitatively [78].

9.2.3.4 Cavity Distribution Analysis

Since the relaxation time of chain conformation is usually much longer than the characteristic time of translational motion of small penetrants, the dynamics of small penetrants are significantly affected by the lipid chain conformation in the membrane interior. Therefore, the analysis of the cavity distribution in the lipid bilayer system will shed light on the dynamics of the small neutral molecules in the membrane interior. The cavity distribution is calculated using 1.5-ns MD trajectories for DPPC and DPhPC membranes and is visualized in Figure 9.8. In the figures, the z-axis is parallel to the bilayer normal, and the bilayer center is located in the middle of the cubic box. The average phosphorus position along the z-axis is depicted by arrows as a guide for the eye in the figure. The isosurface, the equal probability surface of cavity formation at a level of 0.15, appears only inside the hydrophobic region of the lipid bilayers. The cavity probability in the water region ($z > 25$ Å) ($P \sim 0.10$) and that in the interfacial region ($P \sim 0.05$) were significantly lower than the isolevel ($P \sim 0.15$) shown in the figure. The smaller size and high mobility of the water molecule reduce the time-averaged probability of cavity formation in the water region, and the high density of the system atoms at the water/lipid interface explains the lower cavity-formation probability in the interfacial region.

Both DPPC and DPhPC bilayers present high probabilities of cavity formation near the slip plane between two leaflets of each bilayer. However, the three-dimensional cavity distributions in these bilayers differ significantly from each

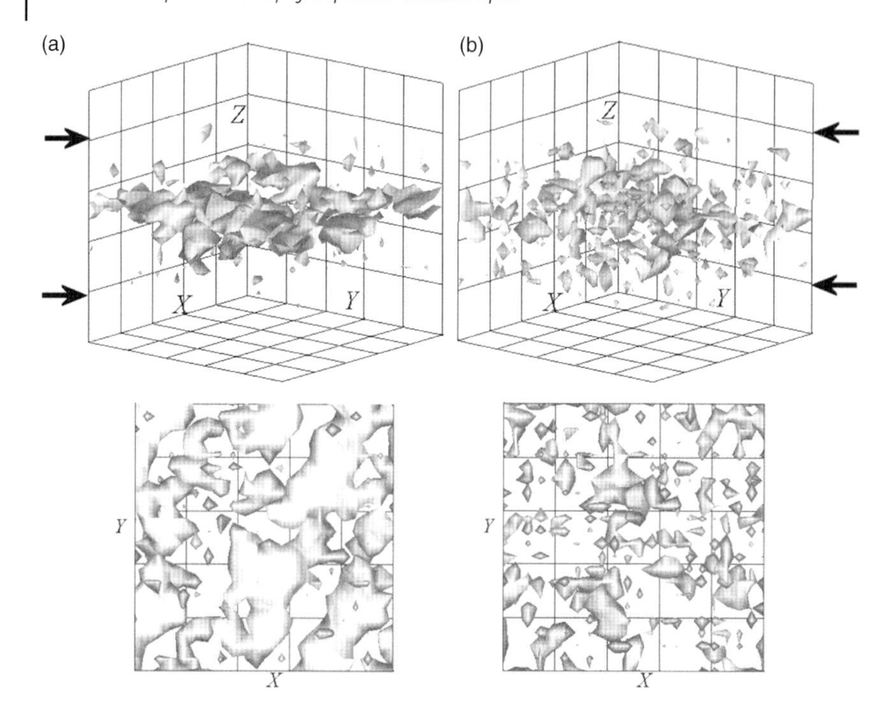

Figure 9.8 Cavity distribution in the lipid bilayers averaged over 1.5ns MD trajectories. The isosurface shows the probability of cavity formation at a level of 0.15. (a) DPPC and (b) DPhPC bilayer systems. The bottom of each is the view from the bilayer normal, z. As a guide for the eye, the grid is drawn at regular intervals of 12.2Å. Arrows denote the average position of phosphorus of lipids.

other. In the DPPC bilayer system, the larger cavity domains are observed near the slip plane. In contrast, DPhPC bilayer demonstrates discrete distribution of small cavities over the hydrophobic membrane domain. Although the appearance of the isosurface depends on the averaging time, the qualitative discussion should be meaningful when we consider the environment for penetrants in a hydrophobic membrane interior.

The difference in cavity distribution suggests that isoprenoid-chained lipid causes (1) a lower mobility of penetrants in the membrane interior, and (2) a lower probability of penetrants clustering in the membrane interior. In the light of the cavity distribution revealed here, let us now discuss the dynamical behavior of small penetrants in the membrane interior, assuming that the dynamics of the penetrants is limited by the cavity distribution in the membrane. Since the discrete distribution of small cavities in a hydrophobic membrane interior would limit the motion of penetrants, the penetrants will be expected to show a restricted, hopping-like motion among the long-lived cavities. The restriction of penetrant motion should be remarkable in the branched DPhPC bilayer. In the DPPC bilayer, the distribution of high cavity-formation probability is spread along the lateral direction near the slip plane, which suggests that penetrants would likely migrate along the plane (x-y plane in Figure 9.8). It is expected that, during the lateral diffusion, the penetrants find a possible pathway along the membrane normal to escape from the membrane region. On the other hand, in DPhPC bilayer, even lateral diffusion of a penetrant will be relatively limited.

Temporal proton-wire (a transient hydrogen-bonded chain of water molecules) formation through the lipid bilayer may be one of the most plausible explanations of abnormally high proton permeability [61, 62, 77] In this hypothesis, the proton is transferred through the hydrogen-bond network that the proton wire has formed. On the basis of the proton-wire hypothesis, let us consider the difference in proton permeability between straight-chained egg yolk PC ($17 \pm 3 \times 10^{-4}$ cm/s) and DPhPC ($3.6 \pm 1.1 \times 10^{-4}$ cm/s) bilayers, which is larger than that in water permeability as seen above, as a result of lipid chain branching. At the slip plane (bilayer center) there exists a shallow free-energy well due to the low density. Therefore, once a small penetrant goes into the region, it will stay there for a while and diffuse along the plane. Our trial MD simulations showed that a single water molecule trapped in the hydrophobic center of a bilayer typically has a lifetime of about 100 ps [72]. During the residence time, once a water molecule meets the other water molecule inside the bilayer, these water molecules will gain electrostatic stabilization. The clustering of the water molecules in the hydrophobic membrane interior could trigger the formation of a proton wire. This suggests that the relatively discrete distribution of small cavities in the DPhPC bilayer reduces the proton conductance significantly. Thus, we consider that the inhibited clustering of water molecules inside the membrane would be responsible for the behavioral difference between proton and water permeability resulting from chain branching.

9.3 Summary

Figure 9.9 shows a brief summary of the difference between the DPhPC and DPPC bilayers elucidated by the MD simulations. The regularly branched (phytanoyl) chain structure of DPhPC shows an enhanced *gauche* conformer in the vicinity of the *tert*-carbon atoms, which in turn leads to a long-lived "bending conformation" of the phytanoyl chain (Figures 9.2 and 9.9). Thus, phytanoyl chains display a less-ordered chain packing in the bilayer membrane as compared to that of DPPC. Namely, in contrast to the preferred parallel alignment of neighboring chains in the DPPC lipid membrane, an enhanced chain crossing or entanglement among neighboring chains is observed in a DPhPC membrane (Figures 9.3 and 9.9). This "entangled chain packing" appears to give rise to a significantly slower chain dynamics of the phytanoyl chain, for example, slower rate of *trans–gauche* isomerization, slower wobbling (rotational) motion of the chain, and even the slower lateral diffusion of whole DPhPC molecules. The slower chain dynamics decrease the dynamics of a small penetrant such as a water molecule in the membrane core significantly. This would explain why the DPhPC bilayer shows a lower permeability,

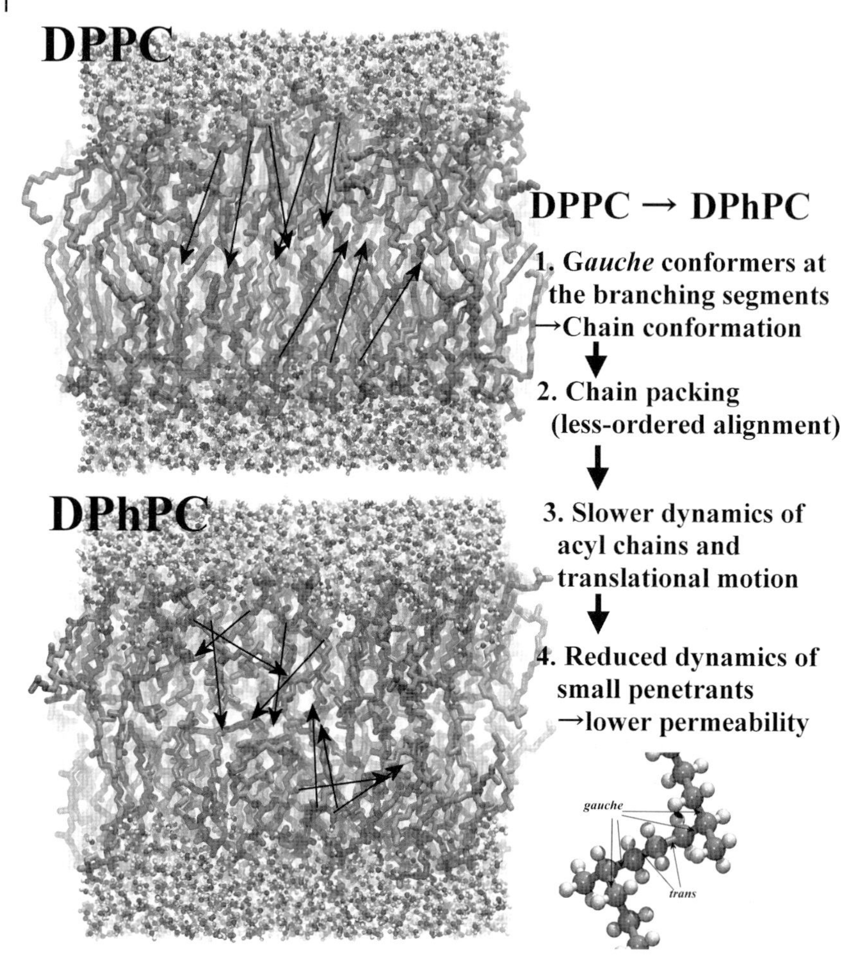

Figure 9.9 Summary of major differences of membrane properties as a result of chain branching. A comparison of DPPC and DPhPC bilayer membranes is made. Allows in the figure are drawn along several selected lipid chain vectors to elucidate the difference of chain packing in these lipid bilayers.

even though the free-energy barrier of water across the DPhPC membrane is slightly smaller than that of the DPPC membrane.

9.4 Future Perspective

Thanks to the continuous increase of available computational resources and better-performing software, MD simulation is now a handy tool to investigate the lipid bilayer membranes. This situation encourages us to design a new molecular structure to generate a synthetic lipid membrane with desirable properties with the help of MD simulations [73, 79]. There has been a significant improvement on the quality of molecular models of lipids in the last two decades, though a further continuous effort for a better potential force field will still be needed [80]. As shown in the present chapter, by using MD simulations, we are now able to evaluate the structural and dynamical properties of lipid bilayer membranes, and furthermore, the membrane permeability. Free-energy analysis is of key importance in characterizing solubility and permeability of molecules in/across lipid bilayers. Taking advantage of a recent development of advanced computational techniques, we are now able to evaluate free-energy profiles of a variety of molecules precisely, though a free-energy calculation of larger molecules to go across the membranes is still a challenging issue and is now actively studied in this field.

Lipid aqueous solution shows rich phase behaviors including not only bilayers but also inverse hexagonal and cubic phases, *etc.* The computational scheme we mentioned in the present paper is applicable, if and only if a self-assembled structure of lipids is known and the atomic initial configuration is prepared to adjust to the structure. It is not straightforward to obtain a reasonable lipid self-assembly from a randomly generated molecular configuration with an all-atomic MD simulation. To exploit such lipid self-assembly on the molecular level sufficiently, a coarse-grained (CG) MD simulation is quite useful [81]. There have been developed several coarse-graining schemes to reproduce several selected important properties to enable a structural prediction of lipid assembly in solvents quantitatively [82–85]. In addition, reverse mapping schemes to obtain AA description from CG configurations have been studied by many researchers, aimed at obtaining a better initial configuration of AA-MD from the predicted self-assembled structure by CG-MD [86, 87]. These technical developments, while there is still room for improvement, will open a way to predict physical properties of lipid self-assemblies under given conditions.

References

1 (a) Birault, V., Pozzi, G., Plobeck, N., Eifler, S., Schmutz, M., Palanche, T., Raya, J., Brisson, A., Nakatani, Y., and Ourisson, G. (1996) *Chem. Eur. J.*, **2**, 789–799; (b) Ghosh, G., Lee, S.J., Ito, K., Akiyoshi, K., Sunamoto, J., Nakatani, Y., and Ourisson, G. (2000) *Chem. Commun.*, **4**, 267–268.

2 (a) Yamauchi, K., Moriya, A., and Kinoshita, M. (1989) *Biochim. Biophys. Acta*, **1003**, 151–160; (b) Yamauchi, K., Sakamoto, Y., Moriya, A., Yamada, K., Hosokawa, T., Higuchi, T., and Kinoshita, M. (1990) *J. Am. Chem. Soc.*, **112**, 3188–3191.

3 Stewart, L.C., and Kates, M. (1989) *Chem. Phys. Lipids*, **50**, 23–42.

4 (a) Morii, H., Eguchi, T., and Koga, Y. (2007) *J. Bacteriol.*, **189**, 4053–4406; (b) Eguchi, T., Arakawa, K., Terachi, T., and Kakinuma, K. (1997) *J. Org. Chem.*, **62**, 1924–1933.

5 Menger, F.M., Chen, X.Y., Brocchini, S., Hopkins, H.P., and Hamilton, D. (1993) *J. Am. Chem. Soc.*, **115**, 6600–6608.

6 Moss, R.A., Fujita, T., and Okumura, Y. (1991) *Langmuir*, **7**, 2415–2418.

7 Fong, C., Wells, D., Krodkiewska, I., Booth, J., and Hartley, P.G. (2007) *J. Phys. Chem. B*, **111**, 1384–1392.

8 Fong, C., Wells, D., Krodkiewska, I., Hartley, P.G., and Drummond, C.J. (2006) *Chem. Mater.*, **18**, 594–597.

9 Milkereit, G., Garamus, V.M., Yamashita, J., Hato, M., Morr, M., and Vill, V. (2005) *J. Phys. Chem. B*, **109**, 1599–1608.

10 Barauskas, J., and Landh, T. (2003) *Langmuir*, **19**, 9562–9565.

11 Hato, M., Minamikawa, H., Tamada, K., Baba, T., and Tanabe, Y. (1999) *Adv. Colloid Interface Sci.*, **80**, 233–270, and references cited therein.

12 (a) Hato, M., Minamikawa, H., Salkar, R.A., and Matsutani, S. (2002) *Langmuir*, **18**, 3425–3429; (b) Hato, M., Minamikawa, H., Salkar, R.A., and Matsutani, S. (2004) *Prog. Colloid Polym. Sci.*, **123**, 56–60.

13 Yamashita, J., Shiono, M., and Hato, M. (2008) *J. Phys. Chem. B*, **112**, 12286–12296.

14 Minamikawa, H., and Hato, M. (2005) *Chem. Phys. Lipids*, **134**, 151–160.

15 Minamikawa, H., and Hato, M. (1998) *Langmuir*, **14**, 4503–4509.

16 Nishikawa, N., Mori, H., and Ono, M. (1994) *Chem. Lett.*, 767–770.

17 Wadsten-Hindrichsen, P., Bender, P., Unga, J., and Engström, S. (2007) *J. Colloid Interface Sci.*, **315**, 701–713.

18 Misiunas, A., Talaikyte, Z., Niaura, G., and Razumas, V. (2004) *Biologia*, **4**, 26–29.

19 Dong, Y.D., Larson, I., Hanley, T., and Boyd, B.J. (2006) *Langmuir*, **22**, 9512–9518.

20 Misiunas, A., Talaikyte, Z., Niaura, G., Razumas, V., and Nylander, T. (2008) *Biophys. Chem.*, **134**, 144–156.

21 Dong, Y.D., Dong, A.W., Larson, I., Rappolt, M., Amenitsch, H., Hanley, T., and Boyd, B.J. (2008) *Langmuir*, **24**, 6998–7003.

22 Salkar, R.J., Minamikawa, H., and Hato, M. (2004) *Chem. Phys. Lipids*, **127**, 65–75.

23 Hato, M., Yamashita, I., Kato, T., and Abe, Y. (2004) *Langmuir*, **20**, 11366–11373.

24 Hato, M., Minamikawa, H., and Kato, T. (2008) *Sugar-Based Surfactants: Fundamental and Applications, Surfactant Science Series*, vol. **143** (ed. C.C. Ruiz), CRC Press, Boca Raton, Chapter 10 and references cited therein.

25 (a) Abraham, T., Hato, M., and Hirai, M. (2004) *Colloid Surf. B*, **35**, 107–117; (b) Abraham, T., Hato, M., and Hirai, M. (2005) *Biotechnol. Prog.*, **21**, 255–262.

26 Smith, P.F. (1988) *Microbial Lipids*, vol. **1** (eds C. Ratledge and S.G. Wilkinson), Academic Press, New York, pp. 489–525.

27 Kates, M. (1990) *Glycolipids, Phosphoglycolipids, and Sulfoglycolipids, Handbook of Lipid Research*, vol. **3** (ed. M. Kates), Plenum Press, New York, pp. 1–122, and reference cited therein.

28 Langworthy, T.A., and Pond, J.L. (1986) *Thermophiles: General, Molecular, and Applied Microbiology* (ed. T.D. Brock), John Wiley & Sons, Inc., New York, pp. 107–135.

29 Dembitsky, V.M. (2005) *Lipids*, **40**, 535–557.

30 Blöcher, D., Six, L., Gutermann, R., Henkel, B., and Ring, K. (1985) *Biochim. Biophys. Acta*, **818**, 333–342.

31 (a) Yamauchi, K., Doi, K., Kinoshita, M., Kii, F., and Fukuda, H. (1992) *Biochim. Biophys. Acta*, **1110**, 171–177; (b) Yamauchi, K., Doi, K., Yoshida, Y., and Kinoshita, K. (1993) *Biochim. Biophys. Acta*, **1146**, 178–182.

32 (a) Gulik, A., Luzzati, V., De Rosa, M., and Gambacorta, A. (1985) *J. Mol. Biol.*, **182**, 131–149; (b) Gulik, A., Luzzati, V., De Rosa, M., and Gambacorta, A. (1988) *J. Mol. Biol.*, **201**, 429–435.

33 Minamikawa, H., and Hato, M. (1997) *Langmuir*, **13**, 2564–2571.

34 Wadsten-Hindrichsen, P., Bender, P., Unga, J., and Engström, S. (2007) *J. Colloid Interface Sci.*, **315**, 701–713.

35 Patel, G.B., and Sprott, G.D. (1999) *Crit. Rev. Biotech.*, **19**, 317–357.

36 Hato, M. (2001) *Curr. Opin. Colloid Interface Sci.*, **6**, 268–276.

37 (a) Larsson, K.J. (1989) *Phys. Chem.*, **93**, 7304–7314; (b) Larsson, K. (2009) *Curr. Opin. Colloid Interface Sci.*, **14**, 16–20.

38 Ganem-Quintanar, A., Quintanar-Guerrero, D., and Buri, P. (2000) *Drug Dev. Ind. Pharm.*, **26** (**8**), 809–820.

39 Shah, J.C., Sadhale, Y., and Chilukuri, D.M. (2001) *Adv. Drug Deliv. Rev.*, **47**, 229–250.

40 Drummond, C.J., and Fong, C. (2000) *Curr. Opin. Colloid Interface Sci.*, **4**, 449–456.

41 Landau, E.M., and Rosenbush, J.P. (1996) *Proc. Natl. Acad. Sci. USA*, **93**, 14532–14535.

42 Lynch, M.J., and Spicer, P.T. (eds) (2005) *Bicontinuous Liquid Crystals, Surfactant Science Series*, vol. **127**, Taylor & Francis, Boca Raton.

43 Briggs, J., and Caffrey, M. (1994) *Biophys. J.*, **66**, 573–587.

44 Briggs, J., and Caffrey, M. (1994) *Biophys. J.*, **67**, 1594–1602.

45 Lutton, E.S. (1965) *J. Am. Oil Chem. Soc.*, **42**, 1068–1070.

46 Qui, H., and Caffrey, M. (1998) *J. Phys. Chem. B*, **102**, 4819–4829.

47 Misquitta, Y., Cherezov, V., Havas, F., Patterson, S., Mohan, J.M., Wells, A.J., Hart, D.J., and Caffrey, M. (2004) *J. Struct. Biol.*, **148**, 169–175.

48 MacKerell, A.D., Jr., Bashford, D., Bellott, M., Dunbrack, R.L., Jr., Evanseck, J.D., Field, M.J., Fischer, S., Gao, J., Guo, H., Ha, S., Joseph-McCarthy, D., Kuchnir, L., Kuczera, K., Lau, F.T.K., Mattos, C., Michnick, S., Ngo, T., Nguyen, D.T., Prodhom, B., Reiher, W.E. III, Roux, B., Schlenkrich, M., Smith, J.C., Stote, R., Straub, J., Watanabe, M., Wiorkiewicz-Kuczera, J., Yin, D., and Karplus, M.J. (1998) *Phys. Chem. B*, **102**, 3586.

49 Shinoda, W., Mikami, M., Baba, T., and Hato, M. (2003) *J. Phys. Chem. B*, **107**, 14030–14035.

50 Nagle, J.F., Zhang, R., Tristram-Nagle, S., Sun, W., Petrache, H.I., and Suter, R.M. (1996) *Biophys. J.*, **70**, 1419.

51 We, Y., He, K., Ludtke, S.J., and Huang, H.W. (1995) *Biophys. J.*, **68**, 2361.

52 (a) Tsuzuki, S., Schafer, L., Goto, H., Jemmis, E.D., Hosoya, H., Siam, K., Tanabe, K., and Osawa, E. (1991) *J. Am. Chem. Soc.*, **113**, 4665–4671; (b) Hafezi, M.J., and Sharif, F. (2007) *J. Mol. Struct.-Theochem*, **814**, 43–49.

53 Shinoda, W., Mikami, M., Baba, T., and Hato, M. (2004) *Chem. Phys. Lett.*, **390**, 35–40.

54 Shinoda, W., Namiki, N., and Okazaki, S. (1997) *J. Chem. Phys.*, **106**, 5731–5743.

55 Brown, M.F., Seelig, J., and Häberlen, U. (1979) *J. Chem. Phys.*, **70**, 5045.

56 Degani, H., Danon, A., and Caplan, S.R. (1980) *Biochemistry*, **19**, 1626.

57 Baba, T., Minamikawa, H., Hato, M., and Handa, T. (2001) *Biophys. J.*, **81**, 3377.

58 Davis, J.H. (1983) *Biochim. Biophys. Acta*, **737**, 117.

59 Saito, H., Araiso, T., Shirahama, H., and Koyama, T. (1991) *J. Biochem.*, **109**, 559.

60 (a) Pastor, R.W., Venable, R.M., and Feller, S.E. (2002) *Acc. Chem. Res.*, **35**, 438; (b) Pastor, R.W., and Feller, S.E. (1996) *Membrane Structure and Dynamics* (eds K.M. Merz and B. Roux), Birkhauser, Boston, p. 3.

61 Venable, R.M., and Pastor, R.W. (2002) *J. Chem. Phys.*, **116**, 2663.

62 Marrink, S.J., Jahnig, F., and Berendsen, H.J.C. (1996) *Biophys. J.*, **71**, 632.

63 Marrink, S.J., and Berendsen, H.J.C. (1994) *J. Phys. Chem.*, **98**, 4155.

64 (a) Allen, M.P., and Tildesley, D.J. (1987) *Computer Simulation of Liquids*, Clarendon Press, Oxford; (b) Frenkel, D., and Smit, B. (2002) *Understanding Molecular Simulation: From Algorithms to Applications*, Academic Press, London; (c) Ulander, J., and Haymet, A.D.J. (2003) *Biophys. J.*, **85**, 3475; (d) Bemporad, D., Essex, J.W., and Luttmann, C. (2004) *J. Phys. Chem. B*, **108**, 4875.

65 (a) Widom, B. (1963) *J. Chem. Phys.*, **39**, 2808; (b) Jedlovszky, P., and Mezei, M. (2000) *J. Am. Chem. Soc.*, **122**, 5125–5131.

66 Mezei, M., and Beveridge, D.L. (1987) Free energy simulations, in *Computer Simulation of Chemical and Biomolecular Systems (Annals of the New York Academy of Sciences*, vol. **482** (ed. W.L. Jorgensen) New York Academy of Sciences, pp. 1–27.

67 (a) Shing, K.S., and Gubbins, K.E. (1982) *Mol. Phys.*, **46**, 1109–1128; (b) Shing, K.S., and Gubbins, K.E. (1983) *Mol. Phys.*, **49**, 1121–1138; (c) Bennett, C.H. (1976) *J. Comp. Phys.*, **22**, 246.

68 (a) Kumar, S., Bouzida, D., Swendsen, R.H., Kollman, P.A., and Rosenberg, J.M. (1992) *J. Comput. Chem.*, **13**, 1011; (b) Carl, D.R., and Feller, S.E. (2003) *Langmuir*, **19** (**20**), 8560–8564.

69 (a) Jarzynski, C. (1997) *Phys. Rev. Lett.*, **78**, 2690–2693; (b) Jarzynski, C. (1997) *Phys. Rev. E*, **56**, 5018–5035; (c) Park, S., Khalili-Araghi, F., Tajkhorshid, E., and Schulten, K. (2003) *J. Chem. Phys.*, **119**, 3559–3566.

70 (a) Darve, E., and Pohorille, A. (2001) *J. Chem. Phys.*, **115**, 9169–9183; (b) Henin, J., Shinoda, W., and Klein, M.L. (2008) *J. Phys. Chem. B*, **112**, 7008.

71 (a) Matubayasi, N., and Nakahara, M. (2000) *J. Chem. Phys.*, **113**, 6070; (b) Matubayasi, N., and Nakahara, M. (2002) *J. Chem. Phys.*, **117**, 3605; (c) Matubayasi, N., Shinoda, W., and Nakahara, M. (2008) *J. Chem. Phys.*, **128**, 195107.

72 Shinoda, W., Mikami, M., Baba, T., and Hato, M. (2004) *J. Phys. Chem. B*, **108**, 9346–9356.

73 (a) Shinoda, K., Shinoda, W., Baba, T., and Mikami, M. (2004) *J. Chem. Phys.*, **121**, 9648–9654; (b) Shinoda, W., Shinoda, K., Baba, T., and Mikami, M. (2005) *Biophys. J.*, **89**, 3195–3202.

74 Shinoda, K., Shinoda, W., and Mikami, M. (2008) *J. Comput. Chem.*, **29**, 1912.

75 Schatzberg, P. (1965) *J. Polym. Sci. C*, **10**, 87.

76 Jorgensen, W.L., Chandrashekar, J., Madura, J.D., Impey, R., and Klein, M.L. (1983) *J. Chem. Phys.*, **79**, 926.

77 (a) Disalvo, E.A., and Simon, S.A. (eds) (1995) *Permeability and Stability of Lipid Bilayers*, CRC Press, Boca Raton, FL; Jansen, M., and Blume, A. (1995) *Biophys. J.*, **68**, 997; (c) Bacic, G., Srejic, R., and Ratkovic, S. (1990) *Stud. Biophys.*, **138**, 95; (d) Lawaczeck, R. (1979) *J. Membr. Biol.*, **51**, 229; (e) Carruthers, A., and Melchior, D.L. (1983) *Biochemistry*, **22**, 5797; (f) Paula, S., Volkov, A.G., van Hoek, A.N., Haines, T.H., and Deamer, D.W. (1996) *Biophys. J.*, **70**, 339; (g) Negrete, H.O., Rivers, R.L., Gough, A.H., Colombini, M., and Zeidel, M.L. (1996) *J. Biol. Chem.*, **271**, 11627.

78 Mathai, J.C., Sprott, G.D., and Zeidel, M.L. (2001) *J. Biol. Chem.*, **276**, 27266.

79 (a) Saito, H., Shinoda, W., and Mikami, M. (2008) *J. Phys. Chem. B*, **112**, 11305; (b) Saito, H., Shinoda, W., and Mikami, M. (2009) *Chem. Phys. Lett.*, **468**, 260.

80 (a) Chiu, S.W., Pandit, S.A., Scott, H.L., and Jakobsson, E. (2009) *J. Phys. Chem. B*, **113**, 2748; (b) Davis, J.E., Raharnan, O., and Patel, S. (2009) *Biophys. J.*, **96**, 385; (c) Hogberg, C.J., Nikitin, A.M., and Lyubartsev, A.P. (2008) *J. Comput. Chem.*, **29**, 2359; (d) Klauda, J.B., Venable, R.M., MacKerell, A.D., and Pastor, R.W. (2008) Considerations for lipid force field development, in *Computational Modeling of Membrane Bilayers* (ed. S.E. Feller), Academic Press, pp 1; (e) Taylor, J., Whiteford, N.E., Bradley, G., and Watson, G.W. (2009) *Biochim. Biophys. Acta*, **1788**, 638; (f) Rosso, L., and Gould, I.R. (2008) *J. Comput. Chem.*, **29**, 24.

81 Klein, M.L., and Shinoda, W. (2008) *Science*, **321**, 798.

82 (a) Shinoda, W., Devane, R., and Klein, M.L. (2007) *Mol. Simul.*, **33**, 27; (b) Shinoda, W., Devane, R., and Klein, M.L. (2008) *Soft Matter*, **4**, 2454.

83 Shelley, J.C., Shelley, M.C., Reeder, R.C., Bandyopadhyay, S., and Klein, M.L. (2001) *J. Phys. Chem. B*, **105**, 4464.

84 Marrink, S.J., de Vries, A.H., and Mark, A.E. (2004) *J. Phys. Chem. B*, **108**, 750.

85 Izvekov, S., and Voth, G.A. (2005) *J. Phys. Chem. B*, **109**, 2469.

86 Santangelo, G., Matteo, A., Muller-Plathe, F., and Milano, G. (2007) *J. Phys. Chem. B*, **111**, 2675.

87 Nielsen, S.O., Ensing, B., Moore, P.B., and Klein, M.L. (2008) Coarse grained-to-atomistic mapping algorithm: a tool for multiscale simulations, in *Multiscale Simulation Methods for Nanomaterials* (ed. R.B. Ross and S. Mohanty), John Wiley & Sons, Inc., New Jersey, pp. 73–88.

10
Structures of Poly(dimethylsiloxane)-Poly(oxyethylene) Diblock Copolymer Micelles in Aqueous Solvents

Masaya Kaneko, Takaaki Sato, Bradley Chmelka, Kenji Aramaki, and Hironobu Kunieda

10.1
Introduction

A-B type diblock copolymers consisting of two incompatible segments often show microphase separation into for example, cubic, hexagonal, and lamellar phases. During such microphase separation, there evolve nanometer-ordered structures that show the specific viscoelastic behaviors. As discussed in several recent review articles [1–5], such phenomena are important in the fields of complex fluids, as well as in self-assembled nanomaterials. The formation of these self-assemblies requires that $\chi N_{AB} > 10.5$, where χ is the Flory interaction parameter between A and B segments, and N_{AB} is the degree of polymerization of the block copolymer species [6]. The addition of a selective solvent to a block copolymer can greatly expand the range of self-assembled compositions and structures that can be prepared [7–9].

For systems that specifically possess poly(oxyethylene) chains as a copolymer block, a variety of solvent-dependent structural configurations can exist. Water, in particular, is a good solvent for poly(oxyethylene) moieties, because of their reciprocal hydrogen-bonding properties [10]. For example, Kunieda *et al.* have reported the phase behaviors of polydimethylsiloxane-poly(oxyethylene) diblock copolymer ($Si_mC_3EO_n$) with different hydrophilic and hydrophobic-chain lengths in water [11]. The type of liquid crystals formed in the water/block copolymer systems is mainly fixed by the volume fraction of the hydrophilic part in block copolymer molecule, f_{EO}. Basically, if f_{EO} is multiplied by 20, the value corresponds to the HLB value in the surfactant field, which determines the spontaneous curvature of amphiphile-film. Lodge *et al.* summarized the phase behavior for block copolymers in solvents of varying selectivity [12]. The full phase diagrams of polystyrene-*b*-polyisoprene were constructed using di-*n*-butyl phtahalate, diethyl phthalate and dimethyl phthalate as solvents. With the increase in the selectivity of solvent, block copolymer tends to form more types of self-assembled morphologies. If poly(ethylene glycol)s (EG_x) are selected as solvents instead of water, the hydrogen-bond donor site is decreased and the formed liquid crystal would be changed. For example, a

Self-Organized Surfactant Structures. Edited by Tharwat F. Tadros

ISBN: 978-3-527-31990-9

lamellar liquid-crystalline phase is dominant in water/$Si_{14}C_3EO_{33.1}$ and ethylene glycol (EG_1)/$Si_{14}C_3EO_{33.1}$ systems, while a hexagonal liquid-crystalline (H_1) phase appears in EG_7/$Si_{14}C_3EO_{33.1}$ system [13]. The type of the formed aggregate of A-B type amphiphilic block copolymer in a selective solvent is dramatically changed depending on the solvent–property and solvent–polymer interacting schemes.

Since the 1980s, phase behaviors of A-homopolymer/A-B diblock copolymer binary systems have been investigated experimentally [14–22] and theoretically [23–25], where the EG_x/$Si_{14}C_3EO_n$ systems that we focus on in this study are also their counterpart. The phase behaviors of binary blends of homopolymer and block copolymer are affected not only by f_{EO}, ϕ, and χN_{AB} but also by $\alpha = N_h/N_{AB}$, where N_h and N_{AB} are the polymerization degrees of homopolymer and block copolymer, respectively. In fact, with increasing N_h, the macro phase separation is enhanced in homopolymer and block copolymer binary blends. Thomas *et al.* found in their small X-ray scattering [26] and TEM [27] investigations that a sphere-to-rod transition of micellar shape takes place at lower concentration with increasing unit number of solvent molecules. However, only a limited number of studies have been reported on the self-assemblies of A-B type diblock copolymer in polar solvents [28].

In this content, we report the change in micellar size and shape in the EG_x/$Si_{14}C_3EO_n$ systems as a function of x, which were systematically investigated by pulsed-field gradient (PFG) ^1H NMR and small-angle X-ray scattering (SAXS). The solvent employed here, EG_x, is commonly used in cosmetic or personal care products to increase the solution viscosity [29]. It is therefore considered that the effect of EG_x on the micellar structure would have important implications in understanding and controlling key properties of these products, such as the emulsion stability and detergency.

10.2
Experimental Section

10.2.1
Materials

The diblock copolymers used in this experiment were polydimethylsiloxane–poly(oxyethylene) diblock copolymer ($Si_{14}C_3EO_{33.1}$ and $Si_{14}C_3EO_{51.6}$), kindly provided from Dow Corning Toray Co Ltd., Japan. C_3 means a propylene residue connecting hydrophobic and hydrophilic parts. The number of subscript indicates the polymerization degree for each part, determined by GPC. The polydispersity indices, M_w/M_n are 1.19 for the $Si_{14}C_3$- part, 1.15 for $EO_{33.1}$ and 1.13 for $EO_{51.6}$, respectively [11]. The purities are 89.4% for $Si_{14}C_3EO_{33.1}$ and 87.9% for $Si_{14}C_3EO_{51.6}$. Ethylene glycol (EG_1) (>99.0%), Diethylene glycol (EG_2) (>99.0%) and Triethylene glycol (EG_3) (>99.0%) were purchased from Tokyo Kasei Kogyo. Poly(ethylene glycol)s of molecular weight 300, 400 and 600 were obtained from Wako Chemicals, and they are abbreviated as EG_7, EG_9 and EG_{14}, respectively, according to the mean polymerization degree of ethylene glycol unit. Poly(ethylene glycol) of molecular weight 900 ($EG_{20.5}$) was purchased from Aldrich Sigma, Japan. The density of each solvent was measured by DSA5000 (Anton Paar, Austria) at 60 °C. All chemicals were used as received without further purification.

10.2.2
Pulsed-Field Gradient (PFG) ^{1}H NMR

Self-diffusion measurements were performed using a Bruker AVANCE 500 NMR spectrometer operating at 500.13 MHz for ^{1}H and equipped with a field-gradient probehead. 5-mm sample tubes were used, with sample volumes of ~400 μL. A separate thin tube made from the end of the glass pipette containing deuterated water (D_2O 99.9%, Cambridge Isotopes) was placed inside the 5-mm sample tubes for signal-locking purposes. The melting points of $Si_{14}C_3EO_{33.1}$ and $Si_{14}C_3EO_{51.6}$ are 45.0 °C and 50.5 °C, respectively. Therefore, samples were kept at 60 °C at least for 15 min to allow thermal equilibration prior to performing the diffusion measurements. Solvent convection currents can interfere with accurate measurement of diffusion coefficients, especially at high temperatures. Here, this was avoided by use of a standard pulsed-gradient double-stimulated-echo (PG-DSTE) ^{1}H NMR technique [30]. From the mono-exponential decay of intensity of the well-resolved proton signal at ~0.45 ppm associated with the $CH_3\{2(CH_3)SiO\}_{13}\{2(CH_3)Si\}(CH_2)_3$- moieties, the self-diffusion coefficient, D, of these surfactant species can be determined. With respect to the signal intensities in the presence and the absence of the pulsed gradient, I and I_0, respectively, the intensity attenuation in PFG-NMR is described by [31]

$$I = I_0 \exp(-kD), \tag{10.1}$$

where $k = \gamma^2 g^2 \delta^2(\Delta - \delta/3)$, γ is the gyromagnetic ratio (=2.67520 × 10^8 rad T^{-1} s^{-1} for protons), g is the field gradient strength, δ is the duration of the applied gradient pulse, and Δ the diffusion time. In all experiments conducted here, the values of δ and Δ were fixed at 7 ms and 200 ms, respectively, and the gradient strength varied from 1.69 T to 32.0 T. Treating the micelles as hard spheres, the hydrodynamic radius R_H can be calculated according to the Stokes–Einstein equation [32]:

$$R_H = \frac{k_B T}{6\pi \eta_0 D}, \tag{10.2}$$

where η_0 is the solvent viscosity, k_B is the Boltzmann constant, and T is the absolute temperature.

10.2.3
Small-Angle X-Ray Scattering (SAXS)

To investigate the size, shape, and internal density distribution of micelles formed by $Si_{14}C_3EO_n$ diblock copolymers in polar solvents, SAXS measurements were performed at 60 °C by using a SAXSess camera (Anton Paar, PANalytical), equipped with a PW3830 laboratory X-ray generator with a long fine focus sealed glass X-ray tube (K_α wavelength of $\lambda = 0.1542$ nm) (PANalytical), which was operated at 40 kV and 50 mA, a multilayer optics, a block collimator, and a semitransparent beam stop. The sample holder made of thin quartz was placed under vacuum and the temperature was controlled with a TCS 120 temperature control unit (Anton Paar). The sample was kept at 60 °C at least for 15 min before performing the measurements. The scattered intensity was detected by an image plate (IP) detector and a recorded intensity was read off by a Cyclone storage phosphor system (Perkin Elmer, USA). All measured data were stored as the function of the scattering vector $q = (4\pi/\lambda)\sin(\theta/2)$, where λ is the X-ray wavelength (Kα; 0.1542 nm) and θ is the total scattering angle, and normalized to the same attenuated primary intensity for transmission calibration. The absolute scale calibration was done using the experimental scattering intensity of water as a secondary standard by referring to $I(q = 0) = 1.633 \times 10^{-2}\,\text{cm}^{-1}$ [33]. The maximum resolution attained in this study was $q_{\text{min}} = \sim 0.08\,\text{nm}^{-1}$, which corresponds to the maximum detectable particle size of ~40 nm.

The scattered intensity $I(q)$ from globular particles is generally given by

$$I(q) = nP(q)S(q) \tag{10.3}$$

where n is the number density of particles, $P(q)$ and $S(q)$ are called the form factor and the structure factor, respectively. $P(q)$ denotes the intraparticle scattering function and is given by Fourier transform of the pair-distance distribution function (PDDF), $p(r)$, as

$$P(q) = 4\pi\int_0^\infty p(r)\frac{\sin qr}{qr}dr \tag{10.4}$$

The real space function $p(r)$ is directly linked to the spatial autocorrelation function (convolution square) of electron-density fluctuations inside the particle,

$$\Delta\tilde{\rho}^2(r) = \langle\Delta\tilde{\rho}^2(\mathbf{r})\rangle = \left\langle\int_{-\infty}^{\infty}\Delta\rho(\mathbf{r}_1)\Delta\rho(\mathbf{r}_1 - \mathbf{r})d\mathbf{r}_1\right\rangle \tag{10.5}$$

where r is the distance between two scattering centers that can be chosen within the single scattering particle. The r-value where $p(r)$ converges to zero gives an estimation of the maximum dimension of the particle, d_{max}, and from the functional shape of $p(r)$, the shape and internal structure of the particle can be deduced. Additionally, when the particle geometry satisfies the spherical symmetry, the radial electron-density profile, $\Delta\rho(r)$, can be calculated by a convolution square root operation.

In an identically diluted nonionic system (<1 wt%), the effects of $S(q)$ can be neglected, that is, the approximation $S(q) \sim 1$ is valid. However, $S(q)$ should be considered for the investigated systems. $S(q)$ is given by the Fourier transformation of the total correlation function, $h(r) = g(r) - 1$, as

$$S(q) - 1 = 4\pi n\int_0^\infty [g(r) - 1]r^2\frac{\sin qr}{qr}dr \tag{10.6}$$

where $g(r)$ is the pair correlation function. $g(r)$ can be interpreted as the probability function to find particles at the distance r from an arbitrarily chosen particle situated at $r = 0$.

All measured SAXS data were analyzed by the generalized indirect Fourier transformation (GIFT) technique with the Boltzmann simplex simulated annealing (BSSA) algorithm [34, 35]. The GIFT calculation is based on the analytical or numerical solution of the Ornstein–Zernike (OZ) equation that describes the interplay between the total ($h(r)$) and direct ($c(r)$) correlation functions;

$$h(r) = c(r) + n\int d\mathbf{r}' c(|\mathbf{r} - \mathbf{r}'|)h(\mathbf{r}'). \tag{10.7}$$

For the investigated systems, the pair potential, $v(r)$, is approximated by a hard-sphere (HS) interaction model,

$$v_{HS}(r) = \begin{cases} \infty; & r < \sigma \\ 0; & r > \sigma \end{cases}, \tag{10.8}$$

where σ is the diameter of a hard sphere, the half of σ corresponding to the interaction radius. To find the solution of Equation 10.7, the further relation among $h(r)$, $c(r)$, and $v(r)$, the so-called closure relation is needed. In this study, within the mean spherical approximation (MSA), the Percus–Yevick (PY) closure relation given by [36]

$$g(r) = e^{-v(r)/k_BT}[1 + h(r) - c(r)] \tag{10.9}$$

was selected for the calculation of $S(q)$. Note that the main purposes of the calculation of $S(q)$ in this study is to suppress the influence of the interparticle scattering contribution on the extraction of $P(q)$ and the corresponding $p(r)$.

10.2.4
Viscosity Measurements

The solvent viscosity was measured by an Automated Microviscometer; AMVn (Anton Paar GmbH, Austria, Graz) that is based on the rolling-ball principle. The diameter of the capillary having an inclination of 30° or 40° and ball were 1.8 mm and 1.5 mm, respectively. The filled solvent in the capillary was kept at 60 °C for 15 min before measurements. All measured values were calibrated by the value of EG_1 (4.95 mPa s) at 60 °C. The accuracy was ±1%, which was taken into account when the diffusion coefficients were converted to the hydrodynamic radii.

10.3
Results and Discussions

10.3.1
Diffusion Coefficients of Micelles for $Si_{14}C_3EO_n$ in EG_x

To study the formed aggregates for $Si_{14}C_3EO_n$ in poly(ethylene glycol) solutions having different molecular weight, the measurement of diffusion coefficients for micelles were carried out by the PFG-NMR technique. The concentration of $Si_{14}C_3EO_n$ in the solution is fixed at 2 wt%. The intensity decays for $CH_3\{2(CH_3)SiO\}_{13}\{2(CH_3)Si\}(CH_2)_3$-peak (~0.45 ppm) at 60 °C are presented in Figure 10.2.

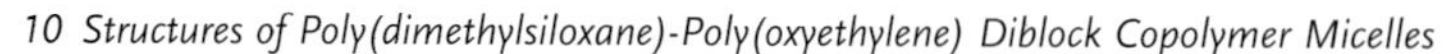

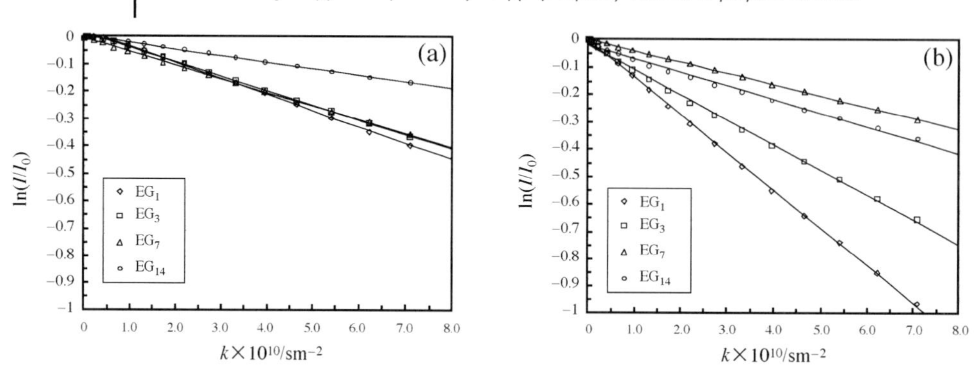

Figure 10.1 Echo decay for the proton in the $CH_3\{2(CH_3)SiO\}_{13}\{2(CH_3)Si\}(CH_2)_3$- moeties as a function of k. (a) the $EG_x/Si_{14}C_3EO_{33.1}$ system; (b) the $EG_x/Si_{14}C_3EO_{51.6}$ system.

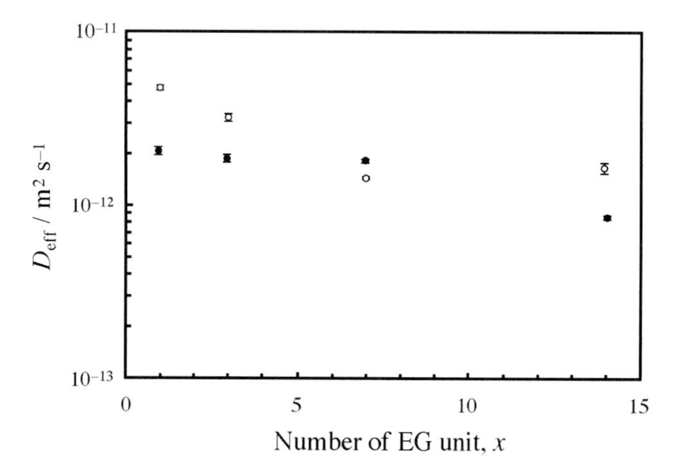

Figure 10.2 The self-diffusion coefficients D_{eff} for micelles as a function of the number of EG unit x at 60 °C in the $EG_x/Si_{14}C_3EO_{33.1}$ systems (filled circles) and the $EG_x/Si_{14}C_3EO_{51.6}$ systems (empty circles).

All the systems show a practically single-exponential decay although the initial slow decay can be seen, presumably due to unreacted silicone chains present as a minor impurity. This denotes both block copolymers form micelles having monomodal size distribution in the present solvents, EG_x. Since $Si_{14}C_3EO_n$ molecule has a long enough hydrophobic chain to give low cmc, a $Si_{14}C_3EO_n$ monomer soluble in the solvent has little contribution to the observed self-diffusion coefficient. The self-diffusion coefficients for micelles in the $EG_x/Si_{14}C_3EO_n$ systems were evaluated according to Equation 10.1 and the results are shown in Figure 10.2. The plot of echo decay has an uncertain distribution depending on taking the different slope in echo decay including first slow decay or not, in Figure 10.1. This factor is considered as an error.

Comparing the self-diffusion coefficients at constant x, the $Si_{14}C_3EO_{51.6}$ systems give a higher self-diffusion coefficient than $Si_{14}C_3EO_{33.1}$ except for $x = 7$ as shown

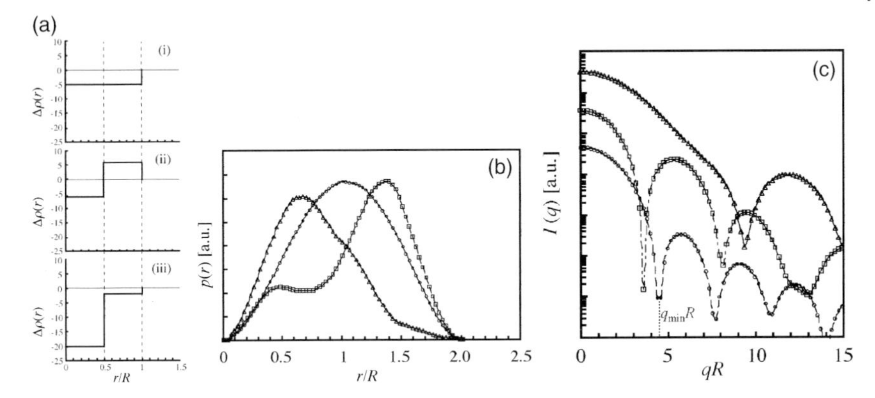

Figure 10.3 The simulated radial electron-density profiles $\Delta\rho(r)$ (a), pair-distance distribution function $p(r)$ (b), and the scattering curves $I(q)$ (c) for a monodisperse spherical particle having (i) a homogeneous electron-density distribution (circle), (ii) the core–shell(corona) structure with negative and positive electron-density fluctuations (square), and (iii) the inhomogeneous electron-density distribution where electron densities of both core and corona are negative (triangle).

in Figure 10.2. Since the hydrodynamic radius is inversely proportional to the self-diffusion coefficient in a given solvent as described in Equation 10.2, the results indicate that the size of micelle in the $EG_x/Si_{14}C_3EO_{51.6}$ systems is smaller than that in the $EG_x/Si_{14}C_3EO_{33.1}$ systems. If the shape of the micelle is spherical in the systems with different copolymer, the size of micelle in the $EG_x/Si_{14}C_3EO_{33.1}$ systems should be smaller than that in the $EG_x/Si_{14}C_3EO_{51.6}$ systems. The obtained self-diffusion coefficients, however, give an opposite trend, indicating that the shape of micelles is different, that is, the formation of more elongated micelles in the $EG_x/Si_{14}C_3EO_{33.1}$ systems, which is reasonable in terms of the generally smaller spontaneous curvature for the shorter hydrophilic chain attached to the same hydrophobic chain. To confirm the difference in shape of the micelles in these systems, we performed SAXS experiments, as discussed in the next section.

10.3.2
Model Calculations of the Scattering Functions

For the better and easy understanding of the SAXS results on the investigated systems systematically shown in the next section, we briefly introduce the models of typical particle systems with different internal electron-density distributions. If the particle has a homogeneous electron-density distribution, the shape of the aggregate can easily be deduced from the shape of $p(r)$ [37]. However, when the electron-density fluctuation inside the particle is inhomogeneous, it becomes somewhat difficult to estimate the exact shape of the aggregate, but instead, $p(r)$ offers the information about the internal structure of aggregates. We simulated $p(r)$ and the scattering curves with the three different types of electron-density profiles described in Figure 10.3a by using the Singlebody MC program [38], and the results are shown in Figures 10.3b and c. The spherical particle having the homogeneous electron density shows the symmetric bell-shape of $p(r)$ and gives $I(q)$ the minimum at $q_{\min}R = 4.49$, where R is the radius of the particle as highlighted by a broken line. In the case of the micelle having a core–shell structure whose signs of the electron-density fluctuations are alternate, $p(r)$ exhibits the local minimum in low-r regime and the inflection points, respectively, just before and after the local minimum approximately correspond to the core-radius and diameter. In addition, $q_{\min}$ giving the first minimum of $I(q)$ is shifted to a low-q value compared to that of the homogeneous particle, depending not only on the ratio of the core and shell thickness but sensitively on the contrast. The spherical particle possessing the inhomogeneous but surely positive or negative electron-density fluctuation offers a small extended tail in high-r regime in $p(r)$ and a lack of the local minimum of $I(q)$ around $qR \sim 4.49$ that exists for the homogeneous spherical particle.

10.3.3
Shape of Micelles for $Si_{14}C_3EO_n$ in EG_x

To investigate the shape of micelles in $EG_x/Si_{14}C_3EO_n$ systems, SAXS measurements were performed at 60 °C. The scattered intensities $I(q)$ are shown in Figure 10.4. The concentration of $Si_{14}C_3EO_n$ in EG_x is fixed to 2 wt%.

The slope of the forward scattering intensities of the $EG_{14}/Si_{14}C_3EO_{33.1}$ and $EG_{20.5}/Si_{14}C_3EO_{33.1}$ systems are markedly steeper than those of the other $EG_x/Si_{14}C_3EO_{33.1}$ systems, which is an indication of the growth of micelle. The change of the scattering curves for the $EG_x/Si_{14}C_3EO_{51.6}$ systems is relatively less pronounced. Further insights into the micellar structures are obtained from the pair-distance distribution functions, $p(r)$. Figure 10.5 shows $p(r)$ calculated as the output of the GIFT analysis.

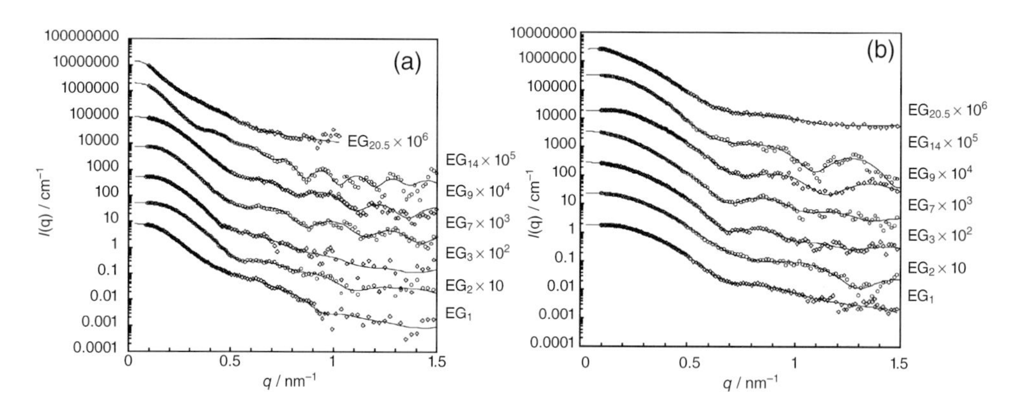

Figure 10.4 The collimation-calibrated (desmeared) SAXS curves for $EG_x/Si_{14}C_3EO_{33.1}$ systems (a), and for $EG_x/Si_{14}C_3EO_{51.6}$ systems (b) at 60 °C on an absolute scale as a function of the magnitude of the scattering vector q. The solid lines represent the fitting with GIFT technique. The arrow indicates $q_{\min}$ calculated from $q_{\min}R$ (= $d_{\max}/2$) ~ 4.49 under the assumption of a homogeneous sphere.

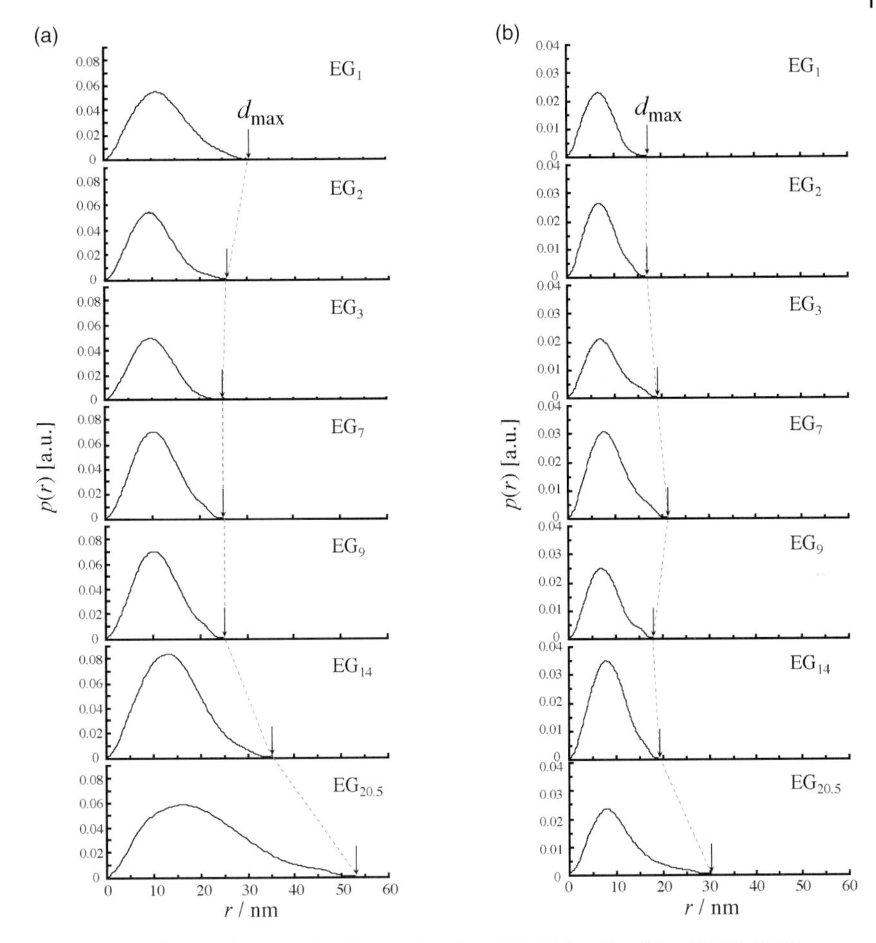

Figure 10.5 The pair-distance distribution functions (PDDFs), $p(r)$, of the $EG_x/Si_{14}C_3EO_{33.1}$ systems (a) and those of the $EG_x/Si_{14}C_3EO_{51.6}$ systems (b) at 60 °C. The concentration of $Si_{14}C_3EO_n$ in EG_x is fixed to 2 wt%.

One can notice that all PDDFs of the $EG_x/Si_{14}C_3EO_n$ systems exhibit small extended tails in the high-r regime. In addition, if the observed particle is assumed to be homogeneous, SAXS curves should give first local minima at $q_{min} = 4.49/R$, where q_{min} is around 0.4 from EG_2 to $EG_9/Si_{14}C_3EO_{33.1}$ systems, and it is around 0.5 from EG_1 to $EG_{14}/Si_{14}C_3EO_{51.6}$ systems. However, there are no local minima at those values in the obtained SAXS curves in Figure 10.4. As pointed out in advance by the MC calculation for typical three model systems (i)–(iii), these observed features underline that the electron-density distribution inside the particle for the $EG_x/Si_{14}C_3EO_n$ systems is inhomogeneous, but different from the conventional core–shell structure of micelles in aqueous media, like poly(oxyethylene) alkyl ethers in water, the electron-density fluctuations of the core and corona from the average (solvent) electron density are sorely negative or positive. Indeed, we confirmed that the electron density of EO_n chain of copolymer (349.6 electrons/nm^3) is slightly smaller than that of the solvent EG (358.2 electrons/nm^3) at 60 °C, and $Si_{14}C_3$ (314.3 electrons/nm^3 at 25 °C) is considerably lower than them. So the EG_x/$Si_{14}C_3EO_n$ systems can be categorized into a counterpart of the model (iii) in terms of the internal electron-density distribution.

The r-value at which $p(r)$ decays to zero, the so-called d_{max}, exhibits the maximum dimension of micellar size including the corona of micelle, which is indicated by the arrows in Figure 10.5. The concentration series of $p(r)$ is consistent with the results of the PFG-NMR in terms of the variation of d_{max} parallel to the change of D_{eff}. In $EG_x/Si_{14}C_3EO_{33.1}$ systems, the micellar size is once reduced when EG_1 is replaced with EG_2. From EG_2 to EG_9, d_{max} is slightly decreased or unchanged. With the further increase in x (EG_{14} and $EG_{20.5}$), the maximum dimension of $p(r)$ is rapidly enhanced, indicating the rapid elongation of micelle. Although, to say exactly, d_{max} in the $EG_{20.5}$ system is beyond the maximum resolution of the SAXS measurements ($\pi/q_{min} \sim 40$ nm) and thus the calculation of $p(r)$ involves a postulated extrapolation, the abrupt growth of micelle is clearly detected. In contrast, different from the peculiar behavior of the $Si_{14}C_3EO_{33.1}$ system, $Si_{14}C_3EO_{51.6}$ forms the smallest micelle in EG_1. With the increase in the number of EG unit, d_{max} monotonously increases up to EG_7. From EG_7 to EG_{14}, the micellar size is decreased or virtually constant. In the $EG_{20.5}$ system, the elongated micelle is observed.

Consistent with the tendency of D_{eff}, d_{max} for the $EG_x/Si_{14}C_3EO_{33.1}$ systems becomes systematically larger than that for the $EG_x/Si_{14}C_3EO_{51.6}$ systems. This finding again rules out the existence of spherical micelles in the $EG_x/Si_{14}C_3EO_{33.1}$ systems. However, interestingly, the $EG_x/Si_{14}C_3EO_{33.1}$ systems show the maximum D_{eff} in EG_7 that exceeds D_{eff} for the $EG_7/Si_{14}C_3EO_{51.6}$. Suppose that the EG_x/$Si_{14}C_3EO_{51.6}$ systems excluding that in $EG_{20.5}$ are nearly spherical, the micelle solely in the $EG_7/Si_{14}C_3EO_{33.1}$ system is also estimated to be spherical-like. In the systems where we observed the elongated structures in SAXS; $Si_{14}C_3EO_{33.1}$ in EG_1, EG_{14}, and $EG_{20.5}$, the shape of micelles is considered to be oblate because a bell-shaped distribution observed in the spherical-like particle becomes gradual and fat, which can be obtained typically for lamellar particles [39]. The fact that neat $Si_{14}C_3EO_{33.1}$ forms a lamellar liquid crystal having a planar interface supports this view because the system approaches the boundary of a macrophase separation on increasing the number of EG unit for the solvent [40]. In the $EG_{20.5}/Si_{14}C_3EO_{51.6}$ system, the r value giving the maximum of $p(r)$ is not largely shifted compared to other EG_x/$Si_{14}C_3EO_{51.6}$ systems in which the spherical-like particles are observed, while the tail alone is extended toward a high-r side. This feature implies that the shape of micelle in the $EG_{20.5}/Si_{14}C_3EO_{51.6}$ system is prolate rather than oblate.

10.3.4
Internal Structures of Micelles

To investigate the internal density distribution inside the micelle, the radial electron-density distribution profile, $\Delta\rho(r)$, was calculated by the deconvolution procedure of $p(r)$ for the circumstantially judged spherical micellar systems; $Si_{14}C_3EO_{33.1}$ from EG_7 to EG_9 and $Si_{14}C_3EO_{51.6}$ from EG_1 to EG_{14}. The results are displayed in Figure 10.6.

On the horizontal axis, r indicates the distance from the center of the micelle as shown in the inserted schematic picture. The deconvolution procedure was not made for $EG_2/Si_{14}C_3EO_{33.1}$ and $EG_3/Si_{14}C_3EO_{33.1}$ systems, because the spherical symmetry does not seem to be valid from the self-diffusion measurements. As the solid arrows in the $\Delta\rho(r)$ profile (Figure 10.6) indicate, the mean radius of the micellar core can be estimated form the point where the electron density starts to change. In the $EG_x/Si_{14}C_3EO_{33.1}$ systems, the length of core (~4.2 nm) is a little longer or almost consistent with the extended length of $Si_{14}C_3$ (~4.1 nm), whereas the hydrophobic chain is mostly extended state in a spherical micelle [43]. In the $EG_x/Si_{14}C_3EO_{51.6}$ systems excluding $EG_{20.5}$, the core radius of the micelle is evaluated to be smaller, if judged from the onset of the change in $\Delta\rho(r)$ at $r \sim 3$ nm, which seems to indicate that the more hydrophilic copolymer ($Si_{14}C_3EO_{51.6}$) has the smaller core size, although the unit numbers for hydrophobic chain in both copolymers are exactly the same, $Si_{14}C_3$-. Thomas *et al.* reported that the core radius is decreased with increasing polymerization degree of polystyrene of block copolymer making a corona of the spherical micelle in the polystyrene–polybutadiene block copolymer and polystyrene binary system [26]. Hence, the core radius is determined not only by N_{CB} but also N_{CA} and N_{hA}, as is described by $l_B \propto N_{CA}{}^{\mu} N_{hA}{}^{\nu}$, in the scaling law, where N_{CA}, N_{CB}, and N_{hA} are the polymerization degree of one block forming micellar corona, that of the other block forming micellar core and that of homopolymer [44]. If $N_{hA} \ll N_{CA}$ as is similar to this system, μ is small and negative. Thus, l_B is decreased with increasing N_{CA}. Several authors attempted to explain theoretically this change in the core size [16, 45]. The copolymer possessing the longer hydrophilic part enables more solvent molecules to be accommodated, which induces the decrease of aggregation numbers. With the decrease in the aggregation number, the micellar core is shrunk. But, to minimize the interfacial energy, the micellar core tends to expand in order not to contact the solvents. These two opposite driving forces determine the radius of a core.

10.3.5 The Change in Micellar Shape

To discuss further the shape and internal structure of the micelles, the hydrodynamic radii calculated from the self-diffusion coefficients according to Equation 10.2 were combined and compared with the SAXS results. The self-diffusion coefficient of a hard sphere at infinite dilution, D_0, is described by [46],

$$D/D_0 = 1 - 2.1\phi_{HS}. \tag{10.10}$$

Here, ϕ_{HS} is the hard-sphere volume fraction. We use ϕ (volume fraction of copolymer) instead of ϕ_{HS} for the calculation of D_0 because the exact volume of micelles is unknown. The concentration of $Si_{14}C_3EO_n$ is ≈ 2 vol% so that the calculated values, D_0, differs from D by ~5% and this difference is noted as an error bar.

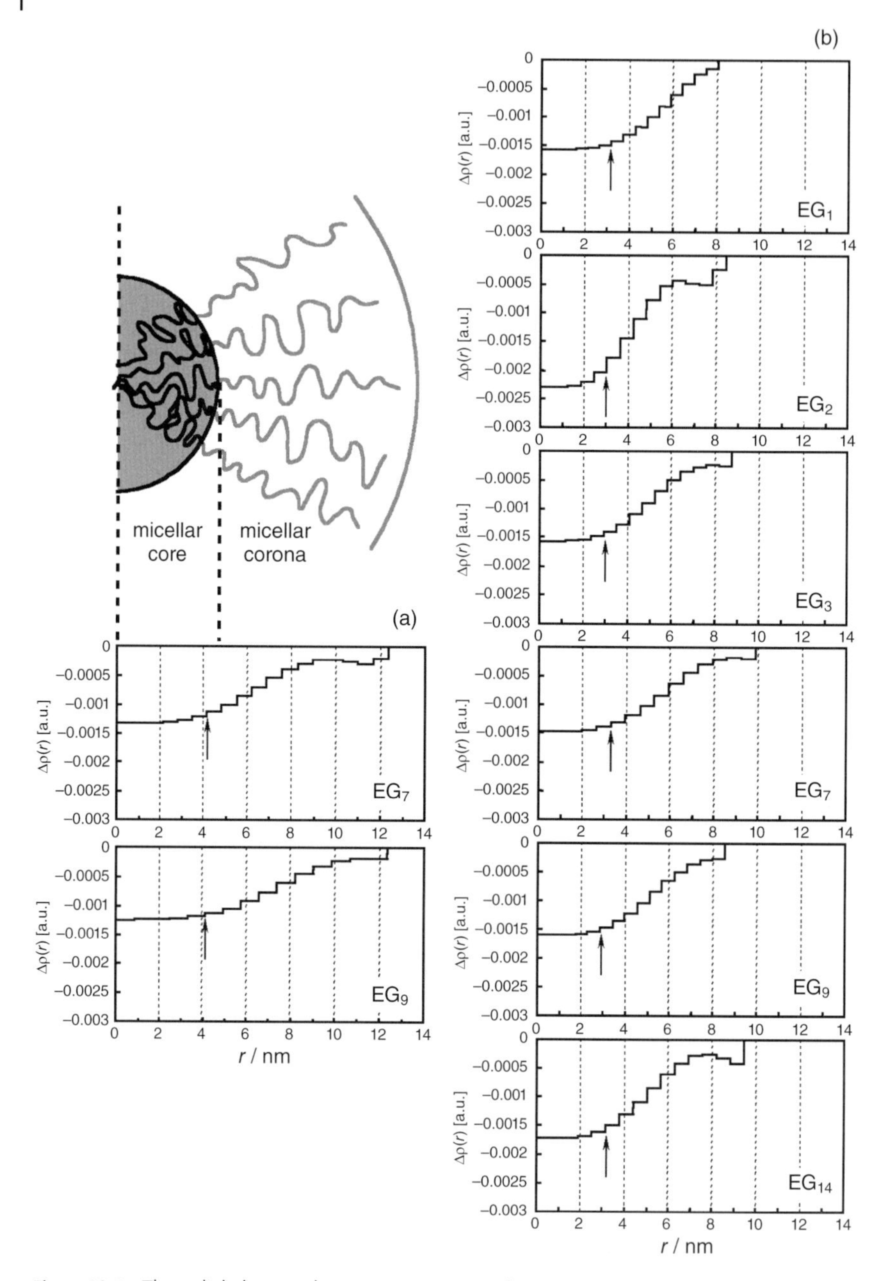

Figure 10.6 The radial electron-density profiles $\Delta\rho(r)$ for the spherical micelles in the $EG_x/Si_{14}C_3EO_{33.1}$ ($x = 7$ and 9) systems (a) and those in the $EG_x/Si_{14}C_3EO_{51.6}$ ($x = 1$–14) systems (b) at 60 °C calculated by a convolution square root operation with the help of the program DECON [41, 42]. The concentration of $Si_{14}C_3EO_n$ in EG_x is fixed at 2 wt%. The schematic picture of the core–corona structure is also inserted.

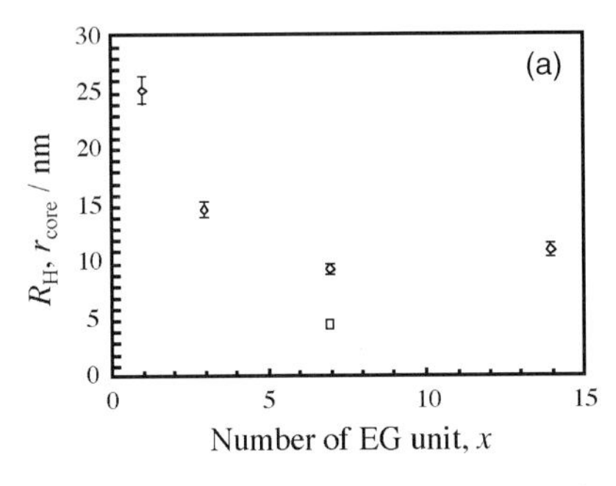

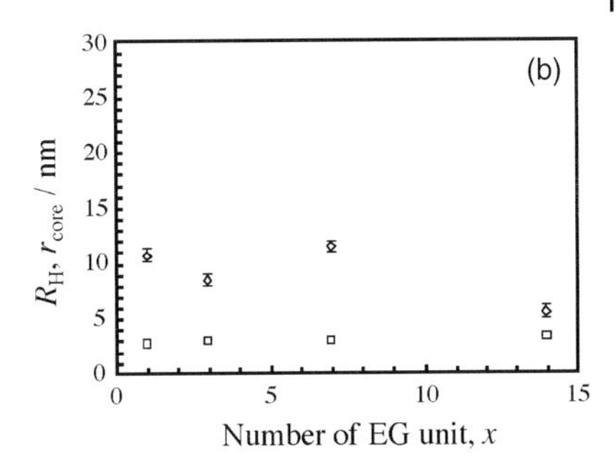

Figure 10.7 The hydrodynamic radii (R_H), diamond, and the radii of the core of spherical-like micelles (r_{core}), square, in EG_x for $Si_{14}C_3EO_{33.1}$ systems (a) and for $Si_{14}C_3EO_{51.6}$ systems (b) at 60 °C as a function of the number of EG unit, x, for EG_x.

Figure 10.7 presents the hydrodynamic radius R_H calculated according to Equation 10.2 using D_0 converted from the experimental D with Equation 10.10 and the radius of core r_{core} obtained from the $\Delta\rho(r)$ profile displayed in Figure 10.6.

In the $EG_x/Si_{14}C_3EO_{33.1}$ systems, R_H is decreased up to $x = 7$ and again increased due to the structural change of micelle, which is in line with the change in $p(r)$. On the other hand, in the $EG_x/Si_{14}C_3EO_{51.6}$ systems, the general trend of R_H shows that the micellar size is slightly increased up to $x = 7$, and the further increase in x makes R_H decrease. This trend is almost compatible with that of $p(r)$. With increasing the EG unit for solvent, the number of OH units is decreased. Thus, the compatibility between EO_n of the block copolymer and EG_x is enhanced, in other words, the solvent selectivity is increased, which promotes the penetration effect for EG_x into the corona of micelle although the size of the solvent molecule becomes large. However, with the further increase in the molecular weight of EG_x, in this studied system, above $x \sim 7$, the number of the swollen solvents is decreased because the large solvents are relatively impenetrable to the corona of a micelle. It is considered that the formed $Si_{14}C_3EO_{51.6}$ micelle in EG_1 is not spherically symmetrical. R_H would be overestimated when their structures are nonspherical. R_H does not always correspond to the half of micellar length extracted from SAXS experiments, probably because the diffusion coefficient is affected by the surface roughness of the micelle. Although the micelle is assumed to be a hard sphere when R_H is computed, the surface of micelle would be uneven due to the entanglement of long EO_n chain or it with solvent. So probably, the R_H value itself may not give the exact size. Nevertheless, the trend of the difference between R_H and r_{core} as a function of x gives a measure of the thickness of a corona for the spherical-like micelle and the trend of its variation depending on the solvents. The thickness of a corona for the micelle is determined by the balance between stretching of the hydrophilic chain by swelling solvent molecules and the configurational entropy that favors the short dense chain [47]. On decreasing the number of solvent molecules penetrating into the corona of the micelle, it becomes favorable for the hydrophilic chain consisting of the micellar corona to shrink and become entangled. Hence, the thinning of the micelle corona could be the reason for the decrease in the R_H as the increase in the EG unit of the solvent.

According to LOW theory [16], the free energy of a single micelle F is described as follows,

$$F = 4\pi l_B^2 \gamma + F_d + F_m. \tag{10.11}$$

The first term is equivalent to the interfacial energy between the silicone core and corona including solvent. First, the contribution of the first term is neglected because the surface tension for EG_1 is $50.2\,mN\,m^{-1}$ and those for $EG_{2\text{-}20.5}$ are $47.3 \pm 0.5\,mN\,m^{-1}$ and they are regarded as being constant. The second contribution is derived from the deformation of the copolymer hydrophobic chains consisting of the core of the micelle. The final term is attributed to the mixing entropy of solvent and poly(oxyethylene) chain in the corona of micelle. Despite the fact that the second contribution, the so-called deformation energy is highly responsible for the shape change of micelle, it is considered that this effect is relatively smaller than the other factors when the mixing entropy is large enough, namely, the large amounts of solvent molecules are incorporated into the hydrophilic corona part. As described previously, the shape change from spherical to nonspherical micelle takes place above $x = 14$ for the $EG_x/Si_{14}C_3EO_{33.1}$ systems and $x = 20.5$ for the $EG_x/Si_{14}C_3EO_{51.6}$ system. When the solvent swelling in the corona decreases, the contribution of F_m in the Equation 10.11 is relatively small, resulting in the deformation energy of hydrophobic chains, F_d, becoming dominant to the total free energy for the micelle. The F_d should be small in the case of nonspherical micelles because the hydrophobic chain is nearly extended length in the spherical micelle [48]. Hence, the transition from spherical to nonspherical shape of micelle could be observed at high x.

10.3.6
Contribution of Interfacial Tension on the Micelle Structure

As was mentioned in the Introduction, Kinning and Thomas *et al.* have reported the micellar growth in the A-B type block copolymer/A-homopolymer system as a function of the molecular weight of A-homopolymer [26, 27]. In the polystyrene–polybutadiene/polystyrene system, the shape of the micelle is changed from globular to elongated structure when the polymerization degree of polystyrene is increased from 20 to more than 300. However, the information on the micellar structure with the homopolymer having much lower polymerization degree, such as EG_1 studied in the present work has not been well known. The micellar structure in the $EG_1/Si_{14}C_3EO_{33.1}$ system is oblate rather than spherical, exceptionally. As is described in the previous section, EG_1 alone gives the slightly higher surface tension among EG_xs. Hence, the interfacial tension between EG_1 and the silicone chain is higher than that between $EG_{2\text{-}20.5}$ and the silicone chain. If the interfacial tension is large, the cross-sectional area for the silicone chain at the interface should be small for less exposure of the silicone chain against solvent, leading to the formation of an oblate or prolate structure, instead of a spherical micelle [11]. On the contrary, $Si_{14}C_3EO_{51.6}$ forms smaller and spherical-like micelles even in EG_1 system, because the longer poly(oxyethylene) chain can hold more solvents, leading to the formation of spherical-like micelle having the positive spontaneous curvature, just as the aqueous systems [49].

10.4 Conclusions

The size, shape, and internal structure of micelles of $Si_{14}C_3EO_{33.1}$ and $Si_{14}C_3EO_{51.6}$ in polar solvents, EG_x were systematically investigated using small-angle X-ray scattering in combination with a pulsed field gradient NMR. With increase in the number of EG unit for a solvent up to $x \sim 7$, $Si_{14}C_3EO_{33.1}$ micelle becomes small due to the structural change of micelle from nonspherical to spherical-like while $Si_{14}C_3EO_{51.6}$ micelle enlarges without the significant geometrical change from the spherical-like micelle owing to the enhancement of the solvent selectivity. As the number of EG units for a solvent is increased above $x \sim 7$, the $Si_{14}C_3EO_{51.6}$ micelle shrinks because the large solvent can penetrate less into the corona of micelle. With the further increase in the solvent chain length, the structure of micelle is changed to the elongated one, which is observed in the EG_{14}, $EG_{20.5}/Si_{14}C_3EO_{33.1}$ systems and $EG_{20.5}/Si_{14}C_3EO_{51.6}$ system, because the deformation energy of the hydrophobic chain of the copolymer is pronounced in the free energy of a single micelle. Although spherical-like micelles tend to be formed in the low molecular weight solvent having shorter chain length, the micelle with oblate shape is formed in the $EG_1/Si_{14}C_3EO_{33.1}$ system owing to the fact that the cross-sectional area of block copolymer at the interface should be small due to relatively high interfacial tension between the silicone core and EG_1.

The micellar structure of these studied $EG_x/Si_{14}C_3EO_n$ systems is mostly determined by the three factors; the interfacial tension between the silicone core and solvent, the solvent selectivity and the penetration effect of the solvent into the corona of micelle.

Acknowledgment

The author thanks Dr. Christian A. Steinbeck for helpful discussions of NMR techniques. This work was supported by Core Research for Evolutional Science and Technology (CREST) of JST Corporation and by the U.S. National Science Foundation Division of Materials Research under award DMR-02-33728 using Central Facilities of the NSF supported UCSB Materials Research Laboratory.

References

1 Klok, H.-A., and Lecommandoux, S. (2001) *Adv. Mater.*, **13**, 1217.
2 Lodge, T.P. (2003) *Macromol. Chem. Phys.*, **204**, 265.
3 Park, C., Yoon, J., and Thomas, E.L. (2003) *Polymer*, **44**, 6725.
4 Hamley, I.W. (2003) *Angew. Chem. Int. Ed.*, **42**, 1692.

5 Shenhar, R., Norsten, T.B., and Rotello, V.M. (2005) *Adv. Mater.*, **17**, 657.

6 Leibler, L. (1980) *Macromolecules*, **13**, 1602.

7 Shibayama, M., Hashimoto, T., and Kawai, H. (1983) *Macromolecules*, **16**, 16.

8 Hanley, K.J., Lodge, T.P., and Huang, C.-I. (2000) *Macromolecules*, **33**, 5918.

9 Lai, C., Russel, W.B., and Register, R.A. (2002) *Macromolecules*, **35**, 841.

10 Kunieda, H., Shigeta, K., Ozawa, K., and Suzuki, M. (1997) *J. Phys. Chem. B*, **101**, 7952.

11 Kunieda, H., Uddin, Md.H., Horii, M., Furukawa, H., and Harashima, A. (2001) *J. Phys. Chem. B*, **105**, 5419.

12 Lodge, T.P., Pudil, B., and Hanley, K.J. (2002) *Macromolecules*, **35**, 4707.

13 Kunieda, H., Uddin, Md.H., Yamashita, Y., Furukawa, H., and Harashima, A. (2002) *J. Oleo Sci.*, **51**, 113.

14 Zin, W.-C., and Roe, R.-J. (1984) *Macromolecules*, **17**, 183.

15 Roe, R.-J., and Zin, W.-C. (1984) *Macromolecules*, **17**, 189.

16 Leibler, L., Orland, H., and Wheeler, J.C. (1983) *J. Chem. Phys.*, **79**, 3550.

17 Hashimoto, T., Tanaka, H., and Hasegawa, H. (1990) *Macromolecules*, **23**, 4378.

18 Tanaka, H., Hasegawa, H., and Hashimoto, T. (1991) *Macromolecules*, **24**, 240.

19 Tanaka, H., and Hashimoto, T. (1991) *Macromolecules*, **24**, 5713.

20 Winey, K.I., Thomas, E.L., and Fetters, L.J. (1991) *Macromolecules*, **24**, 6182.

21 Winey, K.I., Thomas, E.L., and Fetters, L.J. (1992) *Macromolecules*, **25**, 2645.

22 Bodycomb, J., Yamaguchi, D., and Hashimoto, T. (2000) *Macromolecules*, **33**, 5187.

23 Matsen, M.W. (1995) *Macromolecules*, **28**, 5765.

24 Matsen, M.W. (1995) *Phys. Rev. Lett.*, **74**, 4225.

25 Janet, P.K., and Schick, M. (1998) *Macromolecules*, **31**, 1109.

26 Kinning, D.J., Thomas, E.L., and Fetters, L.J. (1989) *J. Chem. Phys.*, **90**, 5806.

27 Kinning, D.J., Winey, K.I., and Thomas, E.L. (1988) *Macromolecules*, **21**, 3502.

28 Riess, G. (2003) *Prog. Polym. Sci.*, **28**, 1107.

29 Jönsson, B., Lindman, B., Holmberg, K., and Kronberg, B. (1998) Surfactant-polymer systems, in *Surfactants and Polymers in Aqueous Solution*, John Wiley & Sons, Ltd, Chichester, p. 219.

30 Jerschow, A., and Müller, N. (1998) *J. Magn. Reson.*, **132**, 13.

31 Stejskal, E.O., and Tanner, J.E. (1965) *J. Chem. Phys.*, **42**, 288.

32 Reif, F. (1965) *Fundamentals of Statistical and Thermal Physics*, McGraw-Hill, New York, Section 15.6.

33 Orthaber, D., Bergmann, A., and Glatter, O. (2000) *J. Appl. Crystallogr.*, **33**, 218.

34 Brunner-Popela, J., and Glatter, O. (1997) *J. Appl. Crystallogr.*, **30**, 431.

35 Weyerich, B., Brunner-Popela, J., and Glatter, O. (1999) *J. Appl. Crystallogr.*, **32**, 197.

36 Klein, R. (2002) Interacting colloidal suspensions, in *Neutrons, X-Rays and Light: Scattering Methods Applied to Soft Condensed Matter* (eds P. Lindner and T. Zemb), Elsevier, Amsterdam, p. 351.

37 Glatter, O. (2002) The inverse scattering problem in small-angle scattering, in *Neutrons, X-Rays and Light: Scattering Methods Applied to Soft Condensed Matter* (eds P. Lindner and T. Zemb), Elsevier, Amsterdam, p. 101.

38 Fritz, G., and Bergmann, A. (2004) *J. Appl. Crystallogr.*, **37**, 815.

39 Glatter, O. (1979) *J. Appl. Crystallogr.*, **12**, 166.

40 Uddin, Md.H., Rodríguez, C., López-Quintela, A., Leisnaer, D., Solans, C., Esquena, J., and Kunieda, H. (2003) *Macromolecules*, **36**, 1261.

41 Glatter, O., and Hainisch, B. (1984) *J. Appl. Crystallogr.*, **17**, 435.

42 Mittelbach, R., and Glatter, O. (1998) *J. Appl. Crystallogr.*, **31**, 600.

43 Israelachvili, J.N. (1992) *Intermolecular & Surface Forces*, 2nd edn, Academic Press, London, Chapter 17.

44 Whitmore, M.D., and Noolandi, J. (1985) *Macromolecules*, **18**, 657.
45 Kao, C.R., and Olvera de la Cruz, M. (1990) *J. Chem. Phys.*, **93**, 8284.
46 Olsson, U., and Schurtenberger, P. (1993) *Langmuir*, **9**, 3389.
47 Milner, S.T. (1991) *Science*, **251**, 905.
48 Rodríguez, C., Shigeta, K., and Kunieda, H. (2000) *J. Colloid Interface Sci.*, **223**, 197.
49 Sato, T., Hossain, Md.K., Acharya, D.P., Glatter, O., Chiba, A., and Kunieda, H. (2004) *J. Phys. Chem. B*, **108**, 12927.

11
Preparation of Mesoporous Materials with Nonhydrocarbon Surfactants

Carlos Rodríguez-Abreu and Jordi Esquena

11.1
Mesoporous Materials: Basic Concepts

Mesoporous materials, with pore sizes between 2 and 50 nm according to IUPAC's classification, have attracted considerable interest in the last decade from both scientific and technological points of view, due to their potential applications in catalysis [1–3], separation [4], optics [5], electronics [6, 7], delivery systems [8, 9] and others [10, 11]. Moreover, mesoporous materials have offered new motivations for the study of surfactant self-assembly, as the fundamental step in the synthesis via sol-gel reaction is the cooperative assembly between surfactants and inorganic species [12]. The first surfactants used in the synthesis were cationic, specifically alkyltrimethyl ammonium halides [13, 14]. Later, nonionic surfactants such as poly(oxyethylene) alkyl ethers [15, 16] were also found to act as structure-directing agents, as well as amphiphilic block copolymers, specially poly(ethylene oxide)-poly (propylene oxide) copolymers, that have been extensively used due to its ability to form materials with a variety of morphologies [17, 18]. The molecular structure of the amphiphiles is of fundamental importance in the characteristics of the supramolecular aggregates to be used as templates or structure-directing agents for the preparation of mesoporous materials. The modulation of the molecular packing parameter and surface curvature of aggregates by selecting the appropriate hydrophile–hydrophobe balance allows the formation of aggregates with different sizes, morphologies and structures that are mimicked by the templated solids [19]. For ethoxylated nonionic surfactants, a correlation has been established between the structure of the mesoporous materials and the ratio between the volume of the hydrophilic head (V_H) and the volume of the lipophilic chain (V_L) of the surfactant molecules [20]. Lamellar mesostructure is formed if the V_H/V_L ratio is between 0.5 and 1.0, hexagonally ordered mesopores are obtained if the ratio is between 1.0 and 1.7; and mesopores with cubic symmetry can be obtained in the range between 1.2 and 2.0. Therefore, the hydrophile–lipophile balance of the surfactant is crucial in determining the structure of the mesoporous material.

Self-Organized Surfactant Structures. Edited by Tharwat F. Tadros

ISBN: 978-3-527-31990-9

In this chapter, we focus in two kinds of surfactants that so far have not been common as structure-directing agents in the synthesis of mesoporous silica, namely, silicone and fluorinated surfactants, having a hydrophobic chain different from that of conventional hydrocarbon surfactants, which imparts peculiar properties to the systems.

11.2 Silicone Surfactants in the Preparation of Mesoporous Materials

11.2.1 General Properties of Silicone Surfactants

Silicone surfactants, bearing siloxane groups as hydrophobes, are substitutes of hydrocarbon surfactants in many applications, due to their particular properties. Siloxane chains are very flexible; the bond angle (Si–O–Si) is significantly wider (~143°) and the bond length (Si–O) (0.165 nm) longer than C–C–C (109°, 0.140 nm) and C–O–C (114°, 0.142 nm) bonds. Therefore, the Si–O bond can rotate freely, and as a result, silicone chains are fluid (i.e., show very low glass transition temperature) even for very high molecular weights [21] and they do not generally show a Krafft point or a gel point in aqueous media [22]. Moreover, silicone chains have low cohesive energy, therefore silicon surfactants can reduce the surface tension down to 20 mN/m, less than the value obtained for most conventional hydrocarbon surfactants. On the other hand, silicone surfactants are less soluble in some organic solvents than their hydrocarbon counterparts, and as a result, silicone surfactants are surface active and able to form aggregates in both aqueous and nonaqueous media, namely, they have the capability of forming reverse aggregates and lyotropic and thermotropic liquid crystals, such as reverse hexagonal and cubic phases, in nonaqueous media [23–25].

Siloxane amphiphilic copolymers are a class of silicone surfactants consisting of a methylated siloxane hydrophobe attached to one or more polar chains such as poly(oxyalkylene). The blocks in the copolymer can be arranged in a linear fashion (as in A–B-type block copolymer) or, as in graft copolymers, one type of blocks can be attached (usually randomly) to a linear, backbone chain of the other type of blocks. Siloxane amphiphilic copolymers are of special interest since their hydrophile–hydrophobe balance can be easily tailored by changing the length of the corresponding blocks [26–28].

Due to the special features mentioned above, siloxane amphiphilic copolymers are used in a wide variety of applications including foam stabilization in plastic (polyurethane) foams, cosmetics, wetting, emulsification, lubrication and in antistatic agents, drug delivery and so forth.

The use of surfactant self-assembly in the synthesis of mesoporous materials has become a very active field of research in the last two decades. However, the literature on the use of silicone surfactants in that field is still scarce.

11.2.2
Mesoporous Materials Obtained Using Silicone Surfactants

Xu *et al.* claimed in a series of papers [29–35] to have obtained not only silica but also titania and zirconia [30] with a lamellar structure (designated as ZSU-L materials) from a silicone amphiphilic graft copolymer composed of a polydimethylsiloxane (PDMS) backbone and a side chain of poly(ethylene oxide)-*b*-poly(propylene oxide) (PEO-*b*-PPO) (see Figure 11.1). Copolymers with different PDMS and PPO block lengths (*n*, *m* and *p* in Figure 11.1) were used. The synthesis was carried out at room temperature, under acidic or neutral conditions [30] using tetraethyl orthosilicate (TEOS) as silica precursor and low surfactant concentrations (1 wt%). Spherical particles (with a size in the range of microns. i.e., 3–7 μm) were obtained by precipitation from the reaction mixture and their inner structure apparently consisted of lamellar parallel silica plates separated by bilayer copolymer aggregates. The silica walls, were found to be amorphous and with a low degree of condensation when compared to SBA-15 or MCM-41 materials [30]. X-ray diffraction measurements gave no evidence of mesopore arrangement. The preparation of transparent silica monoliths and films using the same silicone amphiphile was also reported. However, the lamellar structures observed from the ultramicrotomed particles embedded in a resin extended to the length of the micrometer scale, showing unusually large interlayer spacings (170–330 nm) corresponding to stripes oddly aligned in the same direction for all particles (see Figure 11.2). Such large spacings were attributed unambiguously by Su *et al.* to artifacts arising from the ultramicrotoming [36, 37].

Although the above-mentioned materials may not have a lamellar structure, they do show mesoporosity (see Figure 11.3). The calcined (600 °C) ZSU-L material has a BET surface area of $276\,m^2\,g^{-1}$ and exhibits a type-I isotherm with hysteresis loop of type H_4, associated with narrow slit-like pores. The average pore size estimated by the BJH method was 3.6 nm. Little change was observed in the BET surface area and pore size of the ZSU-L powders calcined at 800 °C, which indicates that the materials have high thermal stability.

Mesoporous silica was also obtained by mixing the silicone surfactant in Figure 11.1 with the cationic surfactant cetyl trimethyl ammonium bromide (CTAB) [34]. A hexagonal array of mesopores was observed, probably templated locally by the CTAB segregated from the surfactant mixture (see Figure 11.4, upper part). The calcined product (designated as ZSU-38) gave a BET surface area of $863\,m^2\,g^{-1}$,

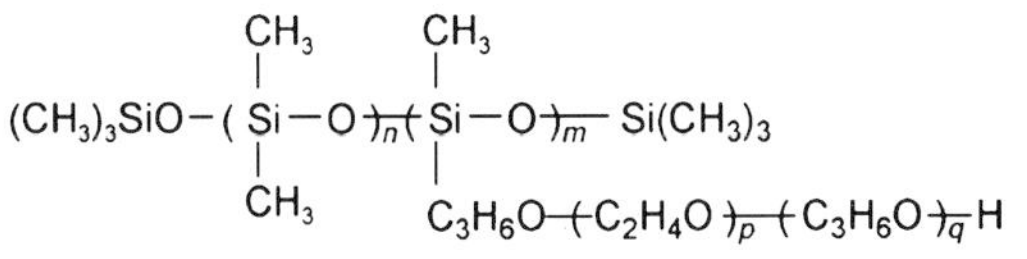

$$(CH_3)_3SiO-(\underset{\underset{CH_3}{|}}{\overset{\overset{CH_3}{|}}{Si}}-O)_n(\underset{\underset{C_3H_6O(C_2H_4O)_p(C_3H_6O)_qH}{|}}{\overset{\overset{CH_3}{|}}{Si}}-O)_m-Si(CH_3)_3$$

Figure 11.1 Molecular formula of the silicone amphiphilic copolymer used in the preparation of silica by Xu *et al.* (adapted from Ref. [29]).

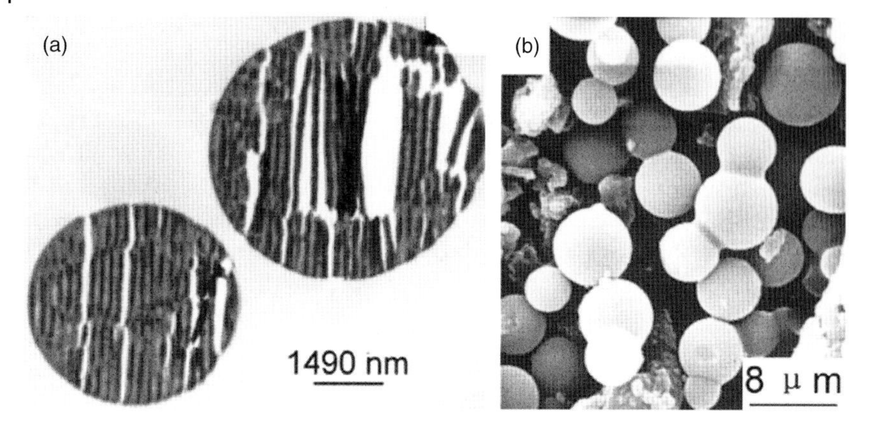

Figure 11.2 (a) TEM image of an ultrathin section of the as-made ZSU-L silica particles formed from silicone surfactant at pH 2. The strips are an artifact coming from the cutting procedure (b) Corresponding SEM image of the as-made ZSU-L silica particles (adapted with permission from Ref. [29]).

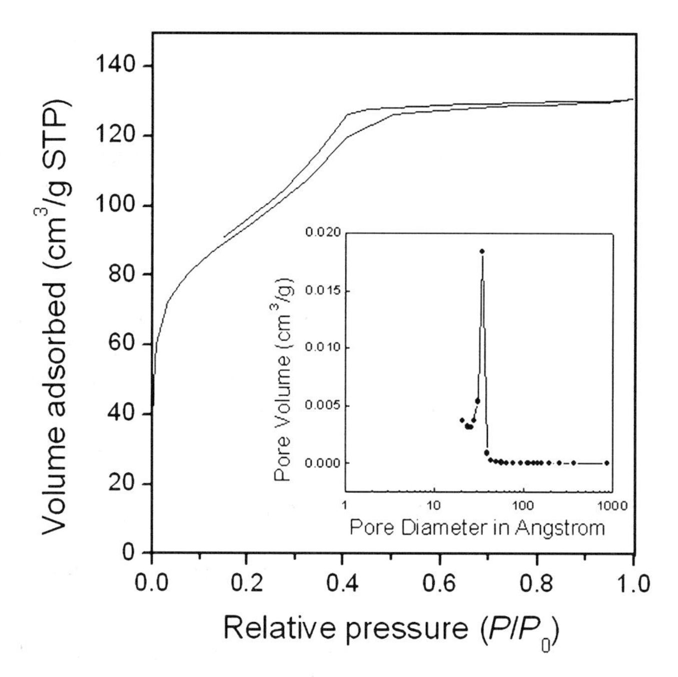

Figure 11.3 Nitrogen adsorption–desorption isotherms for the ZSU-L silica powders calcined at 600 °C for 5 h. The inset is the corresponding Barret–Joyner–Halenda (BJH) pore-size distribution curve calculated from the desorption branch of the isotherm (adapted with permission from Ref. [29]).

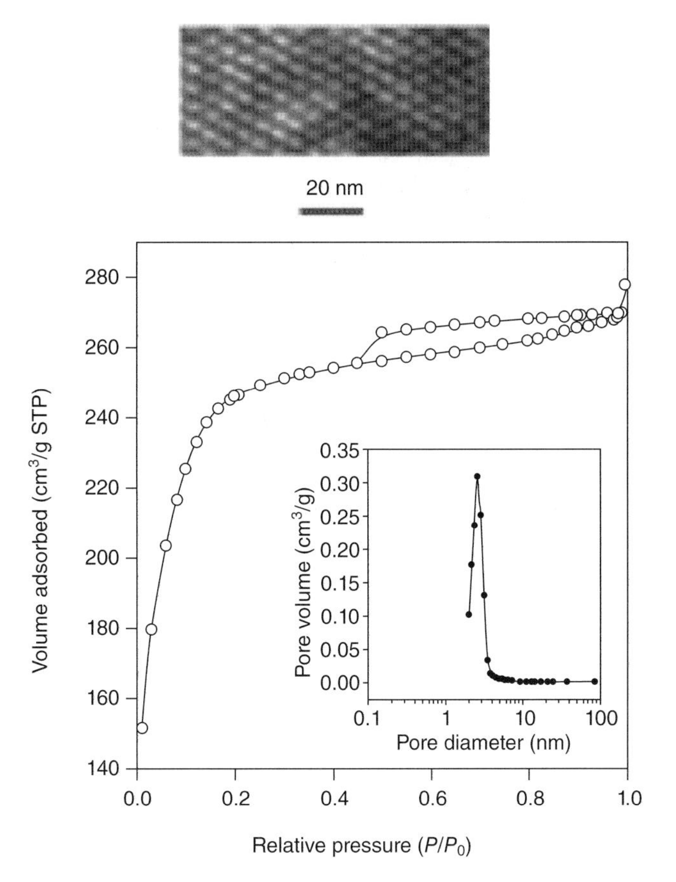

Figure 11.4 Nitrogen adsorption–desorption isotherm for a silica film (designated as ZSU-38) prepared from mixtures of cationic and silicone surfactants. The materials were calcined at 500 °C in air for 5 h. The corresponding BJH pore-size distribution curve calculated from the adsorption branch of the isotherm is also shown in the inset. The hexagonal arrangement of mesopores is shown in the upper part of the figure (adapted with permission from Ref. [34]).

whereas the BJH analysis of the adsorption isotherm with a hysteresis loop gave an average pore size of 2.6 nm (see Figure 11.4, lower part).

Mesoporous silica with hexagonal order (designated as ZSU-4) was also prepared from mixtures of the silicone surfactants (Figure 11.1) and a PEO-PPO-PEO copolymer [33], as evidenced by electron microscopy and X-ray diffraction (see Figure 11.5). The calcined ZSU-4 gave a BET surface area of 841 $m^2 g^{-1}$. The BJH analysis of the isotherm (type IV with a large type-H_2 hysteresis loop) resulted in an average pore size of 8.1 nm.

Very recently, Liu reported [38, 39] on the one-pot preparation of ordered mesoporous carbon with triblock copolymer PEO-PPO-PEO using a diblock copolymer poly(dimethylsiloxane)–poly(ethylene oxide) (PDMS-PEO) (M_w = 3012,

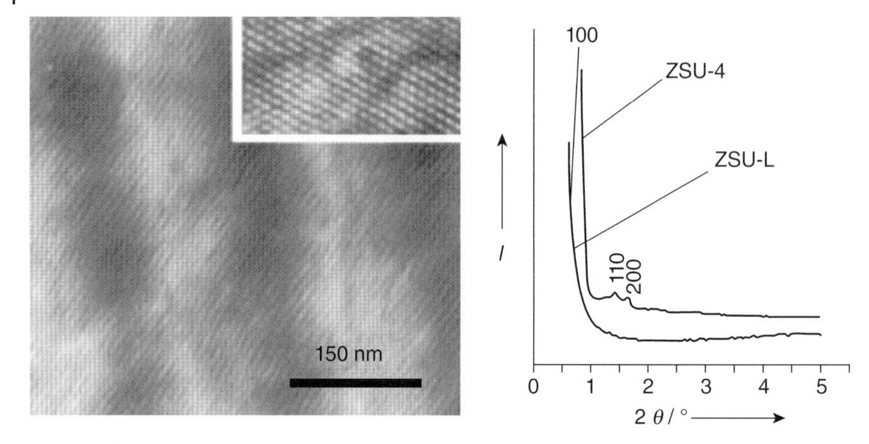

Figure 11.5 TEM images of an ultrathin section of the hierarchically ordered silica mesostructure ZSU-4 obtained with a mixture of silicone surfactant and PEO-PPO-PEO block copolymer as the template (left) X-ray diffraction patterns of the calcined ZSU-4 material (right). The curve of as-prepared ZSU-L material prepared from pure silicone surfactant is also shown for comparison (adapted with permission from Ref. [33]).

DMS_{32}-EO_{20}) as cotemplate. A resol precursor was added to the reaction mixture; such precursor was later carbonized at high temperatures and finally the remaining silica was removed by HF. Carbons with 2D-hexagonal (*p6mm*) and body-centered cubic (space group *Im3m*) mesostructure were obtained (see Figure 11.6). However, silica materials prepared by burning the carbon–silica nanocomposites showed disordered pores with low associated surface areas (below 70 m^2/g). This fact was attributed to the collapse of the silica mesostructure during the combustion that leaves only a small and fragile residue of silica in the pore walls. Experiments showed that the use of PDMS-PEO as cotemplate results in larger pore sizes and reduced framework shrinkage of the materials without evident effect on the specific surface areas. Carbons with a hexagonal array of mesopores had pore sizes ranging from 5.4 to 6.7 nm, and surface areas between 517 and 578 m^2/g, depending on the percentage of the silica content in the carbon–silica nanocomposite after carbonization. On the other hand, for the reported sample of body-centered mesoporous carbon, the pore diameter and surface area were 5.6 nm and 798 m^2/g, respectively.

The systems mentioned so far seem to rely on the structure-directing effect of the nonsilicone surfactant in the mixture. Husing *et al.* [40] synthesized a series of linear A–B-type PEO-b-PDMS block copolymers with different PDMS and PEO length and used them for the first time as templates in a solvent-evaporation-driven synthesis approach to the fabrication of self-assembled mesostructured silica films. The materials were prepared in acidic conditions using tetraethyl orthosilicate as silica precursor. Figure 11.7 shows the diffraction patterns for calcined PEO-*b*-PDMS block copolymer samples prepared with (a) different block copolymer concentrations and (b) different block lengths. For PEO_{12}-*b*-$PDMS_{18}$,

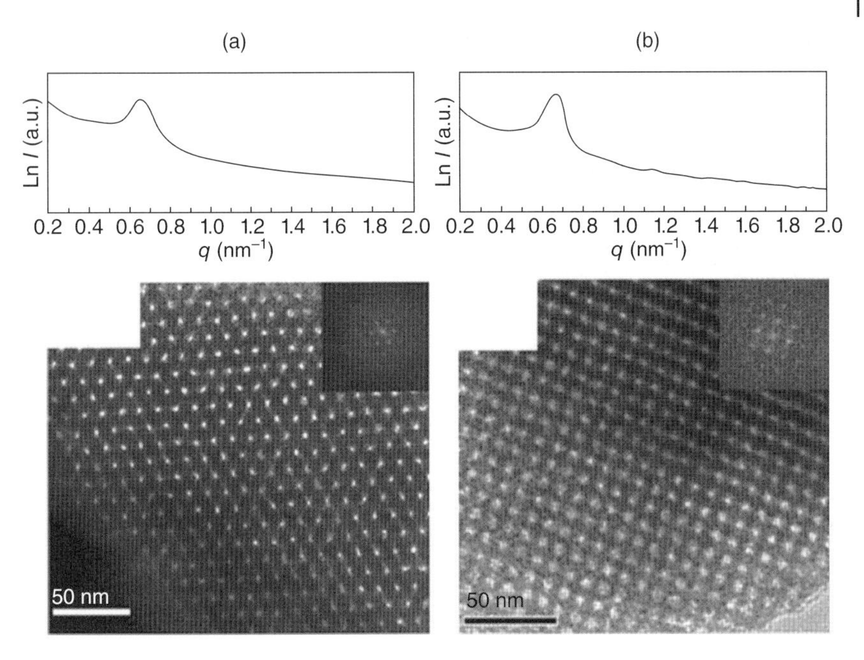

Figure 11.6 X-ray diffraction patterns and corresponding TEM images of mesoporous carbons obtained from mixtures of PEO-PPO-PEO copolymer and silicone surfactant with hexagonal structure, labeled MP-C-8.2 (a) and with body-centered structure (b), labeled MP-C-10 (adapted with permission from Refs. [38] and [39]).

two peaks with relative position ratios of 1 : 2 are observed, which could be assigned either to a lamellar or an hexagonal structure. The higher the concentration (up to 15%) of the surfactant the more pronounced the diffraction peaks are. On the other hand, Figure 11.7b shows that the number and intensity of peaks decrease as the molecular weight of the copolymer increases, namely, structure is lost; this might be due to segregation between the copolymer and the other species present in the reaction mixture. Interestingly, The PEO-*b*-PDMS templated film showed no contraction of the structure upon calcination (the *d*-spacing remains unchanged), contrary to materials prepared with other surfactants.

Transmission electron microscopy (TEM) showed that the samples have a layered structure. The layer spacing obtained from the photomicrograph (Figure 11.8), is about 5 nm, in agreement with the spacing obtained from the XRD pattern (Figure 11.7). However, no porosimetry data have been reported to date for these materials.

Later, Rodriguez *et al.* [41] reported the formation of hexagonal mesoporous silica from a random graft PDMS–PEO copolymer with the molecular formula depicted in Figure 11.9a. The silicone surfactant in this case is hydrophobic and insoluble in water, forming a reverse hexagonal liquid crystal at high surfactant concentrations (see Figure 11.9b). The silica materials were prepared using such reverse liquid crystal as template, which is the first example of reverse templating.

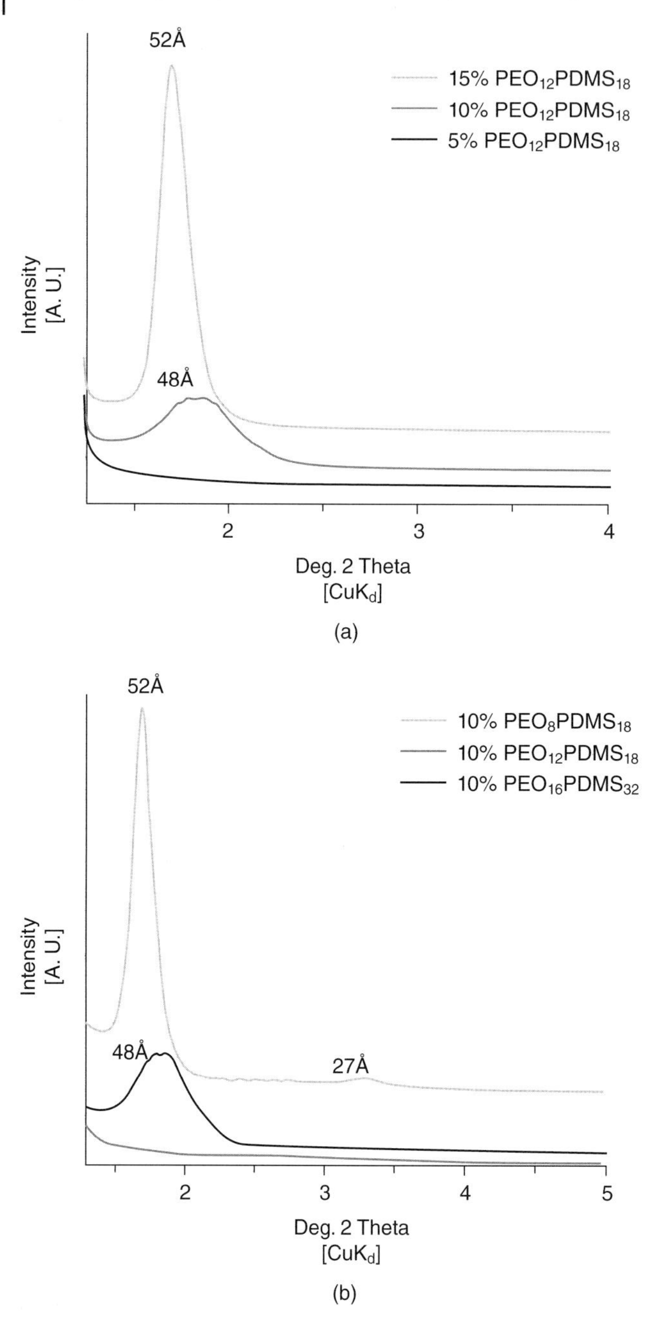

Figure 11.7 X-ray diffraction patterns of different film samples with (a) varying concentration of the template in the film and (b) varying block lengths in the silicone block copolymer (adapted with permission from Ref. [40]).

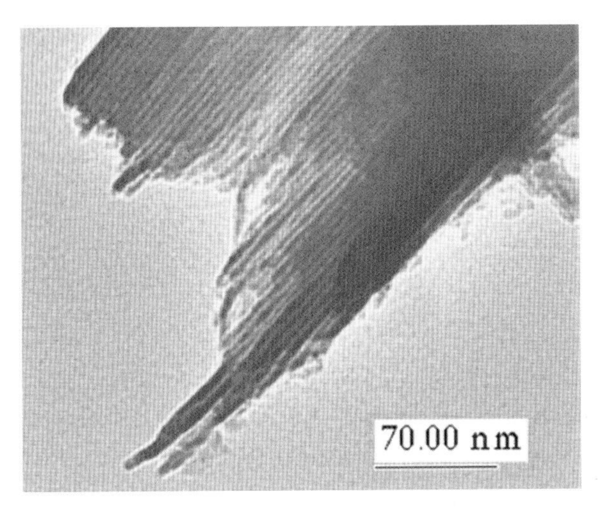

Figure 11.8 TEM image of a calcined PEO_{12}-*b*-$PDMS_{18}$/silica film. (with permission from Ref. [40]).

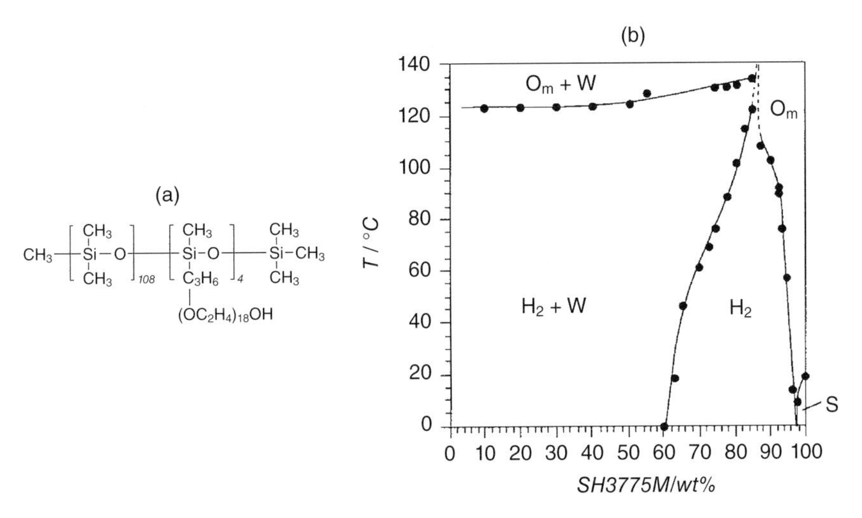

Figure 11.9 (a) Molecular formula of SH3775M (b) Phase behavior of SH3775M in water. H_2 is a reverse hexagonal liquid crystal, O_m is a reverse micellar solution or liquid amphiphile, W is excess water and S is a solid phase (with permission from Ref. [28]).

The SAXS patterns of as-prepared materials showed two peaks in positions that may correspond either to lamellar or hexagonal structures; the smallest one disappeared upon calcination. Low or absent signal intensities from higher–order reflections could be due to rough or diffuse interfaces between the silica and copolymer regions [42]. As in the case of the materials synthesized by Husing *et al.* [40], there is little shrinkage upon calcination of as-prepared samples (see Figure 11.10a),

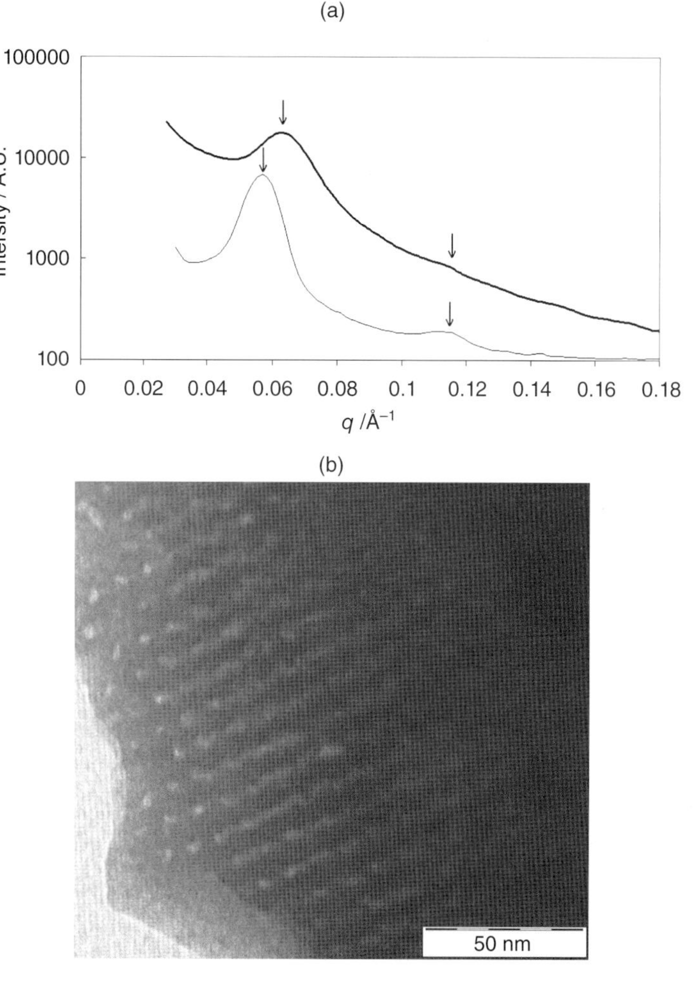

Figure 11.10 (a) SAXS pattern of as-synthesized (lower curve) and calcined (upper curve) materials prepared at SH3775M/TEOS = 1.(b) TEM images of a calcined silica material, prepared at SH3775M/TEOS = 1. (from Ref. [41]).

indicated by the little change in the position of the first peak associated with the Bragg spacing. The electron microscopy image in Figure 11.10b shows unequivocally the presence of hexagonally arranged mesopores (size: 4–5 nm) separated by thick walls (5–6 nm thick).

The pore size was confirmed by sorption measurements that resulted in an isotherm with a hysteresis loop (type IV) typical of mesoporous solids (see Figure 11.11). The estimated BET surface area was 110 m^2/g. This value is relatively low when compared to other mesoporous materials, which can be attributed to the very thick pore walls and/or to the low microporosity of the samples, as PDMS

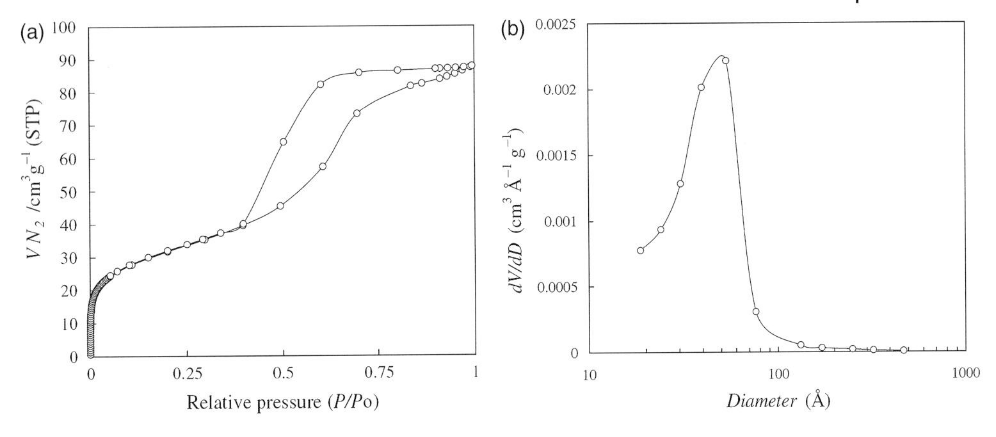

Figure 11.11 (a) Nitrogen adsorption and desorption isotherms of calcined silica prepared using SH3775M. (b) Pore-size distribution calculated by applying BJH equations to the adsorption branch (from Ref. [41]).

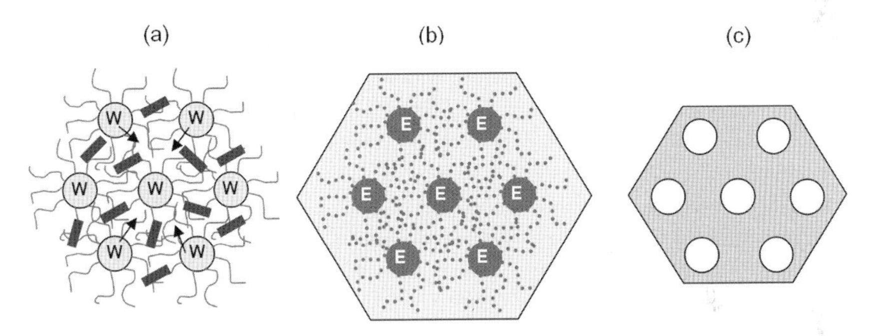

Figure 11.12 Schematics of the proposed structural evolution during silica synthesis from reverse liquid crystals: (a) Initial state. Micelles with a PDMS corona (thin lines) and solubilizing an aqueous phase (W) in their cores are arranged hexagonally, surrounded by TEOS molecules (thick lines). Arrows indicate the possible interfacial diffusion of W (b) After hydrolysis and partial condensation of siloxane moieties, hydrophilic cores containing PEO plus nonreacted water (E) remain in the hexagonal array. PDMS chains (dotted lines) are embedded in the silica matrix. (c) Upon drying and calcination, the structure shrinks and organic groups are burnt off, so that hexagonal mesoporous silica is obtained (adapted from Ref. [41]).

chains are not completely eliminated but seem to contribute to the silica network when calcined, hence not contributing to microporosity in the walls.

Figure 11.12 shows the proposed scheme for the structural evolution during the synthesis, based in a liquid-crystal templating (LCT) mechanism. Initially, the aqueous acid solution is solubilized in the cylindrical reverse micelles forming the reverse hexagonal phase. Nonhydrolyzed TEOS molecules and PDMS chains are mutually soluble and therefore constitute the hydrophobic continuous medium. The hydrolysis is probably driven by interfacial diffusion of the water towards the hydrolysable molecules. After hydrolysis and condensation reactions have progressed in some extent, hydrophilic cores containing PEO plus nonreacted water remain in a hexagonal array, whereas the PDMS chains are embedded in the silica matrix, as suggested by NMR data. Finally, upon calcinations, the condensation reactions progress further and the remaining organic molecules are decomposed, leaving the mesoporous hexagonal matrix. The mechanism sketched here is different from those reported so far for the synthesis of mesoporous silica, in which templating occurs due to the strong interaction between the silicate species and the hydrophilic moieties of the amphiphile.

11.3 Fluorinated Surfactants in the Preparation of Mesoporous Materials

11.3.1 General Properties of Fluorinated Surfactants

The overall properties of fluorinated surfactants are mainly due to the low solubility of fluorocarbons in most conventional solvents. Perfluoroalkanes are more hydrophobic than conventional alkanes with equivalent chain length. Fluorine is an element difficult to polarize due to its high electronegativity. Moreover, the C–F bond is chemically very stable and rigid. Consequently, the fluorocarbon chains are usually rather stiff and bulky, limiting the interactions with other molecules and greatly reducing solubility. Generally, CF_3- and $-CF_2$- groups lead to very weak interaction forces and perfluoroalkanes are insoluble in hydrocarbon-based solvents and only highly soluble in fluorocarbons [43].

Perfluoroalkanes can pack very tightly, as demonstrated by calculating the area per molecule from surface tension measurements (γ–log C plots) or by using Langmuir film determinations [44]. For example, the surface area per molecule at the CMC of CTAB surfactant is $0.47\,nm^2$, whereas the area per molecule of its fluorinated counterpart is only $0.26\,nm^2$. These properties of perfluoroalkanes can be related to their melting temperatures [45], which are higher than the melting points of the corresponding hydrocarbons, except for CF_4. Consequently, the melting points of fluorinated surfactants are also higher than that of the equivalent hydrocarbon surfactants. This increase in melting point is attributed to the stiffness and packing of the fluorinated chains.

Another interesting feature of fluorinated surfactants is their ability to greatly reduce the surface tension of water, even at very small concentrations. Some surfactants, such as sodium perfluorooctanoate, can lower the surface tension at 25 °C down to 15 mN m^{-1}. It should be pointed out that such low tensions can be achieved with a rather short pefluoalkyl chain, because the perfluoalkyl chains are significantly more hydrophobic than the hydrocarbon alkyl chain with the same number of carbon atoms [46]. A remarkable example is sodium fluorobutanoate, with a very short chain length, but still able to form self-aggregates in water [47].

Perfluoroalkyl groups are simultaneously both hydrophobic and oleophobic, and consequently partially fluorinated surfactants are surface active and able to form self-aggregates in both hydrocarbons and fluorocarbons [43, 48–50]. In a hydrocarbon solvent, the perfluoroakyl chain is a lyophobe, whereas the alkyl chain is a lyophile. Consequently, micelles are formed with a fluorinated core. If a perfluorocarbon is used as a solvent, then the perfluoroalkyl chain is the lyophile and the alkyl chain is the lyophobe, forming micelles with a fluorinated shell. Moreover, considering that long-chain alkanes and long-chain perfluoroalkanes are mutually immiscible, partially fluorinated surfactants can be good stabilizers for emulsions consisting of hydrocarbons and fluorocarbons.

Partially fluorinated surfactants have several advantages with respect to their perfluorinated counterparts, due to the low solubility of fully fluorinated chains in hydrocarbons. The hydrocarbon segment of a partially fluorinated surfactant provides solubility in commonly used organic solvents, lowers the melting point of the surfactant, reduces volatility and decreases the acid strength of perfluorinated acids [51].

Fluorinated surfactants can be very interesting for the preparation of mesoporous materials, due to their good self-aggregation properties, allowing a more precise control of the mesostructure and properties of materials. In this context, the use of fluorocarbon surfactants allows the formation of ordered mesoporous materials with a long-range organization [52, 53].

Nowadays, partially or fully fluorinated surfactants are available with a variety of chemical structure and chain lengths. Therefore hydrophilic–lipophilic properties can be controlled and these new surfactant molecules may provide a rich phase behavior, which could be used to obtain well-organized mesoporous materials.

The use of fluorinated surfactants, in the preparation of mesoporous materials will be described in the following section.

11.3.2 Mesoporous Materials Obtained Using Fluorinated Surfactants

The special self-aggregation properties of fluorinated surfactants makes them very interesting for the preparation of mesoporous ordered materials, as has been extensively reported by Stébé and coworkers [54–59]. They have shown that mesoporous silica with a higher long-range degree of organization can be obtained with fluorinated nonionic ethoxylated surfactants $F(CF_2)_nC_2H_4(OC_2H_4)_mOH$, compared to analogous hydrogenated surfactants [52]. As illustration, Table 11.1, shows the main structural properties of mesoporous silica templated with the partially fluorinated surfactant $F(CF_2)_8C_2H_4(OC_2H_4)_9OH$.

The above results demonstrate that partially fluorinated nonionic surfactants, having a fluorinated hydrophobic chain and a nonfluorinated poly(ethylene oxide) polar chain, can be successfully used to obtain mesoporous materials with controlled morphology and pore size. The specific surface areas can be high, similar to MCM-41 materials obtained using hydrocarbon surfactants.

In the conditions reported by the authors [52], very well-ordered 2D hexagonal mesoporous silicas are obtained with a surfactant concentration lower than 25

Table 11.1 Bragg spacings (*d*), Specific surface area (S_{BET}), Pore volume (V_p), and pore diameter, for mesoporous silicas obtained at different concentrations of fluorinated surfactant, $F(CF_2)_8C_2H_4(OC_2H_4)_9OH$. (adapted from results described by Stébé and coworkers [52]).

Surfactant concentration (wt %)	Structure	*d* (nm)	S_{BET} ($m^2 g^{-1}$)	V_p ($cm^3 g^{-1}$ STP)	Pore diameter (nm)
5	Hexagonal	5.3	1044	1.28	4.3
10	Hexagonal	5.3	844	1.28	4.6
15	Hexagonal	5.3	950	1.55	5.0
20	Hexagonal	5.4	831	1.44	6.0
25	Disordered	5.8	744	1.58	7.8
30	Disordered	6.0	660	1.63	9.3

wt%, whereas disordered mesopores are obtained with a concentration higher than 25 wt%. The same authors also showed that the inner diameter of mesopores can be enlarged by swelling the surfactant aggregates with fluorinated oil (perfluorodecalin). Phase behavior and SAXS studies showed that this fluorinated oil was solubilized inside the surfactant aggregates [53].

Antonietti *et al.* described the preparation of mesoporous silica in systems containing mixtures of nonionic fluorinated surfactants and nonionic hydrocarbon block copolymer surfactants [60]. The fluorinated surfactant was $CF_3(CF_2)_{6-16}$ $C_2H_4(EO)_{4-5}$ and the hydrocarbon-based copolymer was poly(ethylene-co-butylene)-block-poly(ethylene oxide).

These two kinds of surfactants are mutually insoluble in aqueous systems, and they can form separated hydrogenated surfactant micelles that coexist with fluorinated surfactant micelles. Consequently, mesoporous silica with a bimodal pore-size distribution can be obtained, because of the coexistence of these two different types of aggregates that template two different pore sizes. An example of bimodal mesoporous silica is shown in Figure 11.13.

The preparation of mesoporous materials with bimodal pore-size distribution has also been described by Stébé and coworkers, using mixtures of $F(CF_2)_nC_2H_4$ $(OC_2H_4)_mOH$ nonionic ethoxylated surfactants and conventional hydrocarbon ethoxylated surfactants [61].

The use of fluorinated cationic surfactants has been studied by Rankin and coworkers [62, 63]. They described the preparation of hexagonally ordered mesoporous silica with rather small pore sizes (pore diameters approximately from 2.0 nm to 2.6 nm), and therefore they have similar properties to silica prepared with conventional hydrocarbon cationic surfactants (such as MCM-41). Thin pore walls could be a disadvantage, since it may lead to low mechanical and hydrothermal stability.

However, other studies with fluorinated ethoxylated surfactants $CF_3(CF_2)_n(EO)_m$, by Xiao *et al.* [64–66] have demonstrated that it is possible to obtain materials that simultaneously have small pore sizes (1.6–4.0 nm), with thicker pore walls (2.5–

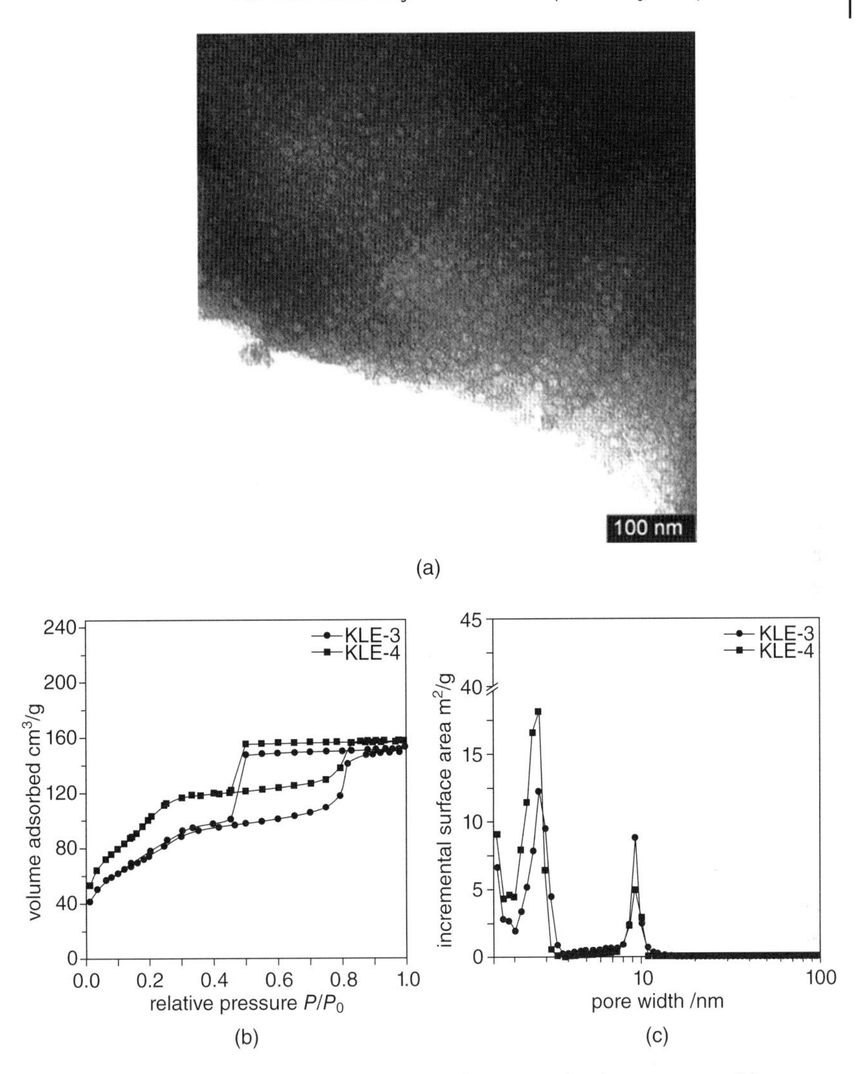

Figure 11.13 Mesoporous materials with a bimodal pore-size distribution prepared from mixtures of fluorinated and hydrocarbon surfactants. (a) TEM image; (b) Nitrogen sorption isotherms, and (c) pore-size distributions. (Reproduced from Ref. [60], with permission).

2.9 nm). It was shown that these silica mesoporous materials have much higher hydrothermal and mechanical stability than those of MCM-41, and this interesting property was attributed to the thick pore walls. The same authors also showed that the pore size can be enlarged by adding a molecule that swells the surfactant aggregates, as shown by the addition of 1,3,5-trimethylbenzene, which also helped to improve the mesoscopic order [65].

Mesoporous materials with cubic symmetry can also be obtained using fluorinated compounds. Tatsumi *et al.* described the preparation of large mesoporous

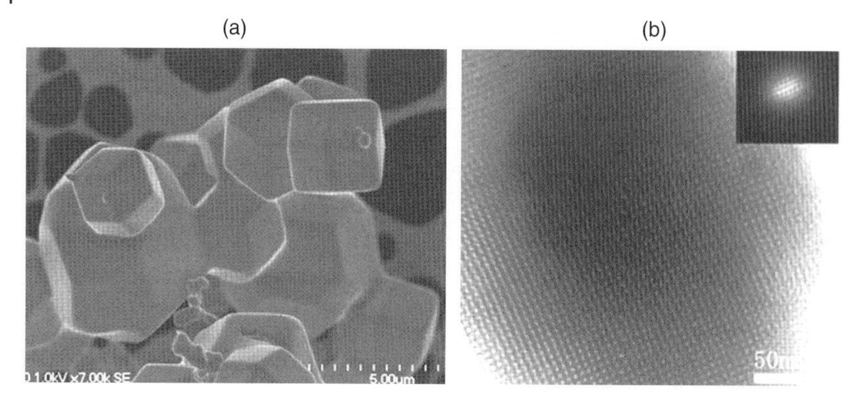

Figure 11.14 Electron microscopy images of a mesoporous silica SEM images for sample prepared by $CF_3(CF_2)_3SO_2NH(CH_2)_3N(CH_3)_3I$, at 100 °C. (a) SEM image, (b) TEM image. (Reproduced from Ref. [67], with permission).

silica single crystals, in rhombododecahedron shape, that have mesopores ordered in Im3m cubic space group [67] (see Figure 11.14). This material can be obtained using mixtures of ionic fluorinated surfactants with short hydrophobic chains ($CF_3(CF_2)_3SO_2NH(CH_2)_3N(CH_3)_3I$ or $CF_3(CF_2)_3SO_3K$), in a mixed system with the block copolymer surfactant $(EO)_{106}(PO)_{70}(EO)_{106}$. This surfactant mixture, consisting of a short-chain ionic fluorinated surfactant and a hyphophilic copolymer surfactant, seems to achieve the appropriate aggregate curvature to obtain a micellar cubic phase.

Lamellar nanostructures can be obtained by hydrolyzing TEOS in acidic mixtures of perfluoroctane sulfonic acid, $C_8F_{17}SO_3H$, and 3-aminopropyltriethoxysilane, APTES. The silica layers in the materials are very thin, and contain a relatively high concentration of attached aminopropyl groups. These hybrid fluorocarbon-silica nanocomposites are thermally stable, hydrophobic and show a low dielectric constant (around 2.8) which is almost independent of frequency [68].

Interesting morphologies can also be obtained by using self-assembly in fluorinated surfactant systems. Rankin and coworkers have reported the preparation of hollow silica particles with ordered mesopores in the shells [69]. These vesicle-like particles were obtained by using perfluorodecylpyridinium chloride as surfactant at high pH. Electron microscopy images of such hollow particles are shown in Figure 11.15. The size of the particles was approximately 100 nm and the size of the mesopores on the shell, calculated from nitrogen sorption isotherms, was 3.6 nm, which was approximately the size of the initial micellar aggregates. The mechanism of formation of this hollow structures is rather complex, since no vesicles were observed in the absence of the silica precursor, and phase-behavior studies indicate the formation of cylindrical micelles in the initial aqueous solutions. It seems that the vesicle formation could be produced

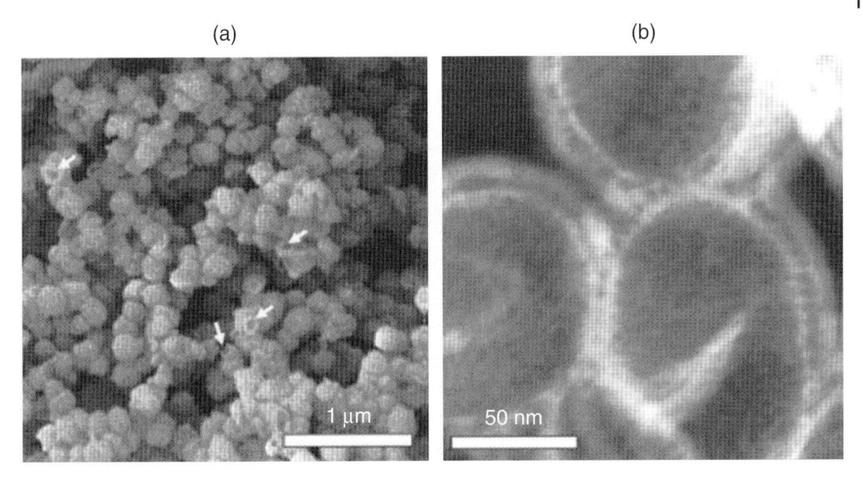

Figure 11.15 Electron microscopy images, (a) SEM and (b) TEM, of hollow particles with mesopores in the shell. The particles were prepared using a cationic fluorinated surfactant (Reproduced from Ref. [69], with permission).

by a phase transition induced by the association of silica species to the surfactant aggregates [69].

It should be pointed out that although most of the research works used silica as a model material, fluorinated surfactants can be used for the preparation of a very wide variety of different mesoporous materials, including TiO_2 [70], aluminophosphates and Fe- aluminophosphates [71], and Fe–Si oxides [72]. Organic functionalizations can also be obtained in direct synthesis processes. An example is the preparation in a one-step precipitation process, by Rankin *et al.*, of vinyl-functionalized porous silica materials using mixtures of TEOS and vinyltriethoxysilane in aqueous solutions of cationic fluorinated surfactants [73].

Kunieda *et al.* described the preparation of mesoporous silica materials, by using novel partially fluorinated surfactants [74], with the general structure $CF_3(CF_2)_7SO_2$–$[CH_3(CH_2)_2]N$–$(C_2H_4O)_nH$. For this type of surfactants, an ethoxylated chain length of 10 EO units produced the appropriate hydrophile–lipophile balance that allowed to obtain well-ordered mesoporous silica with a 2D hexagonal structure [74]. Phase behavior studies showed that this surfactant forms worm-like micelles in aqueous solution. Therefore, there is a correlation between the curvature of the surfactant self-aggregate in water and the morphology of the mesoporous material obtained with the same surfactant.

It is interesting to note that mesoporous silica could be obtained, by cooperative self-assembly and precipitation, even at rather low surfactant concentrations. Figure 11.16 shows the SAXS spectra of different mesoporous silica samples, which were prepared at different surfactant concentrations. The SAXS spectra present a peak sequence $(1:\sqrt{3}:2:\ldots)$, which corresponds to a two-dimensional

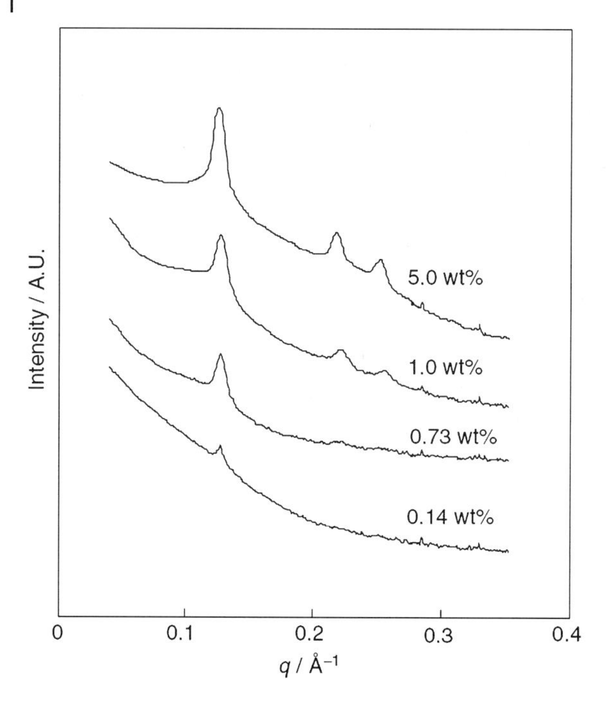

Figure 11.16 SAXS patterns of mesoporous silica samples, at various $C_8F_{17}SO_2[C_3H_7]$ $N(EO)_{10}H$ surfactant concentrations. The synthesis was carried out with 5 wt% TEOS. For improved clarity, the intensity (expressed in arbitrary units) is shown multiplied by arbitrary factors. (Reproduced from Ref. [74], with permission).

hexagonal (p6mm) mesostructure. The repeat distance (intercylinder spacing) was almost constant at 5.8 nm.

With only 1 wt% surfactant concentration, the sequence of the three visible peaks in the SAXS spectra shown in Figure 11.16 ($1:\sqrt{3}:2$) clearly indicates a two-dimensional hexagonal (p6mm) mesostructure. This concentration is lower than those used to obtain SBA-like materials by using nonionic block copolymer surfactants [16].

Low surfactant concentration implies that the number of self-aggregates is limited and therefore the walls that separate adjacent pores becomes thicker. TEM, SAXS and nitrogen porosimetry have shown that mesoporous materials with relatively thick pore walls can be obtained. Representative images are presented in Figure 11.17.

The SAXS spectra confirmed that the sample shown in Figure 11.17 had a hexagonal distribution of cylindrical mesopores. The specific surface, S_{BET} calculated from nitrogen sorption isotherms ($\approx 1000\,m^2/g$), was also comparable to MCM-41 family of materials.

The mesostructure proved to be highly dependent on pH. Hexagonal mesoporous silica was obtained at hydrochloric acid concentrations higher than 0.1 M,

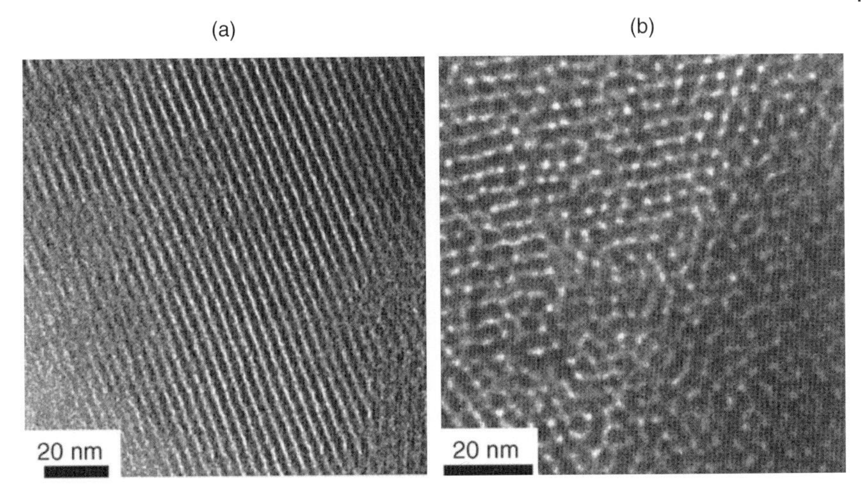

Figure 11.17 Images obtained by TEM of calcined silica particles obtained with $C_8F_{17}SO_2[C_3H_7]N(EO)_{10}H$ surfactant at 2 wt% concentration (TEOS = 5 wt%, HCl = 5 M) concentrations. (a) Longitudinal view, parallel to the mesopores; (b) Transversal view, showing mesopore cross section. (Reproduced from Ref. [74], with permission).

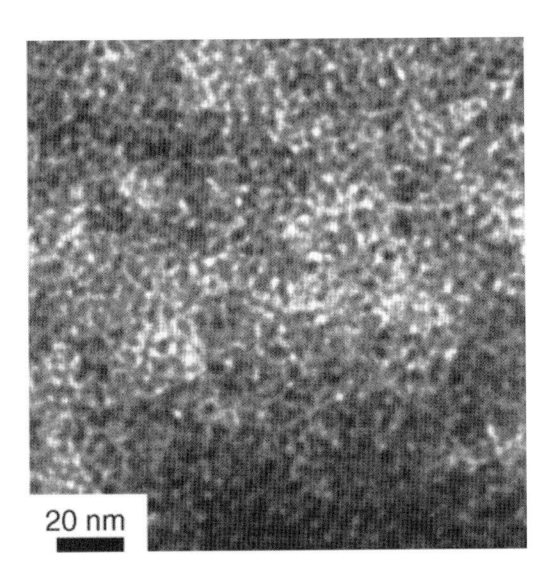

Figure 11.18 TEM image of calcined silica obtained with $C_8F_{17}SO_2[C_3H_7]N(EO)_{10}H$ surfactant at 2 wt% concentration (TEOS = 5 wt%, HCl = 0.0027 M) (Reproduced from Ref. [74], with permission).

Table 11.2 Comparison between the main morphological characteristics of a mesoporous material prepared using the fluorinated surfactant and the typical values obtained for MCM-41 and SBA-15 materials.

Surfactant	Material	Mesostructure	BET area ($m^2 g^{-1}$)	Pore size (nm)	Pore wall thickness (nm)
$C_8F_{17}SO_2[C_3H_7]N(EO)_{10}H$	Described in reference 74	Hexagonal	911	3.5	2.3
CTAB	MCM-41 type [13]	Hexagonal	≈ 1000–1300	≈ 1.5–3.0	≈ 1.0–1.5
Pluronic P123	SBA-15 type [16]	Hexagonal	≈ 800–1000	≈ 4.0–6.0	≈ 2.0–4.0

whereas worm-like disordered mesoporous silica was formed at HCl concentrations lower than 0.1 M. Figure 11.18 shows an example of a mesoporous material with randomly disordered mesopores, obtained by hydrolyzing TEOS at low hydrochloric acid concentration (0.0027 M).

The inner pore diameter was obtained from nitrogen sorption isotherms and the pore wall thickness, in hexagonally well-ordered mesoporous samples, was calculated by subtracting the pore diameter from the repeat distance (intercylinder spacing) obtained from SAXS spectra.

The pore size and the pore wall thickness of mesoporous silica obtained using $C_8F_{17}SO_2[C_3H_7]N(EO)_{10}H$ surfactant are shown in Table 11.2. These results are compared to the typical values described for MCM-41 and SBA-15 silicas, which have a similar mesostructure consisting of hexagonally packed cylindrical mesopores.

Note that the pore walls of the material templated using $C_8F_{17}SO_2[C_3H_7]N(EO)_{10}H$, indicated in Table 11.2, are thicker than those of MCM-41 materials [13]. The pore walls are also thicker than those obtained when using fluorinated cationic surfactants [62, 63], which resemble MCM materials, or when using nonionic surfactants with partially fluorinated alkyl tails [53]. However, the main characteristics of the materials described here are similar to those obtained by using $F(CF_2)_5(EO)_{10}$ or $F(CF_2)_6(EO)_{14}$ fluorinated nonionic surfactants [64, 65].

Thick walls imply a higher robustness, which is illustrated by the material's resistance to hydrothermal treatment. An example is presented in Figure 11.19, showing SAXS spectra of two different materials before and after applying a mild hydrothermal treatment.

Figure 11.19 clearly shows that the mesostructure of MCM-41, with thinner pore walls, collapses during the hydrothermal treatment. However, the mesoporous material obtained using the fluorinated surfactant is more robust and remains practically unmodified after the hydrothermal treatment.

The pore size can be enlarged by the addition of fluorinated oil, which can be solubilized inside the surfactant aggregates. Mesoporous silica with larger pores was obtained by this method in mixtures of the surfactant $C_8F_{17}SO_2[C_3H_7]N(EO)_{10}H$ and the fluorinated oil $(C_3F_6O)_nCOOH$ [75]. Figure 11.20 shows the SAXS spectra and a TEM image of the mesoporous material.

In the absence of polymeric oil, well-ordered mesoporous silica is obtained, as demonstrated by the sharp peaks observed in its SAXS pattern, with a sequence

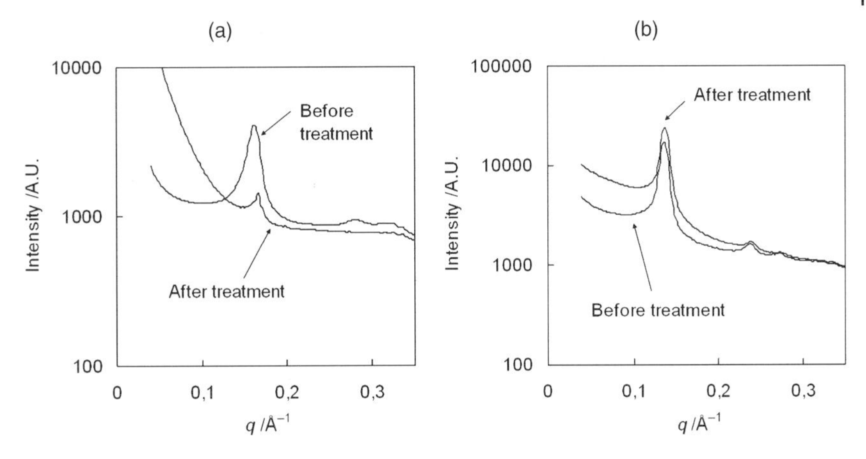

Figure 11.19 SAXS spectra before and after applying a mild hydrothermal treatment, consisting in keeping the material immersed in water at 90 °C for 24 h. (a) MCM-41 silica; (b) silica prepared using $C_8F_{17}SO_2[C_3H_7]N(EO)_{10}H$.

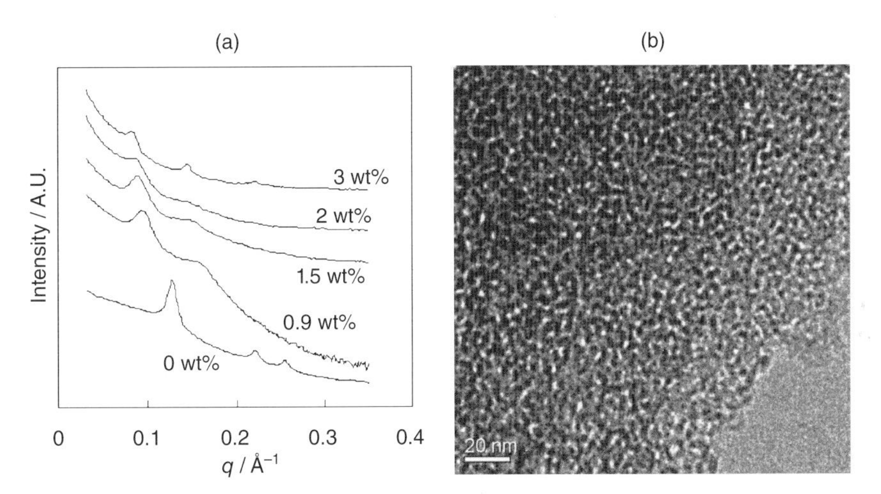

Figure 11.20 Mesostructured silica prepared with mixtures of $C_8F_{17}SO_2[C_3H_7]N(EO)_{10}H$, at 5wt%, and $(C_3F_6O)_nCOOH$ fluorinated oil. (a) SAXS spectra. Concentrations of $(C_3F_6O)_nCOOH$ in the reaction mixture are indicated beside each curve. (b) TEM Image of calcined mesoporous silica prepared at 0.9 wt% oil concentration. (Reproduced from Ref. [75], with permission).

that corresponds to a two-dimensional hexagonal (p6mm) mesostructure. The addition of the fluorinated oil $(C_3F_6O)_nCOOH$ increases the interlayer spacings, indicating larger pores, due to a swelling mechanism.

However, the SAXS spectra also reveal that the mesostructure of the silica materials becomes disordered when adding $(C_3F_6O)_nCOOH$. The diffraction peaks become broad, indicating increasing disorder. TEM imaging indicated

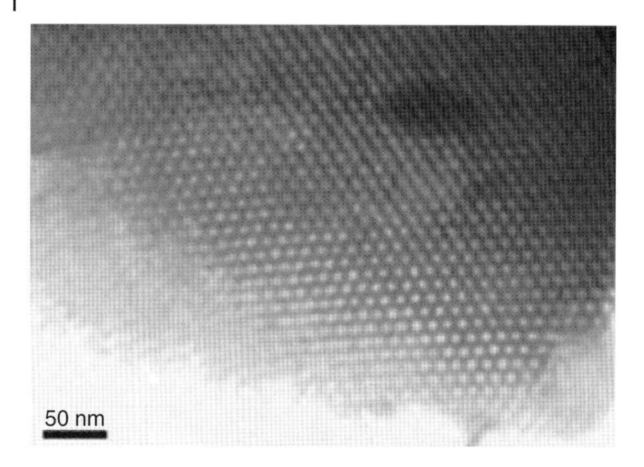

Figure 11.21 TEM image of JLU-20 silica, with approximately 6.5 nm of pore size and 5 nm of pore wall. (Reproduced from Ref. [76], with permission).

worm-like disordered mesopores, as shown in Figure 11.20b. A similar tendency was described by Blin and Stébé when using perfluorodecalin as swelling agent [53].

The preparation of well-ordered mesoporous silica materials, with large pore sizes and very robust pore walls, has been described by Xiao *et al.* [76]. They obtained silica-based materials using $C_3F_7O(CFCF_3CF_2O)_2CFCF_3CONH(CH_2)_3N^+(C_2H_5)_2CH_3I^-$ fluorinated surfactant in mixtures with the conventional triblock copolymer surfactant P123 $(EO)_{20}(PO)_{70}(EO)_{20}$. The synthesis was carried out in strong acidic media and high temperatures (160–220 °C), resulting in a material (designated as JLU-20) with unusual hydrothermal stability [76]. This material has a hexagonal mesoporous structure, with very thick pore walls (thickness ~ 5 nm), which enhance hydrothermal stability. An image of this material is shown in Figure 11.21.

Xiao and coworkers extended that procedure to obtain cubic Im3m symmetry, by the synthesis in mixtures of partially fluorinated and nonfluorinated surfactants at high temperature [77]. More hydrophilic surfactant mixtures are required and mixtures of $C_3F_7O(CFCF_3CF_2O)_2CFCF_3CONH(CH_2)_3N^+(C_2H_5)_2CH_3I$ and triblock copolymer F127 $(EO)_{106}(PO)_{70}(EO)_{106}$ were used. F127 is a block copolymer surfactant more hydrophilic than P123, and therefore the surfactant curvature favors micellar cubic phases.

A different approach, to obtain ordered mesoporous materials, has been described by Stébé *et al.* in a recent work [78]. They used as template the hexagonal liquid-crystalline phase H_1 in the $C_8F_{17}C_2H_4(OC_2H_4)_9OH$/water system. Phase-behavior studies showed that the domain of this H_1 phase goes from 53 to 78 wt % of surfactant, at 20 °C. Therefore, synthesis reactions were carried out in this range of concentration. TMOS was used as silica precursor, which releases methanol as a byproduct. As this alcohol can disrupt mesophases, it was removed during the

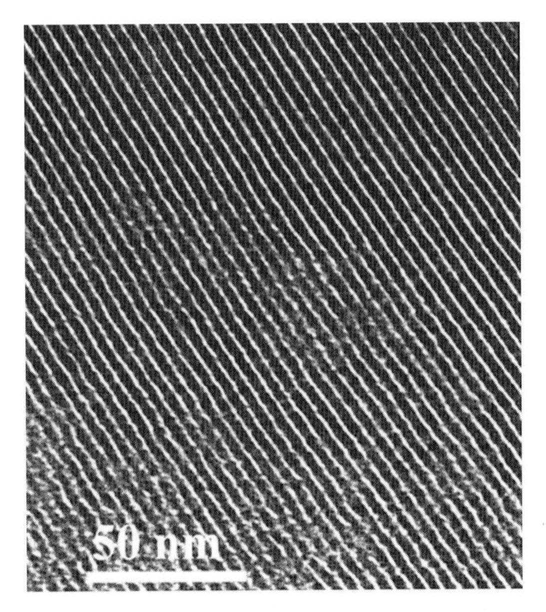

Figure 11.22 Longitudinal TEM view, parallel to the mesopores, of the sample synthesized with 60 wt% of $C_8F_{17}C_2H_4(OC_2H_4)_9OH$, at surfactant/silica molar ratio of 0.145, pH 1.3 and using HCl (aq.). A hydrothermal treatment was performed during 4 days at 80 °C. (Reproduced from Ref. [78], with permission).

synthesis by placing the mixture under vacuum. An example of the resulting material is shown in Figure 11.22.

It should be pointed out that the influence of the alcohols released during the hydrolysis reactions of inorganic precursors is a very important factor in determining the mesostructure. The alcohol can strongly modify the phase diagram of the surfactant in water and, as a consequence, the material pore ordering.

An interesting study is that of Stébé *et al.*, which described the phase behavior of $C_8F_{17}C_2H_4(OC_2H_4)_nOH$/methanol/water systems [59]. Well-ordered liquid crystals can be obtained up to a concentration of methanol in the solution around 8 wt%. If the methanol content in the mixture is further increased, the pore ordering is lost and the recovered materials adopt a wormhole-like structure.

11.4 Summary

Surfactants with hydrophobic chains different from conventional hydrocarbon alkyl chains possess interesting properties as structure-directing agents in self-assembly processes. Both silicone and fluorinated chains are less soluble in many organic solvents than their hydrocarbon counterparts. Moreover, silicone and fluorinated surfactants reduce the surface tension of water more effectively than conventional hydrocarbon surfactants. Consequently, these nonhydrocarbon surfactants can be very appropriate for the preparation of novel mesoporous materials, due to its peculiar self-aggregation properties that allow a precise control of the mesostructure.

Nowadays, silicone and fluorinated surfactants are commercially available with a wide variety of chemical structure and different chain lengths. In addition, novel nonhydrocarbon surfactant systems are being synthesized, which possess rich phase behavior. The molecular structure of surfactants, and specially the hydrophilic–lipophilic properties and the molecular packing parameter, are crucial in determining the surface curvature of the supramolecular aggregates. Consequently, new well-organized mesoporous materials with controlled properties can by obtained by selecting the appropriate surfactant molecules as structure-directing agents (templates).

The use of surfactant self-assembly to precisely control the formation and properties of mesoporous materials has been the focus of a great deal of attention in the last two decades. However, nonhydrocarbon surfactants have not been commonly used as structure-directing agents in the synthesis of mesoporous materials. Silicone and fluorinated surfactants can serve as templates or structure-directing agents for the preparation of mesoporous materials with different structures and morphologies. The use of these kinds of surfactants leads to the improvement of chemical, thermal and electrical properties as compared to those of materials synthesized from conventional hydrocarbon surfactants. Moreover, that improvement can be achieved at reduced surfactant concentrations. Parameters such as pore size and surface area can be tailored by modifying surfactant molecular structure or adding hydrophobic solubilizates. Materials with bimodal pore-size distributions (i.e. two distinct pore sizes) can be obtained by using mutually insoluble fluorinated and hydrocarbon surfactants.

In this context, silicone and fluorinated surfactants, which can impart peculiar properties to mesoporous materials, constitutes a research field that may be exploited to obtain new nanostructured materials for novel applications and many future developments may still arise in this field. The literature on the use of nonhydrocarbon surfactants for the preparation of mesostructured materials remains relatively scarce and many aspects are still unexplored.

References

1 Sayari, A. (1996) *Chem. Mater.*, **8**, 1840.
2 Melero, J.A., Van Grieken, R., and Morales, G. (2006) *Chem. Rev.*, **106**, 3790.
3 Márquez-Alvarez, C., Žilková, N., Pérez-Pariente, J., and Cejka, J. (2008) *Catal. Rev.*, **50**, 222.
4 Kisler, J.M., Dähler, A., Stevens, G.W., and O'Connor, A.J. (2001) *Micropor. Mesopor. Mater.*, **44–45**, 769.
5 Scott, B.J., Wirnsberger, G., and Stucky, G.D. (2001) *Chem. Mater.*, **13**, 3140.
6 Grätzel, M. (1999) *Curr. Opin. Colloid In.*, **4**, 314.
7 Guliants, V.V., Carreon, M.A., and Lin, Y.S. (2004) *J. Membrane Sci.*, **235**, 53.
8 Giri, S., Trewyn, B.G., and Lin, V.S.Y. (2007) *Nanomedicine-UK*, **2**, 99.

9 Slowing, I.I., Vivero-Escoto, J.L., Wu, C.-W., and Lin, V.S.-Y. (2008) *Adv. Drug Deliv. Rev.*, **60**, 1278.
10 Chen, X., and Mao, S.S. (2007) *Chem. Rev.*, **107**, 2891.
11 Ying, J.Y., Mehnert, C.P., and Wong, M.S. (1999) *Angew. Chem. Int. Ed.*, **38**, 56.
12 Antonietti, M. (2001) *Curr. Opin. Colloid In.*, **6**, 244.
13 Kresge, C.T., Leonowicz, M.E., Roth, W.J., Vartuli, J.C., and Beck, J.S. (1992) *Nature*, **359**, 710.
14 Beck, J.S., Vartuli, J.C., Roth, W.J., Leonowicz, M.E., Kresge, C.T., Schmitt, K.D., Chu, C.T., Olson, D.H., Sheppard, E.W., McCullen, S.B., Higgins, J.B., and Schlenker, J.L. (1992) *J. Am. Chem. Soc.*, **114**, 710.
15 Bagshaw, S.A., Prouzet, E., and Pinnavaia, T.J. (1995) *Science*, **269**, 1242.
16 Zhao, D., Huo, O., Feng, J., Chmelka, B.F., and Stucky, G.D. (1998) *J. Am. Chem. Soc.*, **120**, 6024.
17 Zhao, D., Feng, J., Huo, Q., Melosh, N., Fredrickson, G.H., Chmelka, B.F., and Stucky, G.D. (1998) *Science*, **279**, 548.
18 Kirmayer, S., Dovgolevsky, E., Kalina, M., Lakin, E., Cadars, S., Epping, J.D., Fernández-Arteaga, A., Rodríguez, C., Chmelka, B.F., and Frey, G.L. (2008) *Chem. Mater.*, **20**, 3745.
19 Wan, Y., and Zhao, D. (2007) *Chem. Rev.*, **107**, 2821.
20 Kim, J.M., Sakamoto, Y., Hwang, Y.K., Kwon, Y.U., Terasaki, O., Park, S.E., and Stucky, G.D. (2002) *J. Phys. Chem. B*, **106**, 2552.
21 Schmidt, G.L.F. (1973) *Industrial Applications of Surfactants* (ed. D.R. Karsa), Royal Society of Chemistry, London, p. 24.
22 He, M., Hill, R.M., Lin, Z., Scriven, L.E., and Davis, H.T. (1993) *J. Phys. Chem.*, **97**, 8820.
23 Hill, R.M., He, M., Lin, Z., Davis, H.T., and Scriven, L.E. (1993) *Langmuir*, **9**, 2789.
24 Rodríguez, C., Uddin, Md.H., Watanabe, K., Furukawa, H., Harashima, A., and Kunieda, H. (2002) *J. Phys. Chem. B*, **106**, 22.
25 Uddin, M.H., Rodríguez, C., López-Quintela, A., Leisner, D., Solans, C., Esquena, J., and Kunieda, H. (2003) *Macromolecules*, **36**, 1261.
26 Soni, S.S., Sastry, N.V., Aswal, V.K., and Goyal, P.S. (2002) *J. Phys. Chem. B*, **106**, 2606.
27 Lin, Y., and Alexandridis, P. (2002) *J. Phys. Chem. B*, **106**, 10845.
28 Watanabe, K., Kanei, N., and Kunieda, H. (2002) *J. Oleo Sci.*, **51**, 771.
29 Xu, A.-W., Yu, J.C., Cai, Y.-P., Zhang, H.-X., and Zhang, L.-Z. (2002) *Chem. Commun.*, 1614.
30 Xu, A.-W. (2002) *Chem. Mater.*, **14**, 3625.
31 Xu, A.-W. (2002) *Chem. Lett.*, 878.
32 Xu, A.-W. (2002) *Chem. Lett.*, 982.
33 Xu, A.-W., Cai, Y.-P., Zhang, H.-X., Zhang, L.-Z., and Yu, J.C. (2002) *Angew. Chem. Int. Ed.*, **41**, 3844.
34 Xu, A.-W., Yu, J.C., Zhang, H.-X., Zhang, L.-Z., Kuang, D.-B., and Fang, Y.-P. (2002) *Langmuir*, **18**, 9570.
35 Xu, A.-W. (2002) *J. Phys. Chem. B*, **106**, 13161.
36 Yuan, Z.Y., Six-Boulanger, M.F., and Su, B.L. (2003) *Angew. Chem. Int. Ed.*, **42**, 1572.
37 Yuan, Z.Y., and Su, B.L. (2004) *Chem. Mater.*, **16**, 195.
38 Liu, Y.R. (2009) *J. Mater. Sci.*, **44**, 3600.
39 Liu, Y.R. (2009) *Micropor. Mesopor. Mater.*, **124**, 190.
40 Hüsing, N., Launay, B., Bauer, J., Kickelbick, G., and Doshi, D. (2003) *J. Sol-Gel Sci. Technol.*, **26**, 609.
41 Rodríguez, C., Esquena, J., Aramaki, K., and López-Quintela, M.A. (2009) *Micropor. Mesopor. Mater.*, **119**, 338.
42 Melosh, N.A., Lipic, P., Bates, F.S., Wudl, F., Stucky, G.D., Fredrickson, G.H., and Chmelka, B.F. (1999) *Macromolecules*, **32**, 4332.
43 Kissa, E. (1994) *Fluorinated Surfactants and Repellents, Surfactant Science Series*, vol. **97**, Marcel Dekker, New York.
44 Tadros, Th.F. (1980) *J. Colloid Interface Sci.*, **74**, 196.
45 Hudlicky, M. (1962) *Chemistry of Organic Fluorine Compounds*, Macmillan, New York.
46 Schuierer, E. (1976) *Tenside*, **13**, 1.
47 Bernett, M.K., and Zisman, W.A. (1959) *J. Phys. Chem.*, **63**, 1911.
48 Truborg, M.P., and Brady, J.E. (1988) *J. Am. Chem. Soc.*, **110**, 7797.
49 Lo Nostro, P., and Chen, S.H. (1993) *J. Phys. Chem.*, **97**, 6535.

50 Binks, B.P., Fletcher, P.D.I., and Thompson, R.L. (1996) *Ber. Bunsen. Ges.*, **100**, 232.

51 Shafrin, E.G., and Zisman, W.A. (1962) *J. Phys. Chem.*, **66**, 740.

52 Blin, J.L., Lesieur, P., and Stébé, M.J. (2004) *Langmuir*, **20**, 491.

53 Blin, J.L., and Stébé, M.J. (2004) *J. Phys. Chem. B.*, **108**, 11399.

54 Blin, J.L., and Stébé, M.J. (2005) *Micropor. Mesopor. Mater.*, **87**, 67.

55 Blin, J.L., Bleta, R., Ghanbaja, J., and Stébé, M.J. (2006) *Micropor. Mesopor. Mater.*, **94**, 74.

56 Blin, J.L., Henzel, N., and Stébé, M.J. (2006) *J. Colloid Interface Sci.*, **302**, 643.

57 Bleta, R., Blin, J.L., and Stébé, M.J. (2006) *J. Phys. Chem. B*, **110**, 23547.

58 Michaux, F., Carteret, C., Stébé, M.J., and Blin, J.L. (2008) *Micropor. Mesopor. Mater.*, **116**, 308.

59 Zimny, K., Blin, J.L., and Stébé, M.J. (2009) *J. Colloid Interf. Sci.*, **330**, 456.

60 Groenewolt, M., Antonietti, M., and Polarz, S. (2004) *Langmuir*, **20**, 7811.

61 Michaux, F., Blin, J.L., and Stébé, M.J. (2007) *Langmuir*, **23**, 2138.

62 Rankin, S.E., Tan, B., Lehmler, H.J., Hindman, K.P., and Knutson, B.L. (2004) *Micropor. Mesopor. Mater.*, **73**, 197.

63 Tan, B., Lehmler, H.J., Vyas, S.M., Knutson, B.L., and Rankin, S.E. (2005) *Chem. Mater.*, **17**, 916.

64 Meng, X., Di, Y., Zhao, L., Jiang, D., Li, S., and Xiao, F.S. (2004) *Chem. Mater.*, **16**, 5518.

65 Di, Y., Meng, X., Li, S., and Xiao, F.-S. (2005) *Micropor. Mesopor. Mater.*, **82**, 121.

66 Xiao, F.S. (2005) *Curr. Opin. Colloid In.*, **10**, 94–101.

67 Meng, X., Lu, D., and Tatsumi, T. (2007) *Micropor. Mesopor. Mater.*, **105**, 15.

68 Rodríguez-Abreu, C., Botta, P., Rivas, J., Aramaki, K., and Quintela, M.A. (2008) *J. Non-Cryst. Solids*, **354**, 1074.

69 Tan, B., Lehmler, H.J., Vyas, S.M., Knutson, B.L., and Rankin, S.E. (2005) *Adv. Mater.*, **17**, 2368.

70 Pelaez, M., de la Cruz, A.A., Stathatos, E., Falaras, P., and Dionysiou, D.D. (2009) *Catal. Today*, **144**, 19.

71 Du, Y., Yang, Y., Liu, S., Xiao, N., Zhang, Y., and Xiao, F.S. (2008) *Micropor. Mesopor Mater.*, **114**, 250.

72 Du, Y., Liu, S., Ji, Y., Zhang, Y., Liu, F., Gao, Q., and Xiao, F.S. (2008) *Catal. Today*, **131**, 70.

73 Osei-Prempeh, G., Lehmler, H.J., Knutson, B.L., and Rankin, S.E. (2005) *Micropor. Mesopor. Mater.*, **85**, 16.

74 Esquena, J., Rodríguez, C., Solans, C., and Kunieda, H. (2006) *Micropor. Mesopor. Mater.*, **92**, 212.

75 Sharma, S.C., Kunieda, H., Esquena, J., and Rodríguez-Abreu, C. (2006) *J. Colloid Interf. Sci.*, **299**, 297.

76 Han, Y., Li, D., Zhao, L., Song, J., Yang, X., Li, N., Li, C., Wu, S., Xu, X., Meng, X., Lin, K., and Xiao, F.S. (2003) *Angew. Chem. Int. Ed.*, **42**, 3633.

77 Li, D.F., Han, Y., Song, J.W., Zhao, L., Xu, X.Z., Di, Y., and Xiao, F.S. (2004) *Chem. Eur. J.*, **20**, 5911.

78 Zimny, K., Blin, J.L., and Stébé, M.J. (2009) *J. Phys. Chem. C*, **113**, 11285.

12
Worm-Like Micelles in Diluted Mixed Surfactant Solutions: Formation and Rheological Behavior

Alicia Maestro, Jordi Nolla, Carmen González, and José M. Gutiérrez

12.1
Introduction

As known, surfactant molecules self-assemble in aqueous media to form aggregates with different microstructures and shapes depending on concentration, composition, and temperature. Usually, hydrophilic surfactants form spherical micelles in water in the dilute region. The aggregate geometry may be predicted on the basis of the packing of the surfactant molecules in the aggregate or the "critical packing parameter", CP, defined as $v/l_c \cdot a_s$, where v is the volume of the lipophilic chain having maximum effective length l_c, and a_s is the cross-sectional area of the headgroup at the interface. For CP $\leq 1/3$ spherical micelles are expected as a result of a positive high spontaneous curvature of the interface; when $1/3 \leq$ CP $\leq 1/2$, cylinders are expected instead, as a result of a decrease in the mean spontaneous curvature. The sphere–rod transition in the micellar shape can be induced by different ways depending on the surfactant used, such as increasing surfactant concentration, salinity or temperature. In certain conditions, micelles undergo enormous elongation and form very long and highly flexible aggregates, named worm-like or thread-like micelles. Above a system-dependent concentration of the surfactant/s, called the overlap concentration, c^*, worm-like micelles are close enough to each other to entangle and form a transitional three-dimensional network that imparts a strong viscoelastic behavior to the system, in some way analogous to those observed in polymer solutions.

Before Kunieda's group contributions appeared in the field, around 2003, most of the literature reported the formation of viscoelastic worm-like micellar solutions in ionic systems through the presence of strongly binding counterions. These counterions neutralize charges of surfactants, reducing repulsion between headgroups of neighboring molecules in a micelle. This reduction results in a decrease in a_s and the consequent reduction in the CP. For example, works using long-chain cationic surfactants such as hexadecyltrimethylammonium bromide (CTAB) [1–3] and hexadecylpyridinium chloride [4–6] report forming worm-like micelles upon the addition of counterions such as salicylate or bromide ions. There were some reports on significant micellar growth in water–nonionic surfactant systems [7, 8],

Self-Organized Surfactant Structures. Edited by Tharwat F. Tadros

ISBN: 978-3-527-31990-9

but a highly viscoelastic solution in the dilute region had not been reported yet. Around 2003, some of the articles of Kunieda's group started to appear in the field, where worm-like systems with highly viscoelastic behavior in the dilute region in the absence of counterions were obtained. Some of the formulations use surfactants that, due to their natural origin, are biocompatible and environment-friendly, like sucrose alkanoates [9], polyoxyethylene cholesteryl ethers [10], or polyoxyethylene phytosterols [11]. These properties make them suitable additives in the food industry, pharmacy and cosmetics. Kunieda's contribution was very important for several years, and after his demise his group has in some ways taken his heritage and continues publishing in the field.

In this chapter, we present a review of the contribution of Kunieda's group regarding to worm-like micelles formation, the structure and rheological behavior of the micellar solution, through the use of published and unpublished material of his group.

12.2 Worm-Like Micelles: Formation and Rheological Behavior

12.2.1 Mechanism of Formation of Worm-Like Micelles

Kunieda's group reported numerous viscoelastic worm-like micellar systems in the salt-free condition when a lipophilic nonionic surfactant such as short hydrophilic chain poly(oxyethylene) alkyl ether, C_nEO_m, or N-hydroxyethyl-N-methylalkanolamide, NMEA-n, was added to the dilute micellar solution of hydrophilic cationic (dodecyltrimethylammonium bromide, DTAB and hexadecyltrimethylammonium bromide, CTAB) [12–14], anionic (sodium dodecyl sulfate, SDS [15, 16], sodium dodecyl trioxyethylene sulfate, SDES [17], and Gemini-type [18]) or nonionic (sucrose alkanoates, C_nSE [9, 19], polyoxyethylene cholesteryl ethers, $ChEO_n$ [10, 20], polyoxyethylene phytosterol, $PhyEO_n$ [11, 21] and polyoxyethylene sorbitan monooleate, Tween-80 [22]) surfactants. The mechanism of formation of these worm-like structures and the resulting rheological behavior of micellar solutions is discussed in this section based in some actual published and unpublished results, but conclusions can qualitatively be extended to all the systems studied by Kunieda's group.

Surfactants with a big headgroup (or a strong charge) tend to form spherical micelles in water in the dilute region, with a strong positive curvature. One way to increase their CP is to add a cosurfactant in order to reduce the mean cross-sectional area of the headgroup at the interface, $\bar{a}_s$, and, consequently, the mean spontaneous curvature of the interface. Hence, if they are mixed with a lipophilic amphiphile, such as C_nEO_m or NMEA-n, a sphere–rod transition takes place and a viscoelastic micellar system can appear in the dilute region. In rod-like micelles, the cross-sectional area of the endcaps, a_{s1}, is higher than the cross-sectional area of the aggregate at the cylindrical part, a_{s2}. When a mixture of surfactants is used, segregation is expected in such a way that hydrophilic surfactant preferentially

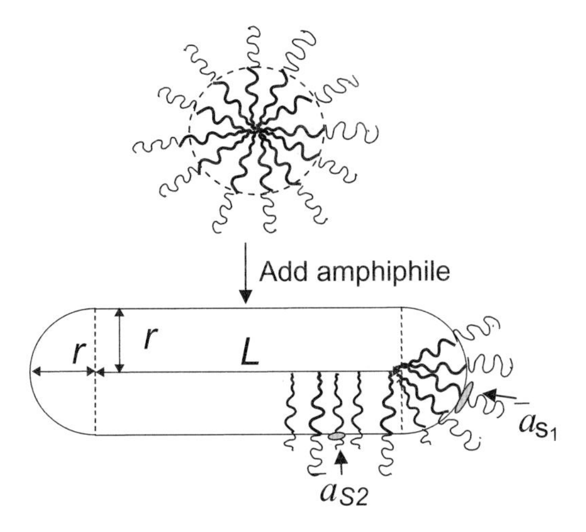

Figure 12.1 Schematic model of a rod-like micelle. *L* is the cylindrical length of the rod-like lipophilic core, *r* is the average lipophilic length, a_{s1} and a_{s2} are the cross-sectional area of endcaps and body of micelles that result in an average cross-sectional area at the hydrophilic/lipophilic interface, $\bar{a}_s$.

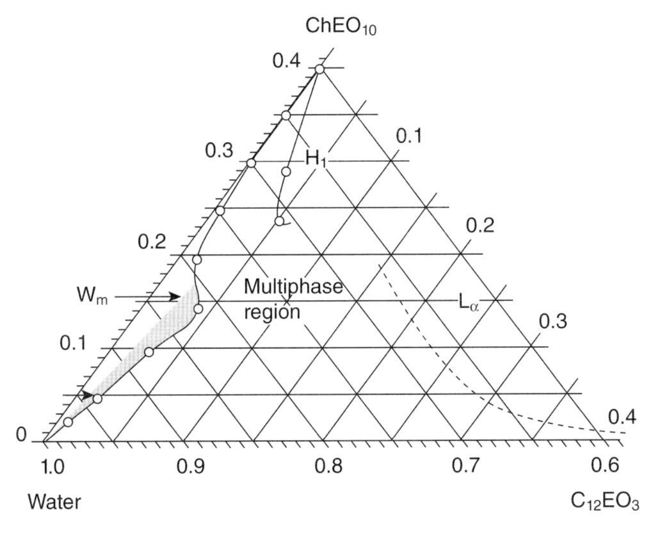

Figure 12.2 Partial phase diagram of the $ChEO_{10}$-$C_{12}EO_3$-water system in the dilute region at 25 °C. H_1 and L_α are direct hexagonal and lamellar liquid phases, and W_m is micellar phase. The region of high viscosity within the W_m phase is tentatively shown by the shaded area (reprinted from ref. [10] with permission from ACS Publications).

situates in the endcaps, and most of the amphiphile resides in the "body" of the micelle (see Figure 12.1).

Figure 12.2 shows the phase behavior of the $ChEO_{10}$-$C_{12}EO_3$-water system at 25 °C in a water-rich region [10]. This phase diagram is presented as an example of the phase behavior of surfactant–cosurfactant–water systems studied. All the systems studied are reported to behave in a qualitatively similar way and, therefore, conclusions exposed here can be extended to the other systems. Successive additions of $C_{12}EO_3$ to the micellar (W_m) phase of the $ChEO_{10}$-water binary system result in a W_m-hexagonal (H_1) phase transition at higher $ChEO_{10}$ concentrations, whereas a W_m–L_α phase transition is observed at lower $ChEO_{10}$ concentrations and is attributed to the decrease in surfactant layer curvature with the incorporation of $C_{12}EO_3$ into the palisade layer of the surfactant aggregate. Inside the W_m region, there is a zone, shown tentatively by the shaded area in the phase diagram, where the samples have very high viscosity with a gel-like behavior. These highly viscous samples are optically isotropic at rest but show shear birefringence, an indication of the formation of entangled worm-like micelles, which can break, disentangle and align in the direction of the flow when shear stress is applied. It will be seen below that, as a consequence, these samples present shear-thinning behavior. The shaded region of the highly viscous micellar phase appears to be extended from the protruded part of the H_1. It seems, then, to be understandable that when a sample in the H_1 region is diluted, although the ordered structure is lost, it is in some way maintained and "rigid and packed cylinders" of the H_1 phase are transformed in "flexible and entangled cylinders" in the W_m phase (worm-like micelles).

While the amphiphile/surfactant ratio is progressively increased from the surfactant–water binary system, as a consequence of the decrease in spontaneous curvature one-dimensional micellar growth occurs that may, then, be considered as a tendency of the system to minimize the excess free energy by reducing the number of endcaps, where the hydrophilic surfactant situates (see Figure 12.1), in spite of the counteracting entropy factor.

The number of endcaps would in principle continuously decrease, due to one-dimensional growth, when the amphiphile–surfactant ratio is increased, and longer worm-like micelles would be expected to be found, resulting in higher viscosity of the system.

Figure 12.3 shows steady-state viscosity of $ChEO_{10}$-$C_{12}EO_3$-water systems at several concentrations of $C_{12}EO_3$ [10]. Results can qualitatively be extended to all the systems studied by Kunieda's group. At low mole fractions X of $C_{12}EO_3$, the system is nearly Newtonian and the viscosity is low. However, above a certain mole fraction, shear-thinning behavior is observed instead, and low-shear viscosity strongly increases with X, up to an X for which the enhancement of the viscosity reaches a maximum. Higher X produces a progressive decrease in viscosity and a return to Newtonian behavior, up to phase separation, where lamellar liquid crystal L_α appears. As stated, the strongly enhanced viscosity and shear-thinning behavior are typical of entangled worm-like micelles, where shear thinning occurs because of alignment of micelles in the direction of the flow. The decrease of viscosity after the X where a maximum viscosity is reached is attributed to branching of micelles, which decreases their length. The tendency of micelles to branching can be explained if one sees branching as the result of a fusion of an endcap of a micelle inside the body of another one. Then, an endcap disappears and the mean interface curvature decreases–we will deep on branching in Section 12.3. Above a certain X, lamellar liquid crystal appears, with curvature zero.

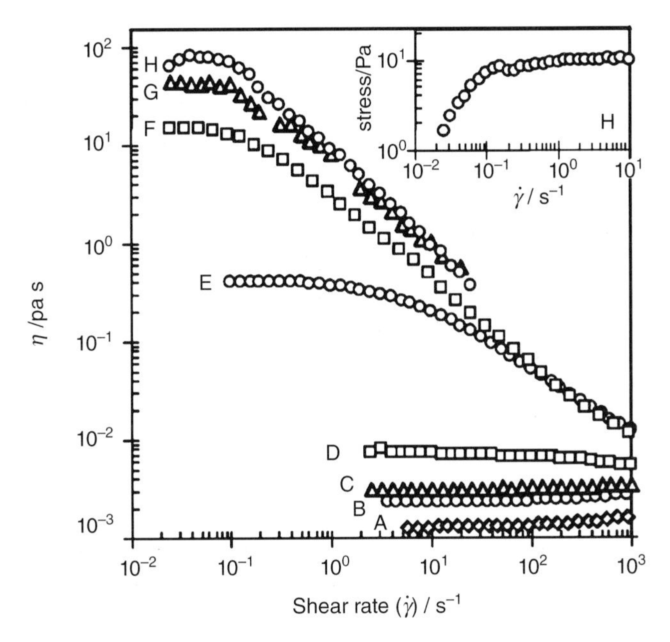

Figure 12.3 Steady-state viscosity *vs.* shear rate for micellar solutions of 0.06 M $ChEO_{10}$-$C_{12}EO_3$ systems at various mole fractions of $C_{12}EO_3$ (X): (A) 0; (B) 0.09; (C) 0.14; (D) 0.17; (E) 0.24; (F) 0.29; (G) 0.36; (H) 0.42 (reprinted from ref. [10] with permission from ACS Publications).

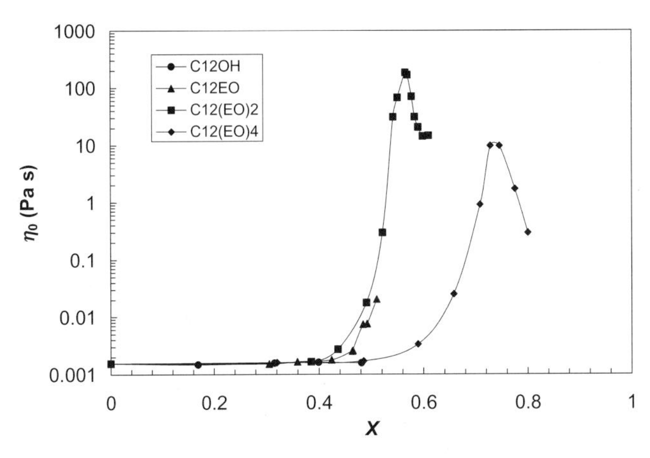

Figure 12.4 Zero-shear viscosity, η_0, *vs.* X (molar fraction of $C_{12}EO_n$) for $ChEO_{15}$ = 0.06 M and $C_{12}EO_n$, with n = 0, 1, 2, and 4.

In Figure 12.4 some until now unpublished results for $ChEO_{15}$-$C_{12}EO_n$ systems are presented. Zero-shear viscosity η_0 is plotted *vs.* mole fraction of amphiphile, X, for $C_{12}EO_n$ with n = 0, 1, 2, and 4. For all systems, tests have been done at X from 0 to X just before phase separation. According (qualitatively) to Figure 12.3,

η_0 is of the order of the viscosity of water in the absence of amphiphile, abruptly increases after an X that depends on n (corresponding to overlapping concentration, c^*) and reaches a maximum, and then decreases again up to phase separation. With $n = 0$ and 1 phase separation into lamellar liquid crystal (curvature zero) appears before an important increase of viscosity can take place. With $n = 2$ the enhancement of viscosity is more important and happens at lower X than with $n = 4$. These results are in agreement with that published elsewhere [10] and with results obtained in other systems where C_mEO_n is added to sucrose alkanoates [9] or SDS [16], as examples. In general, the ability of C_mEO_n to decrease curvature follows the order $C_mOH > C_mEO > C_mEO_2 > C_mEO_3 > C_mEO_4$. It seems coherent as the decrease in mean cross-sectional area will be stronger when an amphiphile with smaller headgroup is added. However, if the decrease in curvature is too strong, lamellar liquid crystal, with curvature 0, appears, and phase separation occurs, before a strong increase in viscosity is observed (see Figure 12.3). In the same way, if the results of ref. [10] are compared with those of ref. [20], it can be seen that when the headgroup of the hydrophilic surfactant is bigger, a greater amount of the amphiphile is required to induce one-dimensional growth, and the viscosity enhancement is more modest, since the initial spontaneous curvature of the interface is higher. Analog results are obtained with sucrose alkanoates [9]; sucrose hexadecanoate watery solutions form stronger worm-like micellar networks than sucrose dodecanoate when C_mEO_n is added.

12.2.2
Rheology of Worm-Like Micelles

As said before, worm-like micellar solutions present viscoelastic behavior due to entanglements between flexible micelles that promote the formation of a three-dimensional transient network analogous to that of a melt polymer network. However, worm-like micelles are dynamic systems that can break and recombine. Because of that, they are also named "living polymers". Cates proposes a model to describe the rheological behavior of these living polymers [23, 24]. The Cates model couples the diffusive disentanglement of micelles due to their reptation along a tube (typical relaxation of polymers) to the kinetic equations describing their reversible breakdown. This implies that two relaxation processes at two different time scales take place, with two characteristic relaxation times, reptation time τ_{rep} and breaking time τ_b. When $\tau_{rep} \ll \tau_b$ the viscoelastic response at low and intermediate frequencies can be fitted to a single Maxwell model with a relaxation time given by $\tau_M = \sqrt{\tau_b \cdot \tau_{rep}}$. Then, the plateau modulus follows the equation $G_0 = \eta_0 / \tau_M$ and the viscoelastic functions storage modulus (G') and loss (G'') modulus can be described by Equations 12.1 and 12.2:

$$G'(\omega) = G_0 \frac{(\omega\tau_M)^2}{1+(\omega\tau_M)^2} \tag{12.1}$$

$$G''(\omega) = G_{0,M} \frac{\omega\tau_M}{1+(\omega\tau_M)^2} \tag{12.2}$$

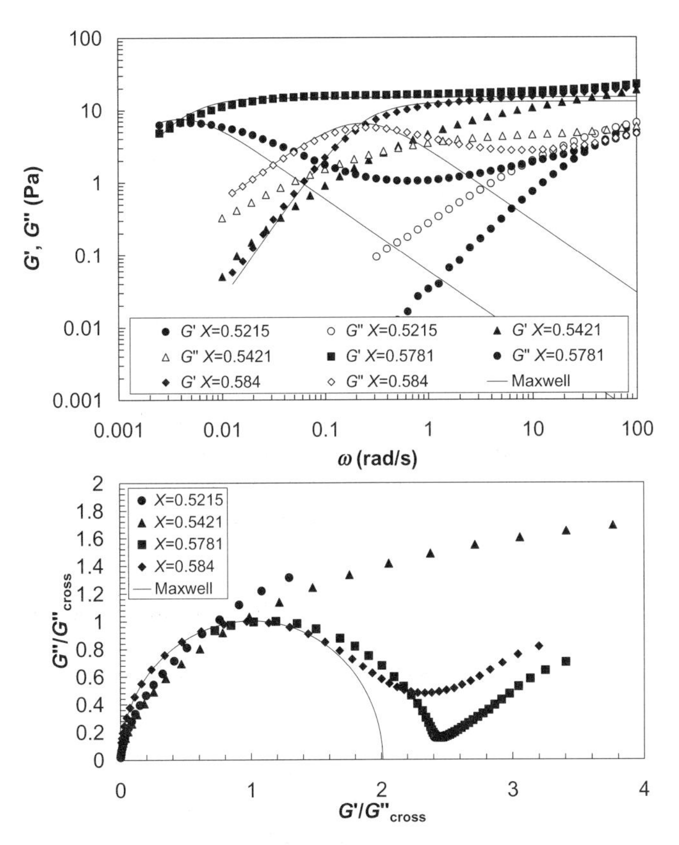

Figure 12.5 Frequency-sweep tests of $ChEO_{15}$ = 0.06 M and $C_{12}EO_2$, for several mole fractions, X. (a) Viscoelastic functions G' and G'' *vs.* frequency; ω. (b) Cole–Cole plot. Lines: fitting to Maxwell model.

where ω is frequency and τ_M is Maxwellian relaxation time that can be calculated as $\tau_M = 1/\omega_{cross}$, being ω_{cross} the frequency for which $G'(\omega) = G''(\omega)$. For Maxwell model, if viscoelastic functions are presented in a Cole–Cole plot ($G''(\omega)/G''_{max}$ *vs.* $G'(\omega)/G''_{max}$) a semicircle is obtained.

Figure 12.5 shows some until now unpublished viscoelastic data for the $ChEO_{15}$/$C_{12}EO_2$/water system, fitted to the Maxwell model if appropriate (lines). Similar results are obtained with other systems [9, 10, 21]. It can be seen that, at low X, G'' is higher than G' in all the range of frequencies tested, indicating a poor viscoelastic behavior. Moreover, the slopes of G' and G'' functions at low frequencies are far from 2 and 1, which they should be if the behavior was described by the Maxwell model. When X increases, micelles grow and entangle, and then the Maxwell model fits the results at low and intermediate frequencies (before the crossing point between G' and G''). At X where a sharp increase in viscoelasticity occurs, the crossing point moves to very low frequencies (too low to be experimentally seen in a time short enough to assure constant composition of the sample), and the zone that can be fitted to the Maxwell model can hardly be seen. After the maximum enhancement, the crossing point moves again to higher frequencies, indicating that samples are losing their viscoelastic properties, and the behavior can no longer be fitted to the Maxwell model. These last results are not shown in the Figure for clarity, but analogous results can be observed in ref. [9].

As can be seen, results at frequencies after the crossing point deviate from the Maxwell model, presenting an upturn in G'' and an increase in G' that are not predicted by the model and have been attributed to additional faster relaxation processes, such as Rouse relaxation [16]. We will discuss this feature further in Section 12.3.

12.2.3 Studied Systems

The general behavior described above was experimentally observed in Kunieda's lab for numerous new systems, and the results were diffused through several papers and congress communications. These works suppose an important contribution in extending the potential applications of new surfactants and cosurfactants, and are now, in the following text, briefly reviewed. The systems studied by Professor Kunieda are reviewed according to his own classification given in his review paper [25].

12.2.3.1 Ionic Surfactant–Cosurfactant Systems

Some works report the formation of worm-like micelles through the addition of C_mEO_n amphiphile to watery dilute solutions of ionic surfactants like sodium dodecyl sulfate (SDS) [16], hexadecyltrimethylammonium bromide (CTAB) [14], dodecyltrimethylammonium bromide (DTAB) [13], and a new Gemini-type surfactant without a spacer chain [18]. Gemini-type surfactants are interesting as it is expected that their biocompatibility will be good compared with a conventional surfactant since their skin penetration may be low due to their high molecular weight. Results obtained with all these ionic surfactants are in accordance with that suggested above, that is, smaller EO chains and bigger hydrophobic chains of the amphiphile favor a faster decrease in curvature when added to surfactant solutions and, consequently, a faster one-dimensional growth of micelles and a higher and faster enhancement of viscoelastic parameters, if the decrease in curvature is not so strong to promote the appearance of lamellar liquid crystal before the formation of worm-like micelles long enough to entangle and form a transient network. In agreement with these results, SAXS data indicate that one-dimensional growth of micelles was obtained with $n = 0$ to 4, with the micelles growing faster at lower n values [16]. When surfactants CTAB and DTAB are compared [13], it is observed that the increase of viscoelastic parameters is more effective for CTAB than for DTAB, as CTAB has a longer hydrophobic group and, as a consequence, micelles in the CTAB–water binary system have a smaller curvature. The contribution of ref. [14] is also worth mentioning. As commercial surfactants are not pure ones, but in fact a mixture of surfactants, the authors considered it interesting to elucidate how the size distribution of amphiphile mixtures affects their effectiveness. They compared the effect of adding $C_{12}EO_{\bar{3}}$, with the same mean $\bar{n} = 3$, but with several controlled size distributions of the headgroup. They observed that the amphiphile was less effective in elongating micelles when the size distribution of the headgroup was wider. They explained this feature for a partial segregation of the $C_{12}EO_n$ molecules, that makes $C_{12}EO_n$ with low n to be preferably located in the cylindrical body of the micelles, while $C_{12}EO_n$ with high n – and, as a result, with higher cross-sectional area – tend to locate in the endcaps. This fact decreases the unfavorable endcap energy and stabilizes endcaps, and higher amounts of $C_{12}EO_n$ are required to obtain the same viscosity enhancement.

The authors also studied the addition of alkanolamides (C8, NMEA-8; C12, NMEA-12; C16, NMEA-16), well known as foam boosters in surfactant aqueous solution, as thickening agents in shampoos and as antistatic and anticorrosion agents in detergents, to watery solutions of SDS [15], sodium dodecyl trioxyethylene sulfate (SDES) [17] CTAB, and DTAB [12]. They observed the formation of worm-like micelles with high viscoelasticity with the same characteristics that formed with the C_mEO_n amphiphiles. The effectiveness of NMEA in decreasing the mean curvature of micelles followed the order NMEA-16 > NMEA12 > NMEA8, since, as said above, a longer hydrophobic tail reduces curvature faster.

12.2.3.2 **Mixed Nonionic Surfactant Systems**

As already stated, several studies have been published using nonionic surfactant systems. The formation of viscoelastic micellar solutions in mixed nonionic systems is interesting in basic research – as the relation between packing constraints of hydrophobic chains and micellar growth would be clarified since the complicated interaction between the counterion and headgroup does not occur – as well as in applications such as cosmetics or pharmacy, where the avoidance of ionic additives is often desirable.

The use and properties of $ChEO_m/C_mEO_n$/water and sucrose alkanoate/C_mEO_n/water systems have been described above. Here, only some additional comments will be done. The authors reached the formation of viscoelastic worm-like micellar networks through the use of $C_mEO_n + ChEO_n$ with n = 10, 15 and 30 [20, 26], having high viscoelastic properties when cholesteryl with smaller endcap was used. A very recent contribution has appeared [20] that states that the temperature sensitivity of these micellar networks can be reduced if the EO_n group of the $ChEO_n$ increases in size. When the $ChEO_m$-C_mEO_n-water systems are studied at progressively higher temperatures, first, a viscosity enhancement was observed at intermediate temperatures, due to stronger one-dimensional growth promoted by dehydration of EO_n groups and the consequent decrease in spontaneous curvature. However, at high temperatures, the viscosity decreased. This was attributed to the formation of branched micelles, that, due to locally negative curvature, are more effective in reducing the mean curvature of micelles than linear ones. The authors find that $ChEO_{30}$ is less sensitive to temperature changes than $ChEO_{15}$, since dehydration of $ChEO_{30}$ must to be stronger to have an appreciable effect on curvature. A paper appeared in 2004 [21] compares results obtained in $ChEO_m$ systems with that obtained in polyoxyethylene phytosterol ($PhyEO_m$) systems. Like the formers, $PhyEO_m$ are surfactants of natural origin and have structures similar to the $ChEO_m$, except that the alkyl chain of the sterol is more branched. Comparison of rheological behavior for both systems when $C_{12}EO_n$ was added under similar conditions stated that in $PhyEO_m$ systems $C_{12}EO_n$ can induce micellar growth more easily, which was attributed to the additional branching of the alkyl chain in the lipophilic moiety of $PhyEO_m$, which promotes a lower curvature.

Some other publications have appeared that find viscoelastic worm-like micellar systems through the addition of C_mEO_n [22] and NMEA-n [17] to commercial surfactant polyoxyethylene sorbitan monooleate (Tween-80) diluted solutions, observing the same behavior as the systems already discussed.

12.3 Deeper Studies of the Surfactant–Cosurfactant Interaction

In this section, some contributions that try to understand surfactant–cosurfactant interactions and their effect in rheological behavior of the systems will be discussed.

As said before, when entangled worm-like micelles are formed, viscoelastic behavior only follows Cates model at low and intermediate frequencies, indicating that other fast relaxation processes exist (see Figure 12.5). Moreover, at cosurfactant–surfactant ratios above the viscosity maximum, further addition of cosurfactant decreases viscosity and viscoelastic functions up to phase separation, where lamellar liquid crystal appears. As said above, this has been related to the fact that, after the maximum in viscosity, the decrease in spontaneous curvature produces branching of micelles before phase separation [9, 10].

In order to go more deeply into the study of the surfactant–cosurfactant interactions, and specially for a better understanding of the structure and behavior of micellar systems after the surfactant–cosurfactant optimum ratio, Kunieda's group obtained and compared rheology and dynamic light scattering (DLS) data in the sucrose hexadecanoate-$C_{12}EO_n$-water system (Maestro *et al.*, 2004). The sucrose hexadecanoate formula ($C_{16}SE$) is shown in Figure 12.6. This nontoxic, biodegradable and biocompatible surfactant aggregates into spherical micelles in water in the dilute region due to the big size of its hydrophilic endcap. Therefore, as in

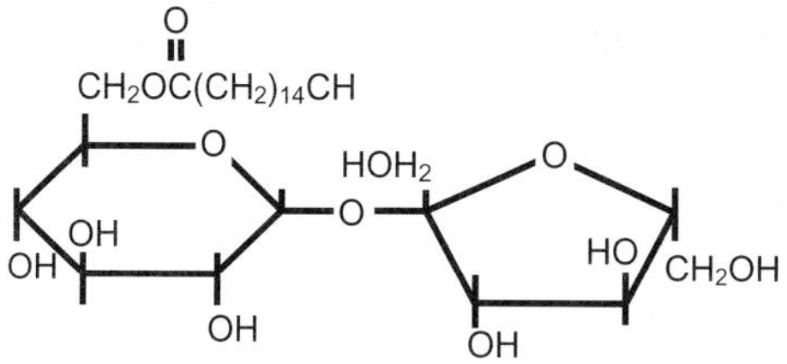

Figure 12.6 Sucrose hexadecanoate.

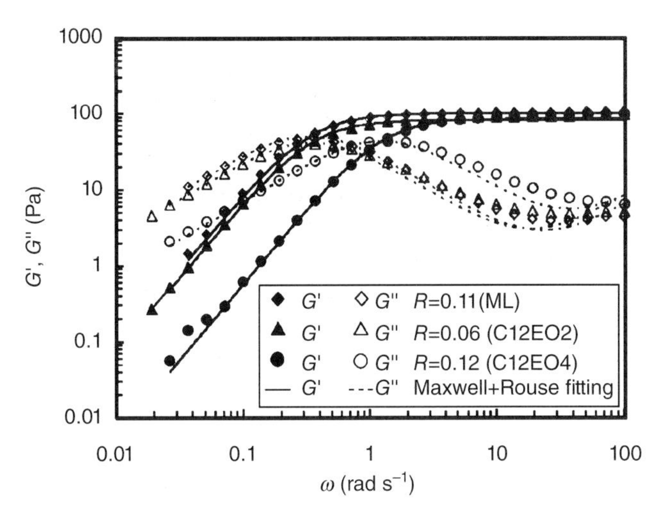

Figure 12.7 Frequency sweep tests and fitting of some C_{16}SE-amphiphile-water systems (90% water) to Maxwell + Rouse models (reprinted from ref. [9] with permission from ACS Publications).

other cases presented above, the addition of an amphiphile with a small endcap would promote the formation of giant worm-like micelles due to a decrease in the interface spontaneous curvature. The phase diagrams of this system are qualitatively analogous to that obtained with other systems (see Figure 12.2 above) [10]. A micellar region W_m is found near the water corner, that separates into $W_m + H_1$ at low concentrations of the cosurfactant and high concentrations of sucroester, and into $W_m + L_\alpha$ at low concentrations of the sucroester. The rheology of this system is likewise analogous to the rheology of the systems already presented, with a strong increase in viscosity and viscoelastic parameters when the amphiphile is progressively added up to a maximum, followed by a decrease until phase separation occurs.

Figure 12.7 shows, as examples, oscillatory functions obtained through frequency-sweep tests at several weight ratios (R) of amphiphile/C_{16}SE, where, in all cases, a 90 wt% water and a 10 wt% total surfactant + cosurfactant was maintained for all R.

Maestro *et al.* (2004) [9] studied the R effect on the formation and disruption of the transient network formed by worm-like micelles, using $C_{12}EO_4$ as the cosurfactant. Oscillatory results at low and intermediate frequencies were fitted to the Maxwell model, as had been done in other publications [9]. The obtained parameters G_0 and τ_M are presented in Figure 12.8 *vs.* ratio R. It can be observed that, when R is increased, G_0 continuously grows, nearly reaching a plateau. On the other hand, τ_M quickly increases up to the R for which the viscosity maximum is found, and then decreases up to phase separation. Kunieda's group found the same tendencies for other worm-like micellar systems, like the water-polyoxyethylene cholesteryl ether $Ch(EO)_m$-$C_{12}(EO)_n$ systems [10]. Figure 12.9 shows some until

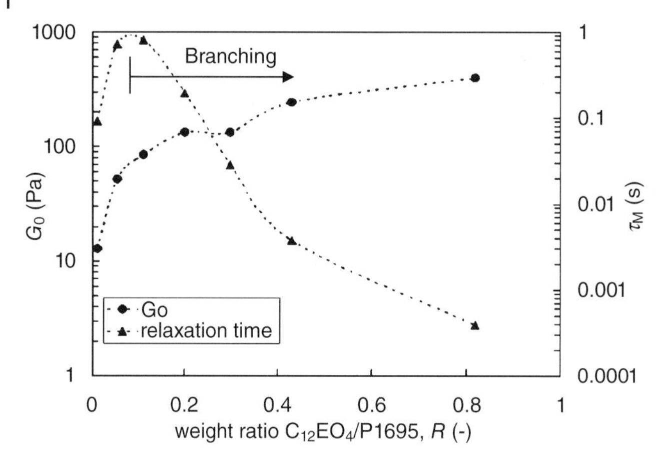

Figure 12.8 G_0 and τ_M *vs.* R for the $C_{16}SE$ = P1695-$C_{12}EO_4$-water system (adapted from ref. [9] with permission from ACS publications).

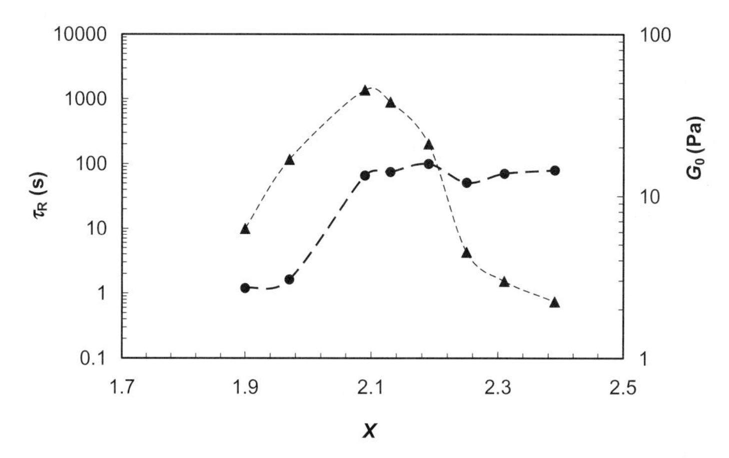

Figure 12.9 G_0 and τ_M *vs.* mole fraction X for the water-Ch(EO)$_{15}$-$C_{12}EO_2$ system. $X = [\text{Ch(EO)}_{15}]/([\text{Ch(EO)}_{15}] + [C_{12}EO_2])$.

now unpublished results obtained with the ChEO$_{15}$-C$_{12}$(EO)$_2$-water system, that follow the same tendencies than that presented in Figure 12.8.

Rehage and Hoffmann [27] propose an increase of the plateau modulus, G_0, when the mean length between entanglements (l_e) decreases, through the expression $G_0 \approx kT/l_e^{9/5}$. Results shown in Figures 12.8 and 12.9 imply therefore that the number density of junctions strongly increases with the cosurfactant–surfactant ratio at low R, up to the maximum in relaxation time (which coincides with the viscosity maximum), and it further slightly increases or reaches a plateau up to phase separation. On the other hand, according to the reptation theory [28], the reptation time is proportional to $\bar{L}^3$, with $\bar{L}$ the average micellar length. As a result, an increase of $\bar{L}$ would imply an increase in τ_{rep} and, therefore, an increase of $\tau_M = \sqrt{\tau_{rep}\tau_{sc}}$. The strong increase of G_0 and τ_M with the cosurfactant–surfactant ratio (R) before the maximum is then related to the elongation of micelles and formation of giant worm-like micelles with progressive longer $\bar{L}$ as a consequence of the decrease in the interface spontaneous curvature when the amphiphile is added, and the subsequent formation of entanglements between them, forming a three-dimensional transient network. After the maximum, the decrease of the relaxation time could be explained by the formation of branched micelles, which has been observed by cryo-TEM for other systems [29]. As a consequence of the decreasing spontaneous curvature when R is further increased, a fusion reaction can occur, where a branch is formed by the fusion of an endcap of a micelle with the cylindrical body of another. The fusion reaction can be energetically favorable because an endcap disappears. Then, the effective length of micelles, $\bar{L}$, decreases, and τ_M is smaller; however, G_0 does not decrease and can, in fact, increase since the number of junctions (now not only entanglements but entanglements + branching points) can yet increase with R. Moreover, when branching occurs the relaxation process may be yet faster since it can involve the fast sliding of connections or branches along the body of micelles at which they are connected, since the surfactant monomers are not chemically connected [30].

As has already been said and can be observed in Figures 12.5 and 12.7, results deviate from Maxwellian behavior at high frequencies. An upturn is observed in G″, which has been attributed in other systems to a transition of the relaxation mode from reptation–scission (Cates model) to breathing or Rouse modes [28, 31]. In the systems presented in this chapter, if the relaxation mechanisms were just living reptation at long times + Rouse relaxations at short ones, results should be fitted by Equations 12.3 and 12.4, where a Rouse relaxation mode has been added to the Maxwell model, subscripts M and R referring to Maxwell and Rouse relaxations, respectively:

$$G'(\omega) = G_{0,M}\frac{(\omega\tau_M)^2}{1+(\omega\tau_M)^2} + G_{0,R}\sum_{p=1}^{N}\frac{(\omega\tau_R/p^2)^2}{1+(\omega\tau_R/p^2)^2} \tag{12.3}$$

$$G''(\omega) = G_{0,M}\frac{\omega\tau_M}{1+(\omega\tau_M)^2} + G_{0,R}\sum_{p=1}^{N}\frac{\omega\tau_R/p^2}{1+(\omega\tau_R/p^2)^2} \tag{12.4}$$

Lines in Figure 12.7 are the best fits obtained by these equations. It can be seen that the model fits the results much better than the single Maxwell model does. Nevertheless, Equations 12.3 and 12.4 predict a well-defined minimum followed by a clear upturn of G''. For this system (and for most of the systems studied by Kunieda's group) G'' does not show a clear minimum, but a very wide one followed by a smooth upturn when frequency is increased, suggesting a wide spectrum of the stress relaxation at high frequencies or several relaxation processes superimposed, apart from living reptation and Rouse relaxation.

DLS experiments done for the C_{16}SE-$C_{12}EO_4$-water system – for which the G_0 and τ_M rheological parameters are presented in Figure 12.8 – reveal the presence of two or three relaxation modes (slow mode, sm, medium mode, mm, and fast mode,

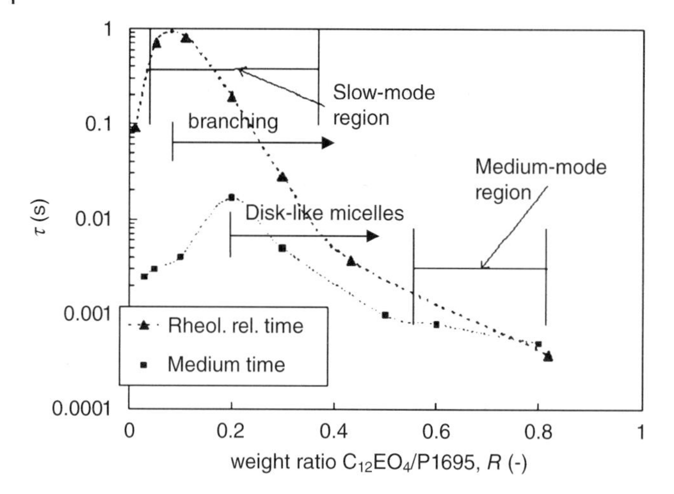

Figure 12.10 Evolution of rheological relaxation time τ_M and DLS medium mode with $C_{12}(EO)_4/C_{16}SE$ = P1695 ratio, *R*. 90% water (adapted from ref. [9] with permission from ACS publications).

fm), depending on *R*. The fm and mm could be the other relaxation processes rheologically observed at high frequencies. The fast and medium modes are diffusive, and the slow mode, nondiffusive. The fast mode, that is present in all the range of *R*, dominates over the mm and the sm at low *R*. It has been related to the movement of small spherical or just slightly elongated free micelles. Therefore, it seems that small micelles coexist with worm-like ones, although their relative importance decreases with *R*. The medium mode, the importance of which increases with *R* up to phase separation and becomes the dominant relaxation mode around $R \approx 0.6$ to 0.8, is related to the movement of bigger aggregates.

The medium mode obtained from DLS and τ_M obtained from rheology experiments are drawn together in Figure 12.10. The mm has been attributed, for low *R*, to the presence of clusters, or small aggregates of worm-like micelles, that are not connected to the whole network and can freely move, since the mm is diffusive. The relaxation time of these clusters increases with *R* up to $R \approx 0.2$, and then decreases, indicating a decrease of the size of these aggregates. However, the relative importance of the mm continuously increases with *R*, and it is the dominant relaxation mode of the system at high *R*, indicating that more and more small aggregates form when more cosurfactant is added. The slow, nondiffusive relaxation mode appears only around the maximum of relaxation time found through rheology. Thus, it is related to the relaxation time of the whole three-dimensional transient network and, afterwards, directly related to τ_M, although its value cannot be accurately calculated from DLS results. Hence, for the determination of this relaxation time, rheology seems to be the better technique.

The authors propose that when the amphiphile is progressively added, the decreasing interfacial curvature promotes the formation and growth of worm-like micelles, and a transient three-dimensional network is formed, related to the slow mode from DLS or the relaxation time τ_M observed by rheology experiments. τ_M increases with R due to an increase in the one-directional length of micelles up to a maximum. Free clusters, formed by noninterconnected small groups of elongated and entangled micelles, also grow in size and number with R. After the maximum in τ_M, around $R \approx 0.1$ for this system, branching occurs, and then τ_M decreases since the network can relax faster by sliding of these branching connections. However, clusters continue growing in size–and number–until $R \approx 0.2$. At $R > 0.2$, these branching points, where the amphiphile is probably locally concentrated due to their negative curvature, begin to separate into free disk-like micelles, which *via* diffusion enter into the medium mode, but have smaller sizes than the clusters present at lower R. Conversely, their number now strongly increases. Hence, the mm starts to go to shorter times, but its relative importance increases. More and more disk-like micelles–that could be the promoters of the lamellar liquid-crystal phase–are formed with further addition of the cosurfactant, up to phase separation. In this region, then, the medium mode is the predominant one.

12.4 Influence of Dissolved Oil in Systems Containing Worm-Like Micelles

Extensive studies have been reported by Kunieda's group regarding the formation of worm-like micelles and micellar transient networks in water–surfactant–cosurfactant systems. However, for applications, it is also relevant to know the effect of additives on systems containing worm-like micelles. It is reported that oils induce a rod–sphere transition in surfactant micellar solutions, leading to a reduction in viscosity [32]. Kunieda's group studied the solubilization of different oils in worm-like micellar solutions [19, 33]. The amount of solubilized oil, its location within the micelle, and its effect on micellar shape and size demonstrated to strongly depend on the nature of the oil and its interactions with the surfactants.

The authors studied the system formed by water, sucrose hexadecanoate, and the cosurfactants $C_{12}(EO)_3$ and monolaurin (ML), and added several oils to the watery micellar solutions. The properties, structure and behavior of the surfactant/cosurfactant/water systems in the absence of oil are analogous to others presented in this chapter. Solubilization limits of decane in the systems studied at 90 wt% and 95 wt% water and several fractions of cosurfactant in the $C_{16}SE$ + cosurfactant mixture (W_1) are presented in Figure 12.11. Results show that the solubility of oil in the system is higher when more surfactant + cosurfactant is present, as it is dissolved in micelles. Moreover, at a constant concentration of surfactant, more oil can progressively be dissolved when W_1 increases as worm-like micelles grow, up to a maximum. After the maximum, the solubility of decane in the system decreases again. However, the maximum of decane solubilization occurs at a W_1 that does not coincide with the maximum of viscosity and τ_M, that is $W_1 \approx 0.1$, but it is higher, which probably suggests that oil accommodates better in branched micelles than in linear ones.

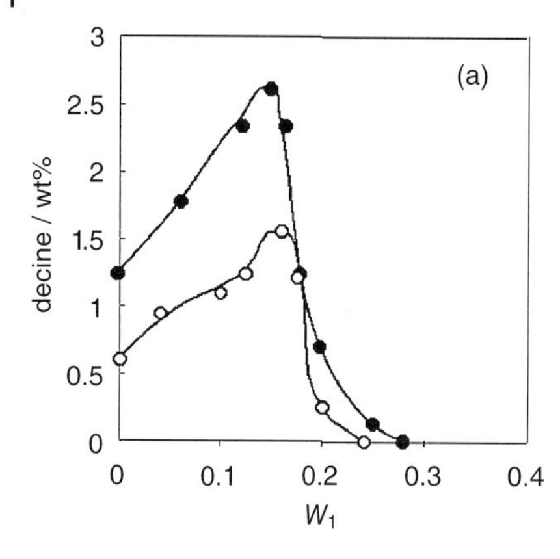

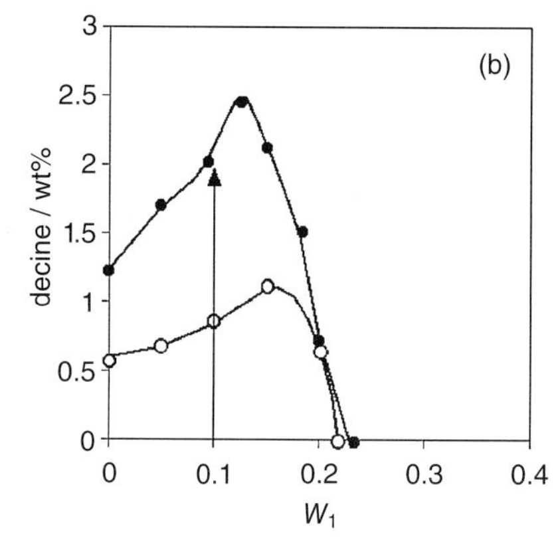

Figure 12.11 Decane solubilization limit in C_{16}SE/cosurfactant/water systems. W_1 is the weight fraction of cosurfactant in the C_{16}SE + cosurfactant mixture. Filled symbols: 90%wt water; open symbols: 95%wt water. (a) $C_{12}(EO)_3$; (b) monolaurin (reprinted from ref. [19] with permission from Elsevier).

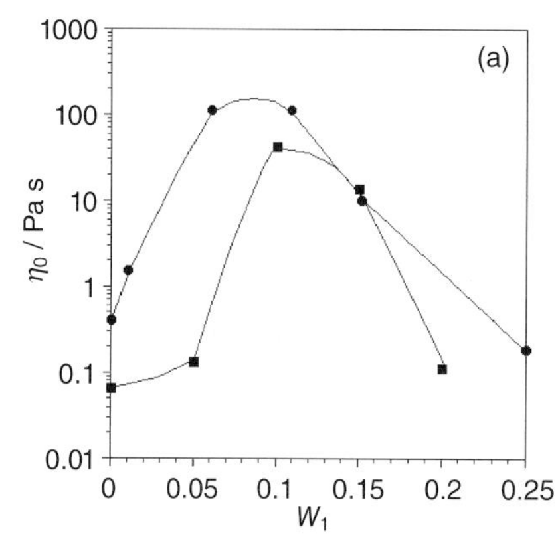

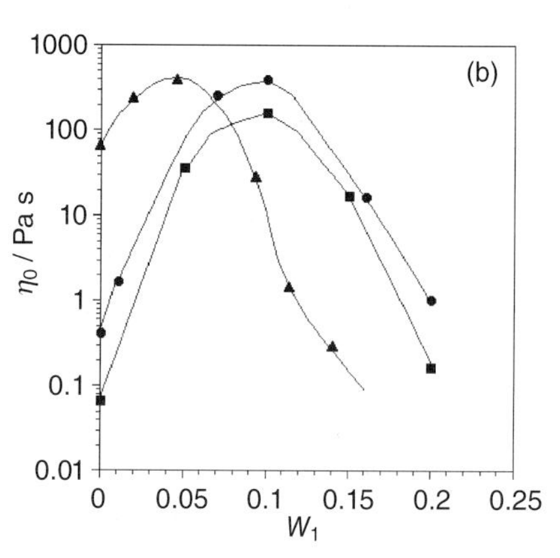

Figure 12.12 Effect of decane on the zero-shear viscosity η_0 as a function of W_1 in (a) C_{16}SE-$C_{12}EO_3$-water systems and (b) C_{16}SE-ML-water systems. The water concentration is 90 wt% (10% total surfactant + cosurfactant). Circles: no decane added; squares: 0.5 wt% added decane; triangles: 0.5 wt% added p-xylene. Lines are only guides to the eye (reprinted from ref. [19] with permission from Elsevier).

The zero shear viscosity (η_0) of the microemulsions formed when oil is added to the micellar solutions is presented in Figure 12.12. When decane is added to C_{16}SE + water systems, the viscosity decreases, yet in the absence of cosurfactant, indicating that also in this system of micelles just slightly elongated there is a decrease in micellar length and interactions between them, according to results already reported [32, 34]. This behavior seems to indicate that decane locates into the core of the micelles, inducing some kind of swelling and an increase in mean interface curvature. Then, micelles become shorter and the interactions between them are weaker. On the other hand, similar to the systems in the absence of oil, the viscosity shows a maximum when plotted against W_1. However, in $C_{12}(EO)_3$ systems a delay occurs in the increase of viscosity with W_1, and the maximum also shifts to higher values of W_1, whereas in ML systems the position of the maximum does not change. It has been attributed to the different solubility of $C_{12}(EO)_3$ and ML in decane. $C_{12}(EO)_3$ is partially soluble in decane and, as a result, at low W_1, most of this cosurfactant is solubilized by the oil in the core of the micelles, and its concentration at the interface is low; therefore, its effect on the shape of micelles and the resulting viscosity is very low, up to a composition where the solubility limit of the surfactant in decane is exceeded. When further $C_{12}(EO)_3$ is added, no more cosurfactant can be dissolved in the limited amount of decane, its interfacial concentration starts to increase, the curvature of the interface decreases and micelles elongate and interact more and more; therefore, the viscosity starts to increase. Contrarily, ML is not soluble in decane and, subsequently, it is readily adsorbed at the interface regardless of the presence of decane. Accordingly, the viscosity starts to increase from the beginning, and reaches the maximum and starts to decrease (due to branching) at the same W_1 than without decane.

When p-xylene is added to $C_{16}SE$ + water samples instead of decane, the effect is very different: This oil causes a large increase in viscosity even in the absence of cosurfactant and, what is more, the position of the viscosity maximum is shifted to lower values of W_1, indicating that less cosurfactant is needed to induce curvature changes and, therefore, less cosurfactant is required for micellar growth, formation of transient network due to entanglements between micelles, and final disruption of this network due to branching. This seems to indicate that p-xylene acts in some way like a cosurfactant, locating near the interface and reducing its spontaneous curvature.

In order to determine the effect of oil solubilization on the structure of micelles, Kunieda's group performed DLS measurements along the arrow drawn in Figure 12.11b. The three peaks obtained in the absence of oil (the diffusive fast and medium modes and the nondiffusive slow mode) yet exist when decane is added. However, the contribution of the slowest mode peak clearly decreases as oil is added, indicating that the addition of this oil induces the disruption of the micellar network, presumably by a decrease of the micelles length, according to previous literature and rheological results presented above.

Finally, Kunieda's group carried out measurements in the concentrated region (40 wt% water), where hexagonal (H_1) or lamellar (L_α) liquid-crystal phase are present depending on composition, in order to detect changes in the microstructural parameters of aggregates when oil (decane or p-xylene) is added, and relate them to those occurring in the adjacent micellar (W_m) phase (where worm-like micelles are formed).

For the hexagonal phase, the radius of the rod-like aggregate lipophilic core (r_L) and the effective cross-sectional area per lipophilic chain (a_s) are calculated by Equations 12.5 and 12.6 [35, 36]:

$$r_L = d\left(\frac{2(\phi_0 + \phi_L)}{3^{1/2}\pi}\right)^{1/2} \tag{12.5}$$

$$a_s = \frac{2\bar{v}_L}{r_L}\left(\frac{\phi_0 + \phi_L}{\phi_L}\right) \tag{12.6}$$

For the lamellar phase, the half-length of the hydrophobic layer (d_L) and a_s are calculated by Equations 12.7 and 12.8 [35]:

$$d_L = d\frac{\phi_0 + \phi_L}{2} \tag{12.7}$$

$$a_s = \frac{\bar{v}_L}{d_L}\left(\frac{\phi_0 + \phi_L}{\phi_L}\right) \tag{12.8}$$

where d is the interlayer spacing of the first diffraction peak, ϕ_L is the lipophilic volume fraction of the surfactant + cosurfactant mixtures, ϕ_0 is the volume fraction of oil and $\bar{v}_L$ is the average molecular volume of the lipophilic part. As shown in Table 12.1, when decane is added there is an increase in the radius of aggregates, r_L, whereas the effective surface area a_s does not change at low oil concentrations. This behavior suggests again that decane is incorporated into the core of aggregates, producing swelling, as it probably happens at higher concentrations of water, when worm-like micelles exist instead of H_1, according to the conclusions derived from rheology and DLS experiments. On the other hand, the increase in micellar radius is smaller for p-xylene, and the effective surface area increases, suggesting that this oil tends to penetrate into the surfactant palisade layer, affecting its curvature, again as DLS and rheology experiments suggested for adjacent diluted systems. As a matter of fact, at higher p-xylene concentrations, there is a transition from the hexagonal to the lamellar liquid-crystalline phase, since there is a decrease in the curvature of aggregates, indicating, once again, the solution of this oil in the interface.

From all these results, one could conclude that long or bulky oils, like decane, tend to solubilize in the cores of aggregates, swelling them and making the cur-

Table 12.1 Structural parameters of liquid crystal phases for several samples in C_{16}SE-ML-water-oil systems. The water concentration is fixed at 40% (rewritten from ref. [19] with permission from Elsevier).

Sample	Phase	d (nm)	r_L, d_L (nm)	a (nm^2)
$W_1 = 0$	H_1	5.44	1.88	0.52
$W_1 = 0.1$	H_1	5.55	1.90	0.50
$W_1 = 0.1$, 1.5% decane	H_1	5.73	2.01	0.50
$W_1 = 0.1$, 1.5% p-xylene	H_1	5.62	1.97	0.51
$W_1 = 0.1$, 6% decane	H_1	6.35	2.37	0.52
$W_1 = 0.1$, 6% p-xylene	L_α	5.38	1.01	0.60

vature more positive. The resulting aggregates are shorter. On the other hand, short and amphiphilic oils like p-xylene tend to solubilize in the surfactant palisade layer, reducing the curvature and increasing the packing parameter by increasing the effective lipophilic volume, and then less cosurfactant is required for the micellar growth and interactions between micelles.

12.5
Conclusion

To conclude, in this work we present a review and explanation of the main contributions of Professor Kunieda in the field of rheology of worm-like micelles. It can be remarked that his works have extended the number and types of surfactant systems able to present worm-like micelles, on the one hand, and have deepened the elucidation of structure and formation mechanism, on the other hand.

Undoubtedly, his contributions constitute a valuable foundation for actual and future works of his colleagues that are continuing his task.

References

1 Kern, F., Lemarechal, P., Candau, S.J., and Cates, M.E. (1992) *Langmuir*, **8**, 437–440.
2 Kim, W.J., Yang, S.M., and Kim, M. (2000) *J. Colloid Interface Sci.*, **232**, 225–234.
3 Innai, S., and Shikata, T. (2001) *J. Colloid Interface Sci.*, **244**, 399–404.
4 Rehage, H., and Hoffmann, H. (1988) *J. Phys. Chem.*, **92**, 4712–4719.
5 Soltero, J.F.A., Puig, J.E., and Manero, O. (1996) *Langmuir*, **12**, 2654–2662.
6 Raghavan, S.R., and Kaler, E.W. (2001) *Langmuir*, **17**, 300–306.
7 Kato, T., and Nozu, D. (2001) *J. Mol. Liq.*, **90**, 167–174.
8 Kato, T., Taguchi, N., and Nozu, D. (1997) *Prog. Colloid Polym. Sci.*, **106**, 57–60.
9 Maestro, A., Acharya, D.P., Furukawa, H., Gutiérrez, J.M., López-Quintela, M.A., Ishitobi, M., and Kunieda, H. (2004) *J. Phys. Chem. B*, **108**, 14009–14016.
10 Acharya, D.P., and Kunieda, H. (2003) *J. Phys. Chem. B*, **107**, 10168–10175.
11 Naito, N., Acharya, D.P., Tanimura, J., and Kunieda, H. (2005) *J. Oleo Sci.*, **54**, 7–13.
12 Acharya, D.P., Hattori, K., Sakai, T., and Kunieda, H. (2003) *Langmuir*, **19**, 9173–9178.
13 Rodríguez, C., Acharya, D.P., Maestro, A., Hattori, K., Aramaki, K., and Kunieda, H. (2004) *J. Chem. Eng. Jpn.*, **37**, 622–629.
14 Engelskirchen, S., Acharya, D.P., García-Román, M., and Kunieda, H. (2006) *Colloid Surf. A*, **279**, 113–120.
15 Rodríguez, C., Acharya, D.P., Hattori, K., Sakai, T., and Kunieda, H. (2003) *Langmuir*, **19**, 8692–8696.
16 Acharya, D.P., Sato, T., Kaneko, M., Singh, Y., and Kunieda, H. (2006) *J. Phys. Chem. B*, **110**, 754–760.
17 Varade, D., Sharma, S.C., and Aramaki, K. (2007) *J. Colloid Interface Sci.*, **313**, 680–685.
18 Acharya, D.P., Kunieda, H., Shiba, Y., and Aratani, K. (2004) *J. Phys. Chem. B*, **108**, 1790–1797.
19 Rodriguez-Abreu, C., Aramaki, K., Tanaka, Y., Lopez-Quintela, M.A., Ishitobi, M., and Kunieda, H. (2005) *J. Colloid Interface Sci.*, **291**, 560–569.
20 Ahmed, T., and Aramaki, K. (2009) *J. Colloid Interface Sci.*, **336**, 335–344.

21 Naito, N., Acharya, D.P., Tanimura, J., and Kunieda, H. (2004) *J. Oleo Sci.*, **53**, 599–606.

22 Varade, D., Ushiyama, K., Shrestha, L.K., and Aramaki, K. (2007) *J. Colloid Interf. Sci.*, **312**, 489–497.

23 Cates, M.E., and Candau, S.J. (1990) *J. Phys. Condens. Matter*, **2**, 6869.

24 Granek, R., and Cates, M.E. (1992) *J. Chem. Phys.*, **96**, 4758.

25 Acharya, D.P., and Kunieda, H. (2006) *Adv. Colloid Interfac.*, **123–126**, 401–413.

26 Ahmed, T., and Aramaki, K. (2008) *J. Colloid Interface Sci.*, **327**, 180–185.

27 Rehage, H., and Hoffmann, H. (1988) *J. Phys. Chem.*, **92**, 4712.

28 Doi, M., and Edwards, S.F. (1986) *The Theory of Polymer Dynamics*, Clarendon, Oxford, UK.

29 May, S., Bohbot, Y., and Ben-Shaul, A. (1997) *J. Phys. Chem.*, **101**, 8648.

30 Bandyopadhyay, R., and Sood, A.K. (2003) *Langmuir*, **19**, 3121.

31 Granek, R., and Cates, M.E.J. (1992) *Chem. Phys.*, **96**, 4758.

32 Penfold, J., Staples, E., and Tucker, I. (2002) *J. Phys. Chem. B*, **106**, 8891.

33 Sato, T., Acharya, D.P., Kaneko, M., Aramaki, K., Singh, Y., Ishitobi, M., and Kunieda, H. (2006) *J. Dispersion Sci. Technol.*, **27**, 611–616.

34 Menge, U., Lang, P., and Findenegg, G.H. (1999) *J. Phys. Chem. B*, **103**, 5768.

35 Kunieda, H., Kabir, H., Aramaki, K., and Shigeta, K. (2001) *J. Mol. Liq.*, **90**, 157.

36 Sjöbom, M.D., and Edlund, H. (2002) *Langmuir*, **18**, 8309.

Index

Self-Organized Surfactant Structures. Edited by Tharwat F. Tadros

ISBN: 978-3-527-31990-9